Crew Resource Management Training

Crew Resource Management Training

A Competence-based Approach for Airline Pilots

Norman MacLeod

CRC Press
Taylor & Francis Group
Boca Raton London New York

CRC Press is an imprint of the
Taylor & Francis Group, an **informa** business

First edition published 2021
by CRC Press
6000 Broken Sound Parkway NW, Suite 300, Boca Raton, FL 33487-2742

and by CRC Press
2 Park Square, Milton Park, Abingdon, Oxon, OX14 4RN

ISBN: 978-0-367-68731-1 (hbk)
ISBN: 978-0-367-68732-8 (pbk)
ISBN: 978-1-003-13883-9 (ebk)

Typeset in Times
by codeMantra

Contents

Preface

The genesis of this book lies in a belief that what we knew in aviation as crew resource management (CRM) was dead. I had been working in the field, in a variety of roles, for more than 20 years and it seemed to me that the things had stalled. I had been fortunate enough to work in a variety of different types of operation, ranging from major flag carriers through to small air taxi companies with one or two aircraft. I had delivered training in North America, Africa, in several of the Gulf States, most of Europe and many former Soviet territories (it seems hard to recall at this distance that the mid-1990s in Europe saw some significant changes in the aviation system). Training was a mandatory requirement, but the quality and quantity of training differed hugely.

In 1995, I was awarded a Winston Churchill Travel Fellowship. My topic was vocational training and the Fellowship afforded me the opportunity to visit different countries to examine why airline pilot training, which was delivered under the same International Civil Aviation Organisation standard, was, in fact, radically different around the world. I visited aviation authorities, flight schools and airlines in Russia, the Ukraine, the USA, Sweden, Germany and the United Kingdom. The key lesson I came away with was one size really does not fit all.

By the early 2000s, CRM was no longer new. Changes to initial pilot training meant that new hire pilots were already familiar with some of the concepts and the language of CRM was commonplace out on the line. My view at that time was that CRM remained a very superficial box-ticking exercise enthusiastically endorsed by a handful of advocates. But it had not really progressed beyond being a cottage industry. Airlines were primarily interested in demonstrating compliance but had little interest in exploring the benefits that might flow from a serious analysis of this thing called 'CRM'. Finally, in conversation with Sidney Dekker, we agreed that the original 'big beasts' – Jim Reason and Bob Helmreich – were no longer working and there were no new names driving the research agenda. Interest was shifting to healthcare where there was some enthusiasm for adopting the airlines' approach to safety and human performance. It was time to move on. At that time, there was no coherent CRM text framed within the European regulatory guidance. I set out to plug the gap. About halfway into writing, an opportunity came up to work for an overseas airline.

As a consultant you are always on the outside looking in: being on the inside gives a very different perspective. Furthermore, I had worked mainly in areas shaped by the European or US regulatory frameworks. It was interesting to work in an area influenced by an Australian mindset. It was interesting to see how ideas have migrated around the world, being adopted and developed in ways that are subtly different. After 8 years, it was time to come home, dust off the stalled manuscript and, frankly, hit 'DELETE'.

My background has been primarily in training analysis and design. I have had some experience of recruitment and selection in both the military and civilian worlds. I have worked on quite specific projects developing training for aircraft equipment, on developing type training courses for different aircraft types and also on more abstract projects like 'leadership' training for military officers and assessing the effectiveness of training systems. Embedded in an airline Flight Operations Department allowed me to see how all of these strands come together but in a context of having to deliver operational capability under financial constraints. It allowed me to move away from delivering CRM as an academic, regulatory requirement to reflecting on what the broader construct might mean in the real world.

Other factors have also shaped the book, not least developments in the broader landscape of aviation training. Training budgets are under intense scrutiny, and there is pressure to reduce the length of training or to find alternative – cheaper – ways to deliver training. Frustration with the perceived lack of effectiveness of training – as manifested in periodic embarrassment to the industry when accidents occur for bewildering reasons – has led to a fashion for 'competency-based training' based on 'evidence'. The industry persists with confusing outcomes with causes and we see periodic fads in training: 'LOC-I'; CFIT; runway overrun; and ALAR. Finally, and fundamentally, the advent of a highly capable, automated aircraft such as the Airbus A-350 signals a change in the nature of the work of an airline pilot. The advent of pilotless flight decks looms, but the role of an airline pilot is in transition just as the requirement for a person at the controls is diminishing. Given all of these factors, any new text on CRM should be forward-looking rather than rehearse old arguments. That is what I intend to do in this book.

Acknowledgements

Finally, I need to offer thanks to the many colleagues and friends in Hong Kong who have, each in their own way, made this a richer book. But I should not forget the thousands of pilots and cabin crew to whom I have delivered CRM over the years, many of them sharing their own experiences of life in aviation with me. I stumbled into CRM by accident but the journey has been worthwhile.

Many colleagues kindly read sections or drafts of chapters and offered comments. Thanks are due to Dr. Nick Dahlstrom, Dr. Mel Todd, Graham Cownie, Andy Craig, Albert Tiong, Adam Johns and Adam Wainwright. Trevor Ng kindly indulged all of my requests for flight data. Matt MacLeod produced the graphics. Prof. Ed Salas generously made suggestions for approaches to teamwork training. Dominique Mineo kindly shared information about the CEFA pilot debriefing tool and Conor McKenna, of Vocavio Technologies, provided information about his company's speech analysis technology. Finally, Dr. Lishuai Li was very patient in educating me about advanced statistical techniques and flight data.

Author

Norman MacLeod has been involved in different aspects of training design and delivery in aviation since 1977. He has designed, delivered and trained trainers in the field of CRM since 1989. He has worked in all types of aviation, including rotary wing, business jets, low-cost carriers and legacy airlines. Most recently, he was the Human Factors Manager for Cathay Pacific Airways in Hong Kong. He has spoken at numerous conferences and published articles in CAT Journal.

List of Abbreviations

AAIB:	Air Accident Investigation Board (UK)
ASRS:	Aviation Safety Reporting System
ATSB:	Australian Transportation Safety Bureau
AQP:	FAA's Advanced Qualification Program
ASRS:	Aviation Safety Reporting System
ATC:	Air Traffic Control
ATQP:	EASA's Alternative Training and Qualification Programme
ATSB:	Australian Transportation Safety Bureau
BEA:	Bureau d'Enquetes et d'Analyses pour le securite de l'aviation civile
CAA:	Civil Aviation Authority (UK)
CASA:	Civil Aviation Safety Authority (Australia)
CBT:	Competency-based Training
CHIRP:	UK Confidential Human Factors Incident Reporting Programme
DDG:	Dispatch Deviations Guide
EASA:	European Aviation Safety Agency
EBT:	Evidence-based Training
EFIS:	Electronic Flight Information System
FAA:	Federal Aviation Administration (USA)
FCOM:	Flight Crew Operating Manual
FLCH:	Flight Level Change
FPA:	Flight Path Angle flight mode
g/s:	Glide Slope
GPS:	Global Positioning System
GPWS:	Ground Proximity Warning System
HDG:	Heading
IATA:	International Air Transport Association
IRA:	Inter-rater Agreement
IRR:	Inter-rater Reliabiity
ICAO:	International Civil Aviation Organisation
ISD:	Instructional Systems Design
LOC:	Localiser
LOFT:	Line Oriented Flight Training
LWB:	Life/Work Balance
MCOQ:	Multiple Choice Objective Question
MEL:	Minimum Equipment List
NASA:	National Aeronautics and Space Administration
NOTECHS:	Non-Technical Skills
NTSB:	National Transportation Safety Board (USA)
ORCE:	Observe, Record, Classify, Evaluate
PAN:	Urgency Call to Air Traffic
SAT:	Systems Approach to Training
SHK:	Swedish Accident Investigation Authority

SME: Subject Matter Expert
TSB: Transportation Safety Board of Canada
VASI: Visual Approach Slope Indicator
VNAV: Vertical Navigation Automation Mode
VR: Virtual Reality
v/s: Vertical Speed

UNITS

ft: foot/feet
kg: kilogram
km: kilometre
kt: knot
lbs: pound
m: metre
nm: nautical mile

Introduction

Behavioural economists recognised a long time ago that the classical 'homo economicus', the rationale being upon which economic theories had been based, probably did not exist. Humans are frail, irrational thinkers subject to a range of biases. Equally, the image of a pilot as steely-eyed daredevil fighting wind and rain to get passengers safely to their destinations has also, possibly, had its day. Pilots are a subset of the population. Their entry into the industry is through a variety of routes, some of which are subject to rigorous selection while others are based simply on access to funding. It is also probably fair to say that there are different flavours of aviation around the world. For example, it was once the case that the bulk of commercial aviation occured within the USA. As a result, an approach to the industry developed inside the continental United States that reflected a degree of introspection arising from having enough to worry about on your own doorstep. Australian aviation is much like that in the United States and both countries can be characterised as large open spaces where air travel offers the easiest solution to the problem of distance. Road and rail networks are poorly developed. The first scheduled, commercial air route was from London to Paris, opened in 1919 and was a continuation of the wartime Royal Air Force route. Of course, England and France were separated by the English Channel, which represented an obstacle, but on mainland Europe, a dense network of road and rail communications made commercial aviation unviable internally. Instead, effort was put into the development of long-haul routes to connect with the outposts of empires. The British pioneered routes through the Gulf to Singapore and on to Australasia, the Germans down to Southwest Africa, and the Spanish and Portuguese to South America. These different historical trajectories resulted in different structures within national aviation systems and, as result, different approaches to the selection, initial training and career development for commercial pilots. Today, an Australian pilot who has spent 5 years hauling cargo between outposts of population looks down on the European pilot who, having completed an 18-month intensive training course, starts their first job as the first officer on a narrow-body jet for a major airline. And yet both systems work and one is no more safe than the other.

Just as pilots differ in their training and background, so do they show all of the foibles and flaws we can see in the average human being. Unfortunately, when you read accident and incident reports, there seems to be an assumption that pilots are a cross between a brain surgeon and a NASA astronaut. Of course, some pilots might identify with that image – and there are, indeed, very many very capable people flying aircraft – but the fact remains that pilots are humans. This book will make use of public domain reports of events but, hopefully, will try to position any analysis in the realms of what was achievable rather than setting impossible expectations.

One current trend in aviation training is a shift towards competency-based training. The idea of 'competency-based' anything is that after training you can be expected to be able to actually do something. Too much of CRM training, historically, has been abstract and divorced from the job of being an airline pilot. In this book, I will look at performance through a competence paradigm and, so, the focus will be on what people do to fly aircraft safely. A competence framework usually includes skills and statements of underpinning knowledge. In fact, the emphasis in this book is making sense of action. I draw on academic theory where it throws light on the *why* things happened. I will attempt to answer the question of why 'homo aeronautics' acted as they did rather than as how some outside observer felt they should have acted. This is not an aviation psychology textbook. This will not be a comprehensive review of current research as it relates to human performance in flight. It is more an ecological examination of pilots at work. Although I started off trying to write a text book, what I now offer is a reflection on safe aviation based on more than 40 years of observing pilots at work.

The aim of this book is to explore what a competence-based approach to training might look like by, first, exploring how the aviation system works and, then, considering the implications for course design. The book is in five parts. First I will introduce the idea of competence, systems and safety. I review some of the key models of accident causation to see how they might inform decisions about training. I then look at more recent, systems-based, approaches. Here I discuss the broader systems concepts I will be drawing on. The second part will look at how individuals do work. I look at how pilots make sense of the world and then make decisions about action. My thesis is that pilot behaviour is goal-directed and involves making decisions about the interventions needed to achieve those goals. I also look at stress and fatigue as inhibitors. The third section looks at the collaborative nature of aviation. In developing my systems concept, I highlight the need to act across interfaces, be they between humans and technology or simply between humans. Interface management is the essence of collaboration. I also look at communication as the driving force of a system. The fourth section looks at functions at the level of the organisation and then at agencies like regulators and investigators. My purpose here is to look at, first, the challenges faced by actors at this level but, more important, how legitimate decisions at these levels can result in unintended consequences. I am interested in cross-scale interactions: how a decision by an authority about a manufacturer's design solution can create the context for failure at the level of an individual crew. Finally, I offer some thoughts on training and assessing competence. It is an attempt to recast CRM training in the light of my analysis.

This book, then, is about how systems are configured to meet operational goals and how people then perform within those systems. In other words, it is about people doing work. The system is, in part, the context of work but can also mean the combination of technologies that form the tools needed to do that work. So, from the start, we see that we need to explore issues at two levels: the level of the organisation and of the operator. It makes no sense to look at why people do things wrong if we do not also look at the context in which erroneous actions take place. In order to understand how people function in systems of production, we also need to understand how people construct acceptable, rational 'realities' that allow them to deliver a performance in uncertain, ambiguous circumstances.

While formalised training to control the technology of aviation has existed for almost 90 years, the need to train humans to work with one another while using the technology was only really appreciated about 25 years ago. To make matters worse, although the training needed to control an aircraft and manipulate its systems is relatively simple to analyse and translate into training courses, the fuzzy behaviours demonstrated by humans working in groups is less readily broken down into convenient training modules. As a result, what we call 'CRM' is an amorphous collection of social skills, not necessarily clearly understood by all involved in aviation and definitely not dealt with very effectively in training.

That said, what do we mean by CRM? The first definition, offered after crew communication was cited in an accident report for the first time, refers to crews making full use of all available resources. As a definition, it establishes an unattainable ideal. Crews cannot possibly know all the resources available to them and lack the mental processing power needed to manipulate the information implicit in the definition in order to achieve an optimum solution to a problem. In many ways, the aim of this book is to answer the question that arises: why not?

CRM has also had to contend with human factors as an overlapping domain. Human factors was initially a subset of engineering that dealt with ergonomic design of devices but, over time, has expanded to include the ways in which humans interact with systems. The failure of CRM to deliver a demonstrable return on investment, especially in terms of an incremental improvement in aviation safety, has driven some to seek newer and better definitions. We are now supposedly in the era of sixth-generation CRM, the realm of error management. We will explore some of these issues more in the later chapters but, for now, it is probably enough to say that CRM seeks to optimise the performance of human operators in a complex system, namely, aviation.

The target audience for this book is, primarily, student pilots in training. I touch on almost all of the subjects covered in a ground school human performance and limitations course but, in many case, not to the depth needed to pass an MCOQ, I am afraid. What I hope the book will do is stimulate an interest in a fascinating subject that should be of direct personal interest to pilots once they get out of ground school. I hope that the book will bring a sufficiently different perspective to also be of interest to the wider aviation community. Because the focus is on the way people do work, the principles are equally applicable to cabin crew, maintenance, ATC and so on.

I have often said that the biggest mistake made in the history of CRM was to call it crew resource management. Even though the 'C' has shifted from 'cockpit' to 'crew' and even 'company', the debate still misses the point. As I hope will become clear, the issue is one of people doing work. How many people are engaged in the task at any specific time is simply an issue of nomenclature. The problems we need to address remain the same no matter how many actors are involved in fouling things up.

1 Why a 'Competence-based' Approach to Crew Resource Management Training?

> The thesis presented in this paper is straightforward: one of the principal causes of incidents and accidents in civil jet transport operations is the lack of effective management of available resources by the flight-deck crew.
>
> **John K. Lauber (1979)**

1.1 INTRODUCTION

In this chapter, I want to address three questions: what do we mean by 'crew resource management' (CRM)?; what might the ultimate goal of any training intervention be?; how do we set about designing training to develop CRM skills? Despite being involved in the design and delivery of CRM since 1989, and was also CRM Instructor Examiner for the UK CAA for part of that time, I have never been convinced that it was a discipline sufficiently distinct from more fundamental approaches to aircraft pilot performance development. Undoubtedly, it is a fascinating field of research and clearly overlaps the broader 'safety' domain but from the narrow perspective of a safety-related training intervention, it remains a muddy concept. As this book unfolds, I hope to make my case more fully. For now, however, I want to set out some broad principles. I start with a brief overview of the CRM concept from its origins, through various iterations to the current position, which is best summarised as 'confused'. Given that the domain, in my opinion, lacks coherence, it follows that we need an analytical framework to help clarify the requirement. In the second part of this chapter, I will propose that the goal of training is to develop expertise and, therefore, I will try to outline what 'expertise' looks like in pilots. As with CRM, generally, I suggest that the current position on training in aviation is equally confused. Therefore, the chapter concludes with an outline of some approaches to training design and a discussion of the challenges to the assessment of performance that flow from my analysis.

I set out in this book to explore CRM through the 'competence' paradigm, by which I mean the development of transferable skills that support expert performance. Unless training achieves some measurable output, it cannot be considered a valid use of the increasingly scarce resources of time and money. Furthermore, by taking the competence route, I hope to show that it becomes very difficult to sustain the technical/nontechnical duality of pilot performance. Therefore, as I develop my arguments in this chapter, the discussion will move away from dealing with CRM, specifically, to a consideration of competence in a more generic sense. Finally, the management of training is, itself, a 'competence'. While the content of this chapter might be of passing interest to line pilots and students, it should be central to the role of anyone involved in the management of the training process.

1.2 A SHORT HISTORY OF CRM

The broad concept of CRM training has been around for over 35 years and, in addition to now being applied in just about every area of aviation, has been adopted by other domains such as maritime and medicine. So, why the need now for a specific competency-based approach to the topic given that it is firmly established as a training requirement? There are two main creation myths attached to 'CRM'. If you are European, CRM started with the Tenerife disaster in 1977 (Project Tenerife), and if you are North American, it was the United Airlines Flight 173 (NTSB, 1979) crash at Portland Oregon in 1978 that was the trigger. The consensus at that time was that aircraft were now so reliable that it was the human that had become the problem. We will see later that this was yet another myth. The actual catalyst for action was undoubtedly the joint NASA/industry workshop – 'Resource Management on the Flight Deck' (NASA, 1980) – in June 1979, at which John Lauber offered the view quoted at the start of this chapter. The workshop resulted in the first course, run by United Airlines, delivering what we would now recognise as CRM training. That course was, in turn, the template for almost all other airlines' initial implementations. Interestingly, as Table 1.1 illustrates, the airline industry has historically tended to seek technological solutions to safety issues. CRM remains one of the few attempts to increase safety through fundamental behaviour change. The problem is that it was never entirely clear what 'CRM' actually was nor if the initiative really worked. Soon after the FAA published its initial proposals, it announced that it was not intending to progress to a full rule-making (Diehl, 2011). The major US airlines objected. As a training intervention, CRM was seen to be of real value.

Despite the concept's longevity, the aviation industry has yet to offer a coherent definition of 'CRM'. Lauber's 'making full use of all available resources' persists but, unfortunately, offers little in the way of guidance to training developers. Attempts

TABLE 1.1
Catalysts for Change

Shock	Approx. date	Response
Mid-air collision	1950s	TCAS
Teamwork failure	Late 1970s	CRM
Controlled flight into terrain	Late 1970s	GPWS
Runway incursions	1990s	Ground radar
Runway over-run	2000	ROPS
Loss of control	2010	Upset recovery training

to elaborate on the concept have generated lists of topics such as communications, teamwork, decision-making and safety culture, which form the basis of curricula around the world. From the outset, 'CRM' appears to have been a generic umbrella label and, over the years, it has been interpreted in a number of different ways. I do not propose to offer my own definition. Instead, I hope to make it clear that 'CRM' is a perspective on, not a subset of, the skills needed to fly aircraft safely. Helmreich et al. (1999a) make the point that 'some have argued that CRM training should ultimately disappear, as it becomes fully integrated into technical training'. While the authors admit to once supporting that position, they were then of the opinion that CRM should remain separate on the grounds that it forms 'a critical and continuing component of a safety culture'. My position is that they were broadly correct in their original analysis. That said, once training has been designed to address real competence, there remains the question of what underpinning knowledge in the broad area of CRM is needed to support performance.

The honeymoon did not last long. Within 10 years, the major US carriers were questioning the cost of training with no discernible reduction in adverse events. Professor Bob Helmreich, at the University of Texas, Austin, had participated in the 1979 conference and his research group was asked to come up with a new approach. Helmreich's team did two things. First, they formulated the 'generational' model of CRM development (Helmreich et al., 1999b). This was another creation myth in that the 'generations' they identified only make sense from a North American perspective. Indeed, an early CRM conference asked if the concept even applied outside of the USA. However, the 'generations' model culminated in 'error management' as the ultimate goal of CRM, which neatly led into the teams' second initiative. The Texas Research Group developed the Line Operations Safety Audit (LOSA) process. Between the first, prototype, LOSA run by Delta Airlines and the second iteration, adopted by Continental Airlines, the 'threat and error' (TEM) conceptual framework had been developed (Klinect, 2005). In the mid-1980s, TEM was adopted by ICAO, a move helped by the fact that Captain Dan Maurino, the ICAO Safety Manager, had been a thesis supervisor for the LOSA/TEM project.

The relationship between TEM and CRM was confused from the outset. Helmreich, himself, recognised that TEM was a data collection process and that CRM was a training intervention aimed at resolving issues identified in the audit. He argued that data were only of use when fitted into a theoretical framework: TEM was that framework. The shift in the centre of gravity from CRM to TEM can be traced, in part to a paper entitled 'Defensive Flying for Pilots: An Introduction to Threat and Error Management' (Merritt & Klinect, 2006). In it, the authors describe TEM as promoting a 'proactive philosophy' and providing 'techniques for maximising safety margins'. In their view, TEM had moved from being a static framework of analysis to a dynamic model of pilot behaviour. They state that 'TEM tools work best when pilots adopt TEM techniques' and 'many of the best practices advocated by CRM can be considered TEM countermeasures'. What was not clear, however, was what constituted a TEM 'tool' or 'technique' nor how a 'practice' differs from a 'countermeasure' in any substantive way. The European Aviation Safety Agency (EASA) pilot training requirements now require TEM to be included in all phases of training such that pilots will demonstrate the 'attitudes and behaviours appropriate to safe conduct of flight, including recognising and managing potential threats and errors'. Apart from cosmetic changes in language, the core of the EASA training requirement remains the same. The latest iteration of the LOSA process has substituted the competence-based training (CBT) initiative's 'competencies' for the original FAA countermeasures, suggesting that 'TEM' is not a fundamental truth and can be changed to suit circumstances.

The confusion between CRM and TEM persists. EASA, in its updated CRM regulations (EASA, 2015) rather trips over itself in trying to reconcile the two initiatives. Of interest, while CRM training persists, no airline (to my knowledge, at least) has managed to develop a 'pure' TEM training course that does anything more than simply define the key concepts of a threat, an error and an outcome. Later in this book, I do draw on data collected in LOSA projects I have managed. The LOSA process is a valuable source of information about normal operations, as was claimed by Bob Helmreich. The benefits that accrue from substituting one acronym for another, however, remain unproven. Despite continuing concerns about the cost-effectiveness of training and the progressive reframing of the issue in a new jargon, the fundamental approach to CRM training has never really changed. The broad themes of an early course are still recognisable today.

Thirty-five years have, however, seen changes in other areas. Our understanding of 'error' as a symptom, not a cause, of failure has changed. New paradigms, such as 'Safety-II' and 'resilience engineering', recognise that human behaviour is adaptive and that we should look more at how pilots respond to change, rather than simply try to prevent the repetition of past failures. The enormous cost of initial and recurrent training, especially in a challenging financial climate, is causing airlines to employ new technologies to deliver training. Finally, advances in aircraft automation have changed the nature of the job of an airline pilot. Today, airline pilots are system managers, more akin to the operator in a power production plant control room. The prospect of autonomous, remotely piloted commercial aircraft is on the horizon and, in all probability, this book is aimed at the last generation of pilots who will travel in the same craft as their passengers. Nonetheless, that generation will still have to contend with the same challenges as their predecessors while, in many cases, carrying larger passenger loads over greater distances. It is probably timely to take stock of the skills they will need.

The fundamental question we need to ask is what are we trying to achieve by exposing pilots to the cluster of ideas that fall under the umbrella of CRM? Furthermore, what do we want pilots to do differently once they have had that exposure? Ultimately, we could say that we want to increase 'safety'. In an industry that is already incredibly safe, as measured in terms of hull loss events and passenger fatalities, this seems a daunting task. Of course, the actual occurrence of and the potential for a future occurrence are not the same things and so risk mitigation might be the goal. In either case, behaviour change must be a candidate target for any intervention. CRM courses have tended to be 'educational' in nature in that they have been rather broad brush and informative rather than requiring trainees to show some incremental improvement in performance. It is not unreasonable to expect to see an output from training that could be measured in behavioural terms but this has never really been the goal. Although performance is assessed in a simulator, there is usually no direct connection between the classroom inputs and the observed outputs. At this point, it is probably fair to say that the historical isolation of CRM as a domain is part of the problem. By requiring ground training to satisfy an externally imposed curriculum, the link between operational need and training design is severed. To restore this link, we need to look more holistically at the problem. To this end, I now want to examine the concept of expertise.

1.3 WHAT DOES EXPERTISE LOOK LIKE?

The Cessna 208 Caravan is a robust workhorse, providing passenger and cargo services in some of the more rugged parts of the world and is the backbone of the parcel feeder networks in the USA. The aircraft's performance in icing conditions is a frequent source of debate among Caravan pilots. The following paragraphs are a set of posts taken from a dedicated Cessna 208 website (caravanpilots.com):

POST 1.1

The a/c has had an icing problem since 1985. I checked out in Feb of 1986, started flying across the Rocky Mountains at 16,000, 17,000 ft, this was before FlightSafety or Pan Am (note – training providers with C-208 sims). I was taught by on old Cessna instructor [...] in Wichita. He told me this a/c will handle ice, shown picture of a 'van with up to 3 inches of ice on the leading edge, the picture was taken by a test pilot upon landing in Minot N.D. So off I went to fly DEN-ASE-GJT, Colorado. I learned fast the a/c is under-powered, and the boots system is marginal at best with low airspeed. I learned to take off with 20 flaps move flaps up at 100 kts. and leave 10 flaps in until reaching cruise airspeed. Ice en route? climb if able, most of time could not. If a/c quit climbing ask for block alt and fly the airspeed if a/c started to descend I would let it lose height as slow as possible on the vsi to the MEA and then declare emergency to the nearest airport. I operate the boots with about 1/4 in and continuous at 1 min intervals. If anyone has to cycle the prop, hang on! It will fill your pants the first time. When we started {recurrent sim training} I was chewed out for not following the flight profiles for t/o, and for ILS approaches. When I have to move the power lever full up I take my foot and "kick it higher". The a/c needs more horsepower and a better boot systems. If Cessna or other operators do not address these issues we got a problem!! Certifying the pilots for flying in ice is not the answer!!!

POST 1.2

The problem was with your route (DEN-ASE-GJT) and your schedule, not the airplane. Go and fly PBI-EYW-TPA a few times on your own schedule and then let us know how you like the airplane and if it's still under-powered. Aside from a few other freight routes in the world with predictable moderate icing, you were flying the worst case scenario every day during the winter. Whomever operated your airplane could have chosen a different aircraft for the job...one with more power yet equal payload and operating costs...let's see, which one would that be? Capt. Slow is right...either leave some freight behind or cancel and stay overnight at a Holiday Inn and wake up refreshed. Icing incidents are solely the fault of the pilot, not the airplane. You are in total control of how much ice gets on your airplane. If you don't believe that, or if for some reason you're not, then there's gonna be trouble.

POST 1.3

Here we go again!!

You are NOT AWAYS in control of how much ice you pick up. Go fly in the Pacific Northwest and experience over an inch of ice in less than three minutes. By the time you can do anything there it is. So, turn around you say? Good-- I just flew through that stuff and now I am supposed to go back into it? I think not. Descend, you say... MEA over the mountains is 14,000. Climb, you say? Yeah, right. As I am already unable to maintain the newly mandated 120kts. that is not an option. And there was only trace forecast and reported along the route.

Pontificating about the Caravan's capabilities should probably be a thing of the past. It is down to skill and luck in the ice and the people that have a certain amount of skill and a lot of experience in the Van are usually the lucky ones. But not always.

The Caravan is a GREAT aircraft if it is kept in a safe environment. I have been flying one for 16 years in the mountains, so I guess I can speak a little about the aircraft. Also, in the last couple years I have elected not to depart more than in the previous 14 years. I'd rather be in trouble on the ground than in trouble in the air and wishing I was on the ground. Been there and done that, but not anymore. And my boss actually supports this new decision making procedure.

So, fly the 'Van safely and wisely and it will get you home every time.

POST 1.4

I have 4000 hours flying the 'van in Alaska. I have had two encounters with severe ice which the plane handled fine. Both times I exceeded one and a half inches of ice on a non-booted plane. Both times I had a forward C of G and way under gross weight. The 'van with a forward C of G has a downloaded stabiliser and it works fine. Knowing what I know now about the van changes how I fly in the ice.

The 'van with an aft C of G has an uploaded stabiliser which has airflow adhesion problems. Thus the vane guides on the stab. Any contamination on the stab when it is uploaded is dangerous and should be avoided.

Now if I am flying the van and I have an aft C of G and pick up ice I deploy some flaps to assure that I have a downloaded stab and land as soon as possible.

In order to make sense of these comments, we need a framework that will allow us to identify the different elements of performance. It is common to categorise behaviour in terms of knowledge, skills and attitudes. Knowledge can be assessed through questioning, either verbally or with a written examination. Skills can be assessed by observing a demonstration of workplace activity. Attitudes are more problematic. In simple terms, attitudes are protobehaviours: they are dispositions to act in a particular way. They are difficult to define and even harder to formally 'train' and assess. This tripartite structure of outputs seems plausible but starts to unravel if we look closer. Skills, for example, can be likened to computer run-time code in that they are the enactment of some sort of knowledge structure. If I need to have a mental representation of the action I am trying to undertake, how is a skill different from 'knowledge'? Without knowledge, I cannot enact the skill. Attitude objectives often read like 'wish lists' of desired behaviours. These labels, of course, are artefacts and the exact mechanism of storage in the brain remains a conjecture rather than an anatomical fact. Modern imaging techniques allow us to study brain activation during various recall tasks but the fact remains that what we call 'learning' is still something of a mystery in terms of storage methods.

It is more useful to consider two classes of knowledge: declarative and process. Declarative knowledge describes attributes about the world while process knowledge describes rules about action. This might sound like we are back to 'skills and knowledge' discussed earlier but this is not the case. Declarative knowledge is usually thought of as concepts and facts but, in relation to expertise, I suggest that declarative knowledge represents sets of constraints that must be satisfied in order for a particular condition to be considered 'true' (Ohlsson, 2011). For example, if asked 'what is the capital of France' I must, first, recall the names of all French cities that I have in my memory – a set of facts – and then select one that satisfies the constraint <IS A CAPITAL CITY>. Of course, 'capital city' is, itself, a construct. Lyon is a large French city known to me but it does not satisfy the 'capital city' constraint. In an aviation context, the requirement to be 'stable' at a specific point on the approach requires a pilot to have a set of constraints in memory that can be applied to the current status of the aircraft. The condition 'stable' can then be declared either true or false. If false, then process knowledge is applied to rectify the situation. Process knowledge comprises rules that guide action. By framing declarative knowledge as a constraint set, we can also reframe 'error' as a form of feedback. The performance deemed 'erroneous' typically violates a constraint, it is discrepant, and should trigger remedial action. We will come back to this when we look at error in more detail in Chapter 6.

Rasmussen (1983) proposes three categories of behaviour: skill-based, rule-based and knowledge-based, generally referred to as the SRK framework. Skill-based behaviour comprises collections of sensorimotor routines that constitute observed action. Hand flying an aircraft requires the complex arrangement of various sensorimotor inputs to the aircraft controls if it is to be conducted successfully. Skill-based behaviour becomes more proficient over time such that, eventually, it becomes almost automatic, requiring little formal conscious control. In effect, the allocation of attention to task monitoring is minimal or even subliminal. Rule-based behaviour comprises collections of <IF-THEN> statements that allow us to select between competing skill-based routines. Rules can be formally prescribed, as in procedures

TABLE 1.2
A Framework of Expertise

Knowledge Constructed solutions to new problems generated by mental models	Improvise equipment configurations	Improvise new techniques or configurations Creating alternative plans	Planned goal denied, improvise new goal
Rule Cue-directed selection between alternative routines	Select configuration in support of operating procedures	Recognising progress towards goal. Selecting alternative procedures or configurations	IF this rule THEN next action ELSE different rule
Skill Sensorimotor control of action	Operation of equipment	Visualisation of path Anticipate selections	Cue generation to corroborate future state
	System Manipulate hardware, software and people to achieve goal	***Manoeuvre*** Direct device in a specific direction. Short-term activity	***Trajectory*** Action intended to achieve future goals Long term

and checklists, or derived from experience. Knowledge-based behaviour is the process of deriving new rules to deal with novel situations. In order to be successful at this level, we need robust, representative models held in memory that can be used to run 'what if' scenarios. The SRK framework represents the actions, or outputs, that flow from the declarative/process knowledge.

In Table 1.2, I have attempted to categorise types of 'knowledge' in a framework that reflects the nature of work. Along the vertical axis, I show Rasmussen's SRK taxonomy. These categories seem to reflect mental storage in that each type of behaviour requires additional cognitive effort as measured in recall response times. Skill-based responses can be recalled from memory faster than knowledge-based responses. The horizontal axis reflects the time available for execution. 'System' reflects actions associated with the immediate configuration of the aircraft, system or device. 'Manoeuvre' reflects action in the near-term while 'trajectory' is a requirement to consider actions needed at some future stage and reflects the fact that an action now could have a consequence in future.

Returning to the Cessna Caravan pilots discussing flight in icing, Table 1.3 contains statements that reflect examples of expertise: knowledge applied to an operational context. Post 1.3 also illustrates what we mean by a possible 'attitude'. The poster comments:

> I'd rather be in trouble on the ground than in trouble in the air and wishing I was on the ground. Been there and done that, but not anymore. And my boss actually supports this new decision making procedure.

What we can see in this statement is a belief that safety is more important than production and, importantly, the poster also shows how positive reinforcement by management influences attitude development.

TABLE 1.3
Caravan Pilot 'Expertise'

Knowledge	The van with a forward c/g has a downloaded stabiliser and it works fine		[L]eave some freight behind or cancel and stay overnight at a Holiday Inn and wake up refreshed
Rule	Now if I am flying [...] I have an aft c/g and pick up ice I deploy some flaps to assure that I have a downloaded stab	I learned to t/o with 20 flaps move flaps up at 100 kts. and leave 10 flaps in until reaching cruise a/s	If a/c quit climbing ask for block alt and fly the airspeed If a/c started to descend, I would let it lose height as slow as possible on the vsi to the MEA and then declare emergency to the nearest airport
Skill	I operate the boots with about 1/4 in and continuous at 1 min intervals		
	System	Manoeuvre	Trajectory

TABLE 1.4
Distribution of Comments

	S	M	T	Total
K	8 (21%)	1 (2.6%)	9 (23%)	18 (47.3%)
R	8 (21%)	6 (16%)	4 (10.5%)	18 (47.3%)
S	2 (5.2%)	0	0	2 (5.2%)
Total	18 (47.3%)	7 (18.4%)	13 (34.2%)	38 (100%)

The Caravan is a relatively simple aircraft in terms of structure and, so, it is not surprising that there are relatively few comments relating to the skills, with just 2 (0.5% of the total) references to operating aircraft systems. In Post 1.1, we see references to the use of the propeller de-icing, leading edge boots and to the operation of the power levers. Of the 38 statements contained in the four posts, most are evenly divided between rule-based and knowledge-based behaviours (see Table 1.4). The single largest category is knowledge/trajectory. This has implications for what we mean by 'CRM competence'. Effective decision-making – a key 'nontechnical' skill – in critical situations depends upon robust knowledge structures.

It is of interest that most of the time, crews work in the four cells that comprise the lower left portion of the framework, highlighted in Table 1.3. The industry model of training is focussed largely on developing proficiency to cope with a set of aircraft 'states' described by this quadrant. Accidents, however, typically involve behaviours associated with the four cells that make up the upper right quadrant. The rule/manoeuvre cell is a transition zone between control and failure. Airline recurrent training, if it is to develop system resilience, should focus on developing the expertise need to control the set of contingencies that occupy the top-right quadrant.

1.4 DEVELOPING EXPERTISE

Airmanship, another much-disputed and ill-defined concept, is the apparently effortless 'reading' of a situation and then adapting to requirements. What we traditionally call 'airmanship' is really knowledge rendered as expertise. Knowledge is stored as scripts and schema (plural schemata). Scripts are canned sets of concepts, facts and rules that guide action in certain situations. I can have a script for eating in a fast food restaurant and another script for eating in a Michelin 3-star restaurant. Both are outlets that provide food but the behaviour expected in each and, therefore, the way of getting fed differs between them. Schemata are mental representations that allow us to control situations and, if necessary, run 'what if' scenarios in order to make choices between alternative courses of action. Some authors differentiate between schemata as structural models and 'frames' as functional models. From a training perspective, effective performance is based on the development of adequate mental representations of the world. It is a more commonplace to use the term 'mental model' in aviation to describe the combination of scripts and schemata. As this is not a psychology textbook, I will use either schemata or mental model to describe the way the brain stores information. To be effective, a mental model must have a number of features, including:

- A representation of the technical system to be controlled,
- A representation of the characteristics of systems and their interconnectivity,
- A representation of the properties of likely disturbances to systems,
- A representation of the criteria to be met in order to be successful (constraint set),
- A representation of the processes to be used in maintaining or restoring control of the system (rule set).

When we consider action in a social, collaborative context, we must add:

- Knowledge of the operational goal we are trying to achieve,
- Knowledge of the procedural actions required to achieve the goal,
- Knowledge of the limitations on possible courses of action,
- Knowledge of the other crew member's understanding of the task,
- An estimation of the other crew member's ability to deliver the required performance.

This elaboration of the mental model concept points to some potential candidate areas of competence but it also starts to break down the traditional divide between soft and hard, technical and nontechnical skills. So, I cannot construct a model of the dynamic environment without a robust understanding of the procedural framework I am working within. I cannot gauge success if I do not understand how the aircraft system will behave. To control activity, I need an understanding of the information available to me in the workspace and, in order to guarantee effective collaboration, I need to understand what information must be communicated to others. Task-related 'hard' skills are enacted in a 'soft skill' context in such a way that any attempt to separate the two becomes arbitrary and, probably, counterproductive.

Two key questions concern us: how do mental models develop and how are they used to manage tasks? To answer the first question, schemata are developed as a result of training or after repeated exposure to similar situations. The problem is that repeated exposure will only result in schemata formation if there is congruence between observed events and the stored explanation. Unfortunately, there is a catch: a schema can be incorrect despite there being congruence. We used to think that the sun went around the earth and the penalty for suggesting otherwise was severe. Apparent congruence must be tested. Schemata are modified by enlargement (adding new components) and by amendment (refining the structure) (Marshall, 1995). From an expertise perspective, enlargement means that, through experience, we develop a richer set of constraints that allow us to detect problems faster while amendment is the process by which we fine-tune our constraint set.

Cognitive competence is considered to be a measure of the completeness of an individual's knowledge about the workspace. Unfortunately, any individuals' 'knowledge' will be incomplete and no two people will hold the same knowledge base about their job. Developments in the field of embodied cognition suggest that, in fact, humans shape 'the world outside' to suit the needs of the task at any moment, partly to reduce cognitive demand. This offers further opportunities for individuals to hold different views of the world in that we activate knowledge in response to a perceived need rather than in some absolute sense.

In work-related domains, training serves to refine existing mental models and develop new task-specific models. Training is supported by company-issued documents, such as manuals. Initial and recurrent training is intended to establish an adequate set of rules and mental models that will suffice for the majority of situations. Manuals elaborate on this fundamental information but gaps remain. In a study of private annotations on paper-based aircraft manuals and checklists, Wright, Pocock and Fields (1998) found that the majority of comments were elaborations on the conduct of the procedure or explanations of the rationale behind the action – the 'how' and 'why' of doing work. Interestingly, the move to electronic checklists, e-flight bags and manuals provided via the internet has removed a source of 'knowledge' – textual annotations – from the flight deck.

Operational experience allows us to consolidate mental models through repetition, enlarge them through exposure to diverse situations and amend them through problem solving. Eventually, we will reach a stage where we can use mental models to reliably extrapolate to novel situations. This is where we start to get close to expertise. Studies of the differences between experts and novices have found that experts do not necessarily possess more 'knowledge' but, instead, it is better organised. Experts also tend to search from the problem to the solution. Problems are decomposed and analysed first before solutions are generated. Novices map solutions onto problems, they engage in pattern matching. Schemata building benefits from constant checking of the behaviour of the world against the expected behaviour, supported by effective diagnosis of any causes of divergence. This process lies at the heart of sense making and is also referred to as metacognition.

Having briefly looked at how mental models are developed, we will now move on to consider at how they are used. At the level of skill-based behaviour, we know that pattern matching with the environment triggers schema containing the actions needed for the current task. Rule-based behaviour, equally, can often be automatically triggered by cues. However, where activity is blocked, rule-based behaviour will require some conscious effort in order to select between alternative possible behavioural responses. In this case, environmental cues are compared with stored cues attached to possible responses. In the domain of knowledge-based behaviour, mental models are used to run mental simulations in order to anticipate which behavioural intervention, in the absence of a relevant rule, will most probably allow us to achieve our goal. From the earlier discussion of expert-novice difference, it seems that experts have richer mental models of the world that can be reconfigured in order to run multiple 'what if' scenarios in order to find an explanation for observed events. Novices, on the contrary, possess collections of responses and select the response that is considered most likely to succeed. We see this in ambiguous situations where crews attempt to apply whichever checklist matches the situation most closely.

In a study of the impact of the loss of experienced staff on safety in the UK rail industry, the authors noted a link between experience and safety (Wright, Turner & Antonelli, 2003). The report makes three telling points in the context of this chapter (p. 38):

> There is evidence to suggest that experience levels are associated with safety performance in the beginning of the employee's tenure. Once the employee has passed through an initial learning period of service, experience ceases to be a major factor in accident causation.
>
> [T]he [accident] rate is likely to be mediated by the types of experience the individual has been exposed to, task difficulty and how familiar they are with specific working styles.
>
> Evidence from studies reported indicates that once employees become more experienced, and older, accident rates can rise. It is likely that this is due to individuals making incorrect assumptions about the safety of present situations based on past experiences.

The report offers clues as to how mental models, shaped by experience, might affect operational performance. On the one hand, inadequate models will contribute to increased error rates among junior staff; but on the other hand, increased risk tolerance (or inadequate risk assessment) can result in errors among more experience staff. The performance of train drivers probably illustrates how skilled performance is dependent upon more than just sound cognitive models. Expertise is the ability to enact a solution within the situational constraints while being able to factor risk into plan execution.

In this section, I have explored ideas around knowledge and expertise. The framework developed here (Table 1.2) offers a route of entry into the underpinning knowledge that supports performance. By applying the framework to an airline's range of operations, more effective training curricula and checking regimes can be developed, requirements that underpin a competence-based approach. Having established the performance expected of crew, the next step is to develop appropriate training interventions. In the next section, I will look at historical developments in training analysis and review some of the current confusion that surrounds the adoption of competence-based approaches to training.

1.5 DEVELOPING TRAINING INTERVENTIONS

Formal pilot training was first codified in 1917 into what became known as the 'Gosport Model', named after the military flight school at the Royal Naval Air Station at Gosport, in the UK. Since then, broadly speaking, aviation has relied on the repetition of manoeuvres and procedures at fixed calendar intervals to validate proficiency. Hours flown and manoeuvres accomplished have long been recognised as a weak guarantee of competence, let alone expertise. Furthermore, initial pilot training courses have been progressively shortened over the years and, increasingly, simulation has been substituted for on-aircraft training. To a large degree, this was driven by economics: pilot training is expensive and there needs to be a cost-benefit cut-off for those, such as the military, who foot the bill.

Historically, the 'apprenticeship' model, or learning by watching others, could almost be described as the default approach to learning. Recent developments in neuroscience and the discovery of the mirror neurone system (MNS) suggest that we are designed by evolutionary processes to learn by copying. This model of repetitive imitation underpinned the medieval craft guild systems found in countries all over the world. Apprentices served under a master craftsman, learning the various tasks and processes involved in the craft over a period of years. 'Graduation' involved producing a piece of work that was then assessed by the guild headmen. The classic instruction to new pilots to 'follow me through on the controls' embodies this approach. Of course, guilds did more than simply train new entrants: they set prices for work, regulated competition, represented their members and provided support to those in need. As a training method, it is restricting and expensive but the fact that it has survived for so long is probably testament to its implicit logic. But, by the 20th century, a need for mass-produced trainees emerged.

Many of the business processes we now take for granted originated in wartime. During the final years of the 1914–1918 war, vocational selection tests were being used to better fit trainees to roles, air gunners and submariners being examples. Selection was seen as a way of making training more efficient by improving course pass rates. Training could also be dangerous. During the Second World War, 10% of UK RAF pilot fatalities were losses in training accidents (Terraine, 1985). By the end of the 1939–1945 war, it was realised that conscript armies could not cope with the level of complex weaponry entering service. New approaches to the design of training were needed. In the late 1950s and early 1960s, the basic building blocks of curriculum development were formulated. Blooms 'taxonomy of learning' (Bloom et al., 1956), Mager's (1961) 'training objectives' and Gagné's (1965) 'conditions of learning' all appeared at this time. The aim was to make training more cost-effective through a better analysis of requirements and organisation of inputs. Known as Instructional Systems Design (ISD) or the Systems Approach to Training (SAT) by the military, who were significant early adopters of this approach, this period is probably epitomised by the textual frame-based 'programmed learning' materials that heavily influenced the design of early computer-based training. The field of psychology at this time was predominantly behaviourist and so training was intended to deliver an observable output. What happened inside the head – cognition – was invisible and, therefore, off limits. Training objectives are still, today, phrased using 'action verbs' for this reason. It is not enough to 'know' something: that 'knowledge' has to be demonstrated in some tangible way so that we can say that learning has occurred.

The Instructional Design (ID) model, by which I mean both ISD and SAT, was applied to military aviation in the 1970s and was introduced to civil aviation in the USA through the Advanced Qualification Program (AQP) in 1990. A much watered-down version of ID was rolled out under the auspices of the then-Joint Aviation Authorities in Europe in the mid-1990s, known as the Alternative Training and Qualification Programme (ATQP). The move to the new training philosophy was not without incident. In 1998, a United Airlines B-747 departing San Francisco suffered an engine failure after take-off (ASRS, 1998). The First Officer (FO), at the controls, mishandled the event and the aircraft came close colliding with the peak of San Bruno Mountain. In March 2003, an American Airlines DC-9-82 landed on the taxiway at Seattle-Tacoma Airport (NTSB, 2004). The subsequent investigations made reference to the introduction of AQP as a possible factor in both events. In neither case could blame be laid at the door of the training design concept but questions were asked of its implementation. In order to encourage airlines to adopt the new philosophy, permission to increase the interval between recurrent training events – and, thus, reduce cost – was granted from the moment an airline undertook to start implementation rather than at some future point after the shift to the AQP philosophy had been evaluated. This 'loss-leader' approach was intended to create a budget to encourage uptake, given that airlines would have been reluctant to invest if they could continue with the old approach, still satisfy the regulator and incur no additional cost. The apparent competitive advantage that would accrue to any airline that could now meet its obligations with crews making just a single visit to the simulator each year, as opposed to at least 2 under existing rules, was what prompted the European authorities to create the ATQP programme.

Although the initial motivation to introduce the military SAT model into commercial aviation through AQP was the belief that a more rigorous approach would offer safety benefits, implementation was problematic. In the USA, the regulation applied to both initial-type conversion training (Indoctrination in the FAA terminology) and also periodic recurrent training (Continuing Qualification). Most airlines that implemented AQP focussed only on Continuing Qualification training, and it is really only the larger carriers that have adopted the concept. The JAA (now EASA) regulation was a skeleton version of AQP and, because of the division of the European regulations into Licensing and Operations, ATQP only applied to that component of training that related to approvals granted by the operator, the equivalent of the US Continuing Qualification. This includes the Operator's Proficiency Check (OPC) and the annual Line Check. Periodic License Proficiency Checks (LPCs) were still required under the Flight Crew Licensing rules but these fell outside of the scope of ATQP, which was therefore hamstrung from the outset. The supposed benefits of delivering training that was both relevant to the operation and innovative ran up against the need to still undertake periodic sets of prescribed manoeuvres.

Competency-based approaches to training emerged in the 1960s. One creation myth is that the USA was humiliated by being beaten into space by the Russians, with their Sputnik, and the blame was laid at the door of the US educational system.

Ironically, one of the first groups to be exposed to the 'competence' approach was US teachers themselves. To be considered 'competent', it was not sufficient simply to have passed a course, you now had to show that you could do the job. The competence movement had a bit of a boost in the 1980s when many of the major global economies started to undergo significant restructuring. Economies that had been based on heavy industries such as steel making, shipbuilding and coal mining were facing fierce competition from the developing world. New industries were emerging and the workforce needed to be redeployed. The challenge was to find a way to redeploy, say, a few thousand unemployed steelworkers or coal miners when what you now needed were call-centre workers or logistics hub staff. The competence movement proposed that you could generalise workplace-specific skills and knowledge into abstract clusters of 'competencies'. Vocational training Lead Bodies were created and tasked with developing competence frameworks for employees in their fields. In the UK, for example, a cabin crew vocational qualification was developed and cabin crew were able to gain a recognised accreditation for the first time. The parallel airline pilot qualification foundered because, in order to be employed as a pilot, you needed a licence issued by the UK CAA, not a vocational qualification issued by an 'industry Lead Body'. Unlike the field of ID, which is a coherent body of concepts and tools, the 'competence' movement lacked any differentiating processes. Instead, it shifted the emphasis to one of looking more at outcomes and workplace performance rather than placing importance on the design of training interventions, or inputs.

1.6 INSTRUCTIONAL DESIGN VS COMPETENCY-BASED TRAINING

Innovation in the training and accreditation of pilots represents a challenge for aviation authorities. Because of the requirement to act as guarantor of the safety of the national aviation network, authorities tend to be conservative and accreditation regimes based on hypothetical 'worst-case' scenarios (engine failure after take-off or on approach in bad weather, for example) have an apparent logic. However, as risks emerge in the operational environment, events get added to the training and testing schedule ('upset recovery', for example). These accretions result in congested training, given that airlines are reluctant to increase resources. The frustration with the 'one-size fits all' model, especially given that many of the events and manoeuvres rehearsed are rare in real life (because of the high reliability of the technology), has prompted a shift towards a competence-based approach in Europe. The incorporation of training models into supranational audit frameworks, such as the IATA Operational Safety Audit (IOSA), means that airlines have been presented with a patchwork of training models they can adopt. Unfortunately, the position is complicated by the fact that these different approaches are not simply substitutes, one for another.

The ID model and the 'competency' concept are probably better seen as overlapping rather than different approaches to developing effective training systems but, before we look at them in more detail, we need to reflect on the nature of airline training. Broadly speaking, initial pilot training takes a novice and turns them into a licence holder. Armed with their licence, new pilots have a body of knowledge about general principles and equipment and have demonstrated their ability to control a small aircraft. The next step is to apply that basic framework of understanding to a specific, usually larger and more complex, aircraft type while learning to operate in a commercial environment. Once established in the industry, pilots then consolidate their skills, broaden their abilities by learning new pieces of equipment and procedures and occasionally apply their skills to a new aircraft type. Airline pilot training, then, involves acquiring new skills and knowledge, extrapolating from that basic foundation to different aircraft types and then, once in employment, consolidating and reinforcing through practice. We should also note that pilots are required to periodically renew and validate their licences and ratings, which are their approvals to operate. The testing required by the state authority to renew a licence or rating is typically delegated to an airline or a flight school. The required internal company-specific pilot validation processes and the requirements of the state in relation to licence renewal and revalidation are often combined into single events. In Europe, for example, a LPC and an OPC can be satisfied with a single simulator profile. This duality of intent, while efficient, can sometimes blur the purpose of an activity.

From this brief discussion, it is clear that initial training for the award of a licence or a rating differs from both the maintenance of skills once in employment and the periodic checking of proficiency through testing. All three areas of interest may fall under the generic umbrella of 'training and checking' but are different processes. The methods we use to design training for new hires will differ from the activities involved in developing in-service consolidation, and it is this distinction that seems to be at the root of much of the misunderstanding that surrounds 'training' as a function in airlines. And just to complicate matters, testing or 'checking' requires yet another approach.

Central to ID is the need for a task analysis (TA) that describes the work done by a trained person. The TA is used to create a set of training objectives that describe the skills and knowledge that underpin a performance. At the end of a training course, we can test against the objectives to see if the trainee has met the output standard. So, for a specific piece of, say, navigation equipment we might want to define some underlying theory, some general characteristics of navigation equipment, the broad functioning of a specific equipment installed in the training aircraft, its principles of operation, utilisation, interactions with other equipments, normal and reversionary modes and actions in the event of failure. With that fundamental knowledge base, the trainee can then gain experience by using the piece of equipment through practical training. Because ID deals with the training inputs needed to bring about change, the TA can often be quite detailed.

A common complaint made about classical ID by advocates of the competency approach is that it is too rooted in task decomposition. It has always been recognised that task analysis methods run the risk of getting bogged down in minutiae and need a 'stop' rule. For example, one such rule would be that if a task element is so simple that most people can do it without training or is so inconsequential that failure or poor performance would have no effect on outcomes then it does not need further analysis. Again, if a piece of information does nothing to support the development of understanding or performance, then it is redundant. However, I think what lies at the root of the problem is that those that make the complaint perhaps do not recognise the difference between entry-level ab initio training and continuing professional development, which is what most airline recurrent training actually is. For an ab initio pilot we clearly need a robust understanding of the skills and the underpinning knowledge need to operate safely. As a result, the task analysis needed is quite fine-grained. For an experienced pilot, on the contrary, we can make assumptions about prior knowledge and simply identify the gap between the existing level of competence and that required to operate in any new context. We do not need a TA but, instead, we need a gap analysis. For periodic checking, I have no need for an analysis of need at all. Instead, I can fall back on company procedures as the frame of reference against which to assess competence.

Another tenet of ID is that the training delivered should reflect the entry level of the trainee. So, having graduated from basic training, our new pilot would not need to be retrained on all the fundamentals of, say, navigation equipments when they start their first job. They would simply need an explanation of the differences between what they already know and what the new systems are capable of.

Designing a 'competence framework' or model still requires an element of analysis. To demonstrate 'competence', an individual still needs an underpinning body of knowledge and a repertoire of skills. Both ID and the competence approach come back to the same fundamental starting point: a description of the job. The difference is the scope of that job or task description. While it will be quite detailed and fine-grained for initial training, for airline recurrent training it can be a description predicated on a set of declared assumptions, most of which are contained in company documents.

Another commonality between ID and 'competencies' that seems to cause some confusion is the concept of 'training to proficiency'. In the ID approach, this is interpreted as exposing the student to as many repetitions as are needed to demonstrate achievement plus some additional attempts to consolidate. By definition, faster learners will progress through the material more rapidly than slower learners. 'Competence'-based approaches require that trainees do not progress to the next stage of the course, or be signed off as competent, until they have demonstrated mastery of the current learning task. The difference between these two interpretations is that the ID approach is a rule for assigning time to achieve proficiency, a resource allocation rule, whereas the competence-based usage is really a constraint on progress: a hurdle or gate. They are the same idea but, ironically, one that is difficult to implement in aviation. Airline recurrent training is usually based around a 'crew' of two pilots who are scheduled together for an event. Allowing a free flow model of training that adapts to the pace of the learner would seem to be at odds with the need to assign a 'crew' to the training device, if one pilot is of a higher ability than the other. True training to proficiency would require scheduling to be open-ended, with implications for subsequent rostering of crew. That said, implicit in both uses of the concept is that the trainee should be expected to actually show that not only have they learnt something but that they can actually do something to a desired standard.

The ID model requires that the standard of performance expected of a competent person is defined. For an entry-level course, it is quite easy to specify the output standard. For example, I can either name the parts of a jet engine or I cannot.

But how can I set a standard for, say, 'communications' or 'decision-making'? Coupled with this is the issue of how to conduct assessment. I will cover the technical aspects of these issues in Chapter 11 but, for our purposes here, it is important to understand that both ID and the 'competence' approach have a requirement to assess and the problems inherent in judging performance are common to both. As an aside, we do have a secondary problem. In many jurisdictions, the assessment of individuals' CRM skills in a classroom setting is actually expressly forbidden in regulations. If a competence-based approach requires trainees to demonstrate mastery, then this restriction on assessment reduces our opportunities to establish actual performance and, as a consequence, the effectiveness of our training.

This brings us to one final point. The classical ID model assumes that training will be evaluated. In short, evaluation requires us to take steps to establish that the training system actually works. For novice trainees, evaluation is typically covered by course satisfaction forms ('happy sheets'), examinations and flight tests and, finally, success in gaining a licence. In the case of operational continuation training, we are really looking at how training contributes to successful performance in the real world: training transfer. The goal here is to estimate the added value that is represented by training. I was once told that the CRM programme I was running for an airline was, actually, no more than the cost of compliance: the regulator demands it so we have to deliver. The value of the course was its ability to satisfy a regulatory requirement.

This brief overview has hopefully indicated the scope of a function – training – that is crucial to safety. Quite often, an accident report will state that the crew were properly licensed and often reference is made to attendance at their most recent refresher event. Unfortunately, the content of training, the quality of its design and delivery, and the effectiveness of any measurement systems are never referred to. In short, the fact that the crew's training record is up to date is of little relevance to the quality of their performance: we know that some boxes were ticked but know nothing of the competence of the crew.

To summarise, AQP is a conventional ID model. Over time, different regulatory frameworks have incorporated technological innovations and broader industry developments. So, while ATQP cherry-picked bits of AQP, it also included a version of Line Operational Safety Audits (LOSA) and required the routine use of flight data as a form of quality control – initiatives that had come into play since the original framing of the AQP regulation. More recently, evidence-based training (EBT), and its conjoined twin CBT, has been endorsed by International Civil Aviation Organization (ICAO) and adopted by EASA. The broad framework comprises some elements of the ID model but, now, with the emphasis placed firmly on assessing outputs. The use of behavioural markers in assessment is a requirement that has blossomed since AQP/ATQP was first launched. Once again, innovation has simply been the accommodation of existing developments.

The real problem we face in airline training is that we need to develop a programme of training that is aimed at already competent pilots but the tool kit we use was really designed for training novices. A competence approach, just as with ID, cannot get away from the fact that we need some sort of representation of the job as a starting point. We will discuss the idea of resilience in more detail in Chapter 3 but, in relation to training, the resilience engineering paradigm offers a telling critique of current thinking. Woods (2006) describes two components of a 'textbook' competence model. First, there must be a representation of the variability and uncertainty faced by an operator and second, there must be an understanding of how the strategies, plans and countermeasures provided by the organisation will handle these. Competence, from this perspective, represents a model of control and our training system, therefore, needs to be based around the skills of 'control' applied to a valid model of the operational environment. This insight offers a path through the ID/competence dichotomy and the implication is quite radical. While an ID approach might be sufficient to get a novice to the stage where they can function in the workplace, the real goal is to train individuals to cope with workplace variability. What we define as 'competence' will not be sufficient if performance is brittle under conditions of uncertainty. Consequently, the assessment of competence should aim to identify how well integrated are the pilot's knowledge structures and how the pilot can apply those structures to novel situations. What follows from this is that training regimes designed to satisfy a regulatory requirement, while compliant, might not be fit for purpose. Airlines will probably need the flexibility to design training that meets the demands of their operating environment while accommodating the needs of their pilot cadre to sustain competence. The role of the regulator will change from one of verifying compliance, a task increasingly delegated to third-party audits, to one of evaluating effectiveness. Finally, airlines, themselves, will probably need more expertise in the field of training development.

1.7 UNDERSTANDING 'SUCCESS'

The current popularity of 'Safety-II' thinking is, in part, a consequence of its shift in focus from failure to look, instead, at success. Without a doubt, it is a fundamental truth that there are more successful performances delivered on a daily basis in aviation than failures. But what does 'success' actually look like? I would argue that success flows from rich constraint sets that allow for early detection of departure from the desired task status and effective rule sets that restore acceptable performance before events become anomalies. Again, the upper right quadrant in Table 1.3 is the space occupied by expert performance. This leads to a final aspect of expertise: knowledge must be abstract and generative. Knowledge is abstract when it becomes detached from specific contexts (the left hand column of Table 1.2) and is rendered as generic sets of constraints and rules. Knowledge can then be transported across contexts. Knowledge is generative when it can be used to construct new rules to control novel situations (the right hand column of Table 1.2).

To conclude this section, I want to consider where knowledge resides in networked systems. Because of differences in training and experience, and given the limitations on memory, it would be impossible for any individual to have perfect knowledge. Knowledge is distributed between individuals and between humans and technology. For example, is: a pilot will insert a flight level or altitude into the mode control panel (MCP) and the autopilot will maintain that level until instructed to do otherwise. The aircraft now has 'knowledge' of how high it is supposed to be. The pilot might momentarily forget the assigned level but a quick check of the MCP will provide the information. Human and technology share knowledge. Here is another, more complex example. In June 2013, a Boeing 747-8 cargo aircraft was flying from Dubai to Frankfurt, cruising at FL320, when the outboard aileron section on the starboard wing unlocked and motored to the fully up position (personal communication). The ailerons on the B-747-8 are operated by fly-by-wire and the outboard section locks as the aircraft accelerates through 210 kts after take-off. The aircraft's autopilot had sufficient control authority to stay engaged but the starboard inboard aileron was operating in the opposite sense to the outboard section in order to maintain that control. The increased drag caused by this configuration meant that there was now insufficient fuel to make Frankfurt and so a diversion would be required. More important, however, was that this was a configuration outside of the design scope of the aircraft: it should not have been possible for this to happen. The outboard aileron was not supposed to be able to unlock above 210 kts and, accordingly, there were no checklists or procedures to deal with the situation. Equally, the future behaviour of the control surface and any subsequent effects on aircraft handling were not known. The crew called the company Operations Centre.

The call was initially passed to the Duty Operations Manager, a B-777 pilot, who passed it on to a colleague who had led the team introducing the B-747-8 into service when he had been the B-747 Chief Pilot. He placed a call to a Boeing Vice-President, who he knew was visiting Frankfurt. At the time, it was the middle of the night in Seattle, Headquarters of the Boeing Aircraft Company, but it was breakfast time in Frankfurt. Permission was sought to wake up the B-747-8 design team in Seattle and get them working on the problem. In the Operations Centre, the Duty Engineering Manager contacted a colleague who worked for his previous employer and was based at Istanbul Airport, which was under the planned flight path to Frankfurt. He warned him that the aircraft might be diverting and requested support if needed.

In Seattle, the aircraft control systems design team convened and began brainstorming the problem. A second, Structures, team was set up as it was not known if the airframe could sustain the loads imposed on it in this unplanned configuration. There now followed an iterative cycle of possible solutions being proposed and debated by the teams in Seattle, the Management Pilot and the Engineering Manager, most of which were rejected. Finally, it was decided that, if the speed was reduced to below 210 kts, then the system might relock. In order to do that, the aircraft had to descend. The current position was over high terrain but 45 minutes further along track was the Black Sea. The flight crew were briefed on the plan and began discussing the various ways the aileron might respond and what they would have to do to maintain control. Finally, over the Black Sea, the crew descended to FL190, reduced speed and reset the system successfully.

This example is interesting because, in terms of 'knowledge', no single individual knew what the cause of the problem was and, therefore, what the answer might be. A virtual team was formed comprising people at several locations, spread across the globe. A solution was created which was not technically 'correct'. It was a guess as to what might solve the problem. It was arrived at by debate and counter argument. It was created through iterative communication cycles. In this case, 'knowledge' was actually an approximation, not a truth, and it was created in the space between individuals, not inside the head of any one person. In some ways, this is the epitome of what we call CRM.

1.8 CONCLUSION

This first chapter has set the scene for a subsequent discussion of what 'competence' might mean in the context of CRM. I have suggested that CRM has been poorly defined since its introduction. It has grown by accretion as new topics have been added but, at no stage, has the industry taken stock of what the concept really means. If we are to make progress, we need a valid description of performance that is sufficient to cope with operational demands. Effective knowledge structures are created through training and reflection on experience. The mechanics of developing interventions to facilitate these processes is the province of training design. The current approach to CRM, specifically, and training, generally, while satisfying compliance requirements, is probably failing to fully prepare crew to cope with the demands of a challenging environment.

The key point that emerges from the discussion of expertise is that a successful performance is, first, directed towards an operational goal and, second, requires the integration of different categories of knowledge. An implication of this is that what we cannot separate CRM from traditional 'technical skills'. The two categories of performance fit hand in glove, the one unable to work without the other. To maintain a separation probably undermines any meaningful approach to future training design. Fundamentally, performance in the workplace is directed at maintaining control and it is this idea of 'control' in a dynamic system that underpins the rest of this book.

Effective training underpins resilient system performance. Before moving on to looking at aviation as a system, I want to offer one last thought. All learning requires reflection. We compare what was once known with what we now know. As one of the Caravan pilots says: 'Knowing what I know now about the 'van changes how I fly in the ice'. In CRM, we learn by reflecting on how our own performance supported or inhibited the achieving of the operational goal. Self-awareness is probably the most important 'competence' we need to develop in CRM.

REFERENCES

Aviation Safety Reporting System (ASRS). (1998). ACN 406810.

Bloom, B. S., Engelhart, M. D., Furst, E. J., Hill, W. H., & Krathwohl, D. R. (1956). *Taxonomy of Educational Objectives: The Classification of Educational Goals. Handbook I: Cognitive Domain.* New York: David McKay Company.

Diehl, A. (2011). Human performance training: Successes and failures in civil aviation. In *Proceedings of the 16th Aviation Psychology International Symposium.* Dayton, OH.

European Aviation Safety Agency. (2015). Annex II to ED Decision 2015/022/R.

Gagné, R. M. (1965). *Conditions of Learning*. Holt, Rinehart and Winston. New York.

Helmreich, R.L., Klinect, J.R., & Wilhelm, J.A. (1999a). Models of threat, error, and CRM in flight operations. In *Proceedings of the Tenth International Symposium on Aviation Psychology* (pp. 677–682). Columbus, OH: The Ohio State University.

Helmreich, R.L., Merritt, A.C., & Wilhelm, J.A. (1999b). The evolution of crew resource management training in commercial aviation. *International Journal of Aviation Psychology*, 9(1), 19–32.

Klinect, J.R. (2005). Line Operations Safety Audit: A Cockpit Observation Methodology for Monitoring Commercial Airline Safety Performance. Unpublished PhD thesis. University of Texas. Austin, TX.

Mager, R.F. (1961). *Preparing Objectives for Programmed Instruction*. San Francisco, CA: Fearon.

Marshall, S.P. (1995). *Schemas in Problem Solving*. Cambridge University Press: Cambridge.

Merritt, A. & Klinect, J.R. (2006). Defensive Flying for Pilots: An Introduction to Threat and Error Management https://flightsafety.org/files/tem_dspt_12-6-06.pdf

NASA. (1980). Resource management on the flight deck. In *Proceedings of a NASA/Industry Workshop June 26–28, 1979*. Conference Publication 2120. Ames Research Center Moffett Field, California.

NTSB. (1979). Aircraft Accident Report No: AAR-79-7 United Airlines, INC. Mcdonnell-Douglas, DC-8-61, N8082U Portland, Oregon. 28 December 1978.

NTSB. (2004). Incident Report No: SEA03IA046. Accidental landing on taxiway, Douglas DC-9-82, March 14, 2003.

Ohlsson, S. (2011). *Deep Learning: How the Mind Over-rides Experience*. Cambridge.

Project Tenerife. http://www.project-tenerife.com/engels/ accessed 7 August 2020

Rasmussen J. (1983) Skills, rules, and knowledge; signals, signs and symbols, and other distinctions in human performance models. *IEEE Transactions on Systems, Man, and Cybernetics* 13(3): 257–66

Terraine, J. (1985). *The Right of the Line: The Royal Air Force in the European War 1939–1945*. London: Hodder & Stoughton.

Woods, D. (2006). Essential characteristics of resilience. In Hollnagel, E., Woods, D., & Leveson, N. (Eds.), *Resilience Engineering: Concepts and Precepts*, pp. 21–33. Aldershot, UK: Ashgate.

Wright, P.C., Pocock, S., & Fields, R.E. (1998). The prescription and practice of work on the flight deck. In *Proceedings of the Ninth European Conference on Cognitive Ergonomics*. Limerick.

Wright, M., Turner, D., & Antonelli, A. (2003). The impact of experience erosion on safety. Final Report for Rail Safety & Standards Board. Report No: 20-T153-GSBr-55-TRT5-I1. 9 December 2003.

2 Thinking about Failure

2.1 INTRODUCTION

Aviation has a long tradition of investigating failure and taking remedial action where necessary. If crew resource management (CRM) is to support safety initiatives, then it would be useful to better understand how safety margins are eroded and what behaviour-based countermeasures can be developed. In this chapter, I want to look at the mainstream historical approaches to safety investigation, and in the next chapter, I will outline a new framework of analysis. I start with an overview of some safety concepts, and then, I will briefly look at a number of broad approaches to understanding accident causation to see how they throw light on why things go wrong. The first thing we will discover is that it is difficult to be specific about causation, about why something happened the way it did. Why is it that the same set of circumstances on one day will have no observable effect and yet, on another day, will result in an adverse event? It is clear that consequences shape our view of specific events. We are more likely to think negatively about the performance of the actors if we know that the outcome was a failure of some sort. Hollnagel (2009) observes, though, that success and failure can both flow from the same performance, the implication being that we need a change of perspective. By default, investigation tends to look at what the crew failed to do or did wrong. What we often overlook is that the crews were actively trying to solve problems as things unravelled around them. We need ways to explore why performance was insufficiently robust such that safety was put in jeopardy. The following two examples illustrate the, often banal, nature of failure.

On 16 December 2004, a Shorts 360 was substantially damaged, and the two crews were seriously injured when the aircraft failed to fly after an attempted go-around (TSB, 2005a). The crew was carrying motor vehicle parts from Toledo, Ohio to Oshawa, Ontario. The runway at Oshawa is 4000ft long. Because of some inaccuracies in flying the approach, the aircraft touched down a third of the way along the runway. The Shorts 360, given the aircraft weight at the time of landing, required at least 4100ft in which to stop. However, the runway was completely covered with wet, slushy snow. Under these conditions, the aircraft stopping distance was at least 7400ft. Notwithstanding aircraft handling and performance issues, the crew should never have attempted to land at Oshawa in the first place. All of the landing constraints were known before take-off. When it was clear that the aircraft was not going to stop, the crew attempted to get airborne again, striking the trees just outside the airfield boundary.

On 10 September 1999, a Beech 1900 landed at Williamstown, New South Wales, Australia (ATSB, 2001). As the crew taxied the aircraft towards the terminal, they were presented with indications of an electrical failure and

> also a low fuel pressure warning. The electrical problem was dealt with using the checklist but, because they were close to the terminal, they took no action over the fuel warning. As they turned onto their parking position, the First Officer (FO) saw flames coming from the underside of the right-hand engine nacelle. Both engines were shut down, and the fire bottles discharged but to no effect. Ground personnel dealt with the fire using hand-held extinguishers. Subsequent investigation revealed that, at some unknown time, maintenance work involving the removal of cable looms in the wing had been undertaken. When the looms were replaced, cable ties were used to attach the wiring to the aircraft structure. However, special collars normally used to prevent the looms from chafing against the metal structure had been left out. Over time, the landing light cable had been rubbing against the fuel pipe in the area behind the right-hand engine. The cable insulation had finally worn sufficiently to allow electrical arcing between the wires and the fuel pipe, causing the pipe to burn through and ignite the fuel. The electrical failure indication on the flight deck was caused by the short-circuit, and the low fuel pressure warning was the result of the holed pipe. Ironically, the crews' attempts to use the fire bottles were frustrated because the bottles were electrically operated, and the power supply had been destroyed by fire. However, the bottles work in a protected area around the engine, and the fuel pipe runs outside of that area; the fire bottles, had they been working, would have done nothing to control the fire.

These two examples clearly illustrate conditions that, with hindsight, we would call 'unsafe'. The Shorts 360 outcome was clearly both predictable and preventable based on the information available at the time to both the dispatchers and the crew. The Williamstown event is more complex, involving both third parties (the maintenance event) and decisions about aircraft design (the functioning and the operation of the fire extinguishing system). If CRM is really about sustaining safety or, more accurately, mitigating risk, then we ought to be able to identify actions that would be required, or even expected, of the flight crew to prevent an adverse outcome. Based on the discussion in Chapter 1, we should be able to identify deficiencies in crew knowledge structures or their execution of skill-, rule- and knowledge-based behaviours in order to maintain control.

We might argue that the crew of the Shorts 360 should have demonstrated a professional 'attitude' towards pre-flight planning. After all, they should never have attempted to land at an airfield where the runway was known in advance to be too short. The risk of a runway overrun was present before they even took off. Unfortunately, such a critique of the crew is one based on hindsight. Because we know the outcome, it is easy to be critical of their actions. Nonetheless, there are questions I would have liked to have asked of both the dispatcher and the captain, especially, relating to planning and operational management that were not answered by the investigation report.

The crew of the Beech 1900, on the other hand, could not possibly have known what was wrong with their aircraft. The fire could have occurred at any point (and on almost any flight) whenever the landing light was turned on, and it is only by huge good fortune that the consequence of the failure could be readily dealt with. Ironically, by choosing not to act on the 'low fuel pressure' checklist, the crew inadvertently mitigated their problem. One of the actions on the checklist was to turn the fuel boost pumps on, which would have fed fuel to the fire at a higher rate. Part of the problem we face in making sense of events is that processes and technologies combine in ways we fail to understand or anticipate. These two relatively simple examples illustrate the range of factors that shape outcomes. They show how humans do work in complex systems that include technology, infrastructure and work undertaken by others removed in space and time. The fact that failure is usually the observable end point of sequences of events involving many factors is an unfortunate cliche in aviation. From the Oshawa report, it is not clear how often aircraft had been dispatched to an unsuitable runway but with no consequence. Again, it was not clear how long the unsafe condition was present on the Beech 1900 or if any subsequent maintenance activity had simply overlooked the missing spacers. These examples suggest that we are often unaware of the fragile state of normal operations.

2.2 WHAT IS SAFETY?

Lord Cullen, reporting on the Piper Alpha disaster in the North Sea oil field (Cullen, 1990), recommended that all safety-critical industries declare safety to be their first priority. Since that time, organisations have regularly claimed that 'safety is the cornerstone of our business' or 'safety is our number one priority'. It is now a mantra, but few managers can explain what it means in practice, tending to fall back on the absence of accidents as a measure of safety. For most people, safety is an abstract concept. If we want to bring about both qualitative and quantitative change, then we need to find more tangible expressions of safety enactment (MacLeod, 2005). One generally accepted definition of safety is the 'freedom from unnecessary risk' (Blockley, 1992). Three aspects of safety arise from this definition: first, safety is not a property in its own right but, instead, is the absence of some other factor, risk; second, risk is recognised as ever-present; third, safety is characterised by our exposure to risks deemed additional to those considered 'acceptable'. I will explain these terms in the next section, but I want to briefly discuss another, which is 'reliability'. All organisations depend upon predictable, reliable processes but, as Leveson (2011) makes clear, reliability and safety are not the same thing. A process can be considered reliable in that it delivers a consistent outcome, but the manner in which the process is undertaken can be fundamentally unsafe. This is another reason why the absence of something tells us nothing about the safety status of an operation.

2.3 THE BUILDING BLOCKS OF SAFETY THINKING

In the Bible, specifically in the Gospels, you can read about two Roman soldiers sitting at the foot of the cross rolling dice to see who gets to keep Jesus' clothes. The soldiers were playing what later became known as the game of chance. The game involved rolling dice and betting on which numbers came up. The possible combination of numbers on the face-up dice was called the 'risk'. In order to make the game more fun, certain combinations of numbers were deemed of no value, and the name of the game was changed to 'hazard'. Chance, risk and hazard are terms in common usage today but with very different meanings. Chance refers to a random, unavoidable event. A hazard is a source of a problem. Risk, meanwhile, has grown in complexity as a concept. It retains its earlier association with a probabilistic distribution of outcomes, but in a safety context, those outcomes are more firmly linked to disadvantage or harm. In fact, safety management processes now define risk as a combination of the probability of an event and the severity of any outcome. Hazard and risk warrant closer examination.

A hazard is, quite simply, a potential source of a problem. In golf, a sand trap or bunker is a hazard: hit your ball into a bunker and you need to be a pretty good player not to drop a shot. Objects are hazardous if they have the potential to do harm or frustrate us in our pursuit of a specific goal. However, entities only become hazardous when they come into juxtaposition with our course of action, that is, once they interfere with our route towards our intended goal. If my golf ball lands beside the bunker, rather than in it, then the bunker is no longer a hazard (unless I am a really bad golfer).

The following event shows the range of hazard types that can exist in the workplace. The cabin crew on an aircraft reported a sticking toilet door in business class and requested maintenance support when they landed. A crew change was planned during the turnaround, and as the new crew prepared the aircraft, a man, dressed in a business suit, turned up asking for information about the toilet door. One of the new crew, only vaguely aware of the reported fault, directed him to the problem toilet and the man then used a lubricant spray to get the door moving freely. Shortly after take-off, several passengers complained about strong fumes at the front of the cabin. One cabin attendant and two passengers collapsed with respiratory problems, and the aircraft had to divert. It was subsequently found that the man who fixed the door was an office worker employed by the company that provided engineering support to the airline. He had taken the message, but there were no maintenance technicians available at that time, so he decided to see what he could do. He found a can of what he thought was a suitable lubricant in a locker but did not notice the health warning on the label. The spray was heavy-duty oil for use on external aircraft hinged surfaces. It was highly toxic and should never be used in confined spaces and, even then, breathing equipment must be worn. Spraying the toilet door hinges was guaranteed to cause problems (source: ASRS).

We can identify three archetypical hazard states in this story (Cox & Cox, 1996). First, the lubricant was a substance that could cause harm to humans who are exposed to it. The first class of hazard, then, is objects or substances (think of the sand trap or a mis-directed golf ball landing in a crowd of spectators). However, the lubricant was formulated for a specific purpose. When used under the correct conditions – in the outdoors by a trained engineer equipped with breathing apparatus – the hazard is controlled. The second class of hazards, then, are situations where hazards are a part of the designed process but are properly managed. Flight at altitude in a pressurised aircraft is a hazardous condition. A failure to maintain pressurisation will expose passengers to hypoxia. The risk is managed through the design of the aircraft's pressurisation system and also by providing emergency oxygen masks in the event of a loss of cabin pressure. Finally, an untrained person, simply trying to help, used the spray in circumstances for which it was never intended. Our actions can be hazardous. This third category is the domain of human error and will be discussed in a later chapter. Hazards, then, exist as entities, contexts and actions.

Moving on to 'risk', although the term simply describes a possible distribution of outcomes, in safety management it has become attached to a negative outcome: risk has 'consequence'. Going back to the sand trap on a golf course, as I drive off from the tee, the 'risk' is a function of my ability to control the trajectory of the ball. There are any number of places the ball could come to rest, and not all of them are bad. As I get better at playing golf, the 'negative' risk – landing in the sand trap, in the rough or out of bounds – should reduce. Conversely, the 'positive' risk – landing slap in the middle of the fairway or even a hole-in-one – should increase. There are several problems with this formulation of risk. First, safety management systems use tools based on some understanding of frequencies of events. In engineering terms, it is possible to record the mean time between a specific component failure. That information could then be used to construct preventive maintenance schedules that offer protection against component failures during an operational flight. However, in the case of human performance, such data are probably unknowable. Furthermore, in complex systems with multiple agents, it is impossible to quantify probabilities in any meaningful sense. Failures involving two or more components require us to be able to calculate combinatorial probabilities, something we are not very good at. Probabilistic approaches to safety are probably irrelevant in the context of CRM.

Humans tend to deal with risk in a subjective sense based on a general understanding of an issue and any previous exposure to events. As a result, we construct lay understandings of the distribution of risk (Tversky & Kahneman, 1974). We saw in the previous chapter that risk assessment by railway workers varied with age and experience. The severity of an event will also influence the way we construct a 'value' of the risk attached to the event. In effect, we tend to think that 'nasty stuff' is more commonplace than 'not so nasty stuff' when, in fact, the opposite is usually true. Table 2.1 shows the responses of 100 airline pilots to questions about the importance of workplace hazards. The data offer indirect evidence of the distribution of crews'

TABLE 2.1
Pilot Responses to Hazard Questionnaire (*n* = 100. Scale 0 = low; 8 = high)

Condition or Event	Average	Range
Failure to correctly follow procedures	6.8	6–8
Inadequate manuals	5.2	4–7
Handling aircraft outside published standards	6.9	4–8
Regulations open to interpretation	7.1	6–8
Inadequate training	6.6	4–8
Failure to regularly inspect airport ground facilities	5.4	3–6
Failure to correctly inspect airframe	7	5–8
Failure to correctly maintain aircraft powerplant	6	4–8
Incorrect aircraft loading	7	6–8

views on risk. The range of scores reflects, in part, individual beliefs about the significance of each hazard. For example, airport inspections are of less interest to pilots than airframe inspections. That said, these pilots worked for the UK operator and flew mainly within Europe or to the USA. Pilots operating in, say, Indonesia or the Philippines might have a very different view. The score will also be influenced by the individual pilot's previous encounters with the specific hazard. The item 'handling aircraft outside of published standards' could reflect what pilots have seen others do in the past. Risk, then, seems to be something that is socially constructed. It is an abstract entity. Unfortunately, it is a fundamental to aviation.

That individual crew members perceive risk differently is illustrated by this crew of a Boeing 757 (source: ASRS). Company procedures required the aircraft to be established in the correct landing configuration by 1000 ft on the final approach. However, delays in handover between air traffic controllers resulted in the aircraft being about 4000 ft above the profile at the start of the descent. The captain, who was flying the aircraft, began the descent with the power at idle and with the speed brakes out. The landing gear was lowered, and the descent was flown at 250 kt kts with the gear and speed brakes extended. As the aircraft slowed for the final intercept, the captain called for flaps as the aircraft progressively slowed through the various flap extension limiting speeds. The flight continued with the speed brakes out, eventually slowing to the point at which landing flaps could be selected, which was achieved at about 15 kt above the target speed for the approach. The landing was normal and the rollout, on the airport's longest runway, did not call for any unusual braking. It was only after the touchdown that the captain realised that the speed brakes had been out the whole time. After they parked the aircraft and shut the engine down, the captain asked the FO for his input on the events that had taken place. The FO commented that he knew what the captain was doing and did not have a concern about the outcome because of the long runway. The captain suggested that a go-around was probably a better option. The captain later said that he subconsciously calculated that the speed would not be a factor because of the long runway but had, perhaps, not accounted for the 30-kt tailwind during the descent below 5000 ft agl. In this incident, two observers participated in the same event and but had different perspectives on the risks associated with their course of action. What is interesting is that the captain appears to have had an expectation of how the FO was viewing the progress of the flight. He was anticipating a call to 'go-around', which never came. This example also shows how other external factors – the unexpected wind and the performance of air traffic control (ATC) – shape events. I suggested earlier that probabilistic models of risk start to unravel in the dynamic world of live operations. Risk is probably better seen in terms of a set of influencing factors or attributes of the situation that can shape outcomes.

Hazards and risks, then, are constants in aviation. If we now think about competence, safety flows from our ability to complete tasks within set performance criteria. If the training system has been designed effectively, our performance should allow us to manage the risks associated with any exposure to hazards. Unfortunately, in order to accommodate the variability encountered in the workplace, performance has a degree of discretion – workers can elect to act in different ways to achieve the same goal. Performance has two components: first, the extent to which a workplace solution will satisfy performance goals and, second, the ability of the worker to deliver the performance as required by the task situation. Safety is a function of tools, tasks and operators coming together to achieve goals. Inadequacies in any of these components will increase the risk of task failure, possibly resulting in loss and harm. Historically, there has been a tendency to treat safety as a structural component open to quantification and management. There is a contradiction here in that, on the one hand we are treating safety as something that can be engineered into an organisation and, at the same time, we are suggesting that safety is an artefact of other activities. Aviation involves organisations and people working in collaboration to meet production targets. Those various actors are kept in a dynamic but stable equilibrium. Systems become unsafe when the control being exercised over the component parts fails or is degraded in some way. As a result, margins of performance are eroded, and the probability of exceeding a limitation is increased.

Safety, then, is not a condition in which an organisation exists or some property that can be assessed but, rather, it is a reflection of the competence of workers in that system. It flows from the degree to which workers successfully negotiate solutions to workplace problems. In the rest of this chapter, I want to examine a specific case study from the perspective of different approaches to accident causation. I want to see how different models can elaborate on the types of behaviour that support competent performance in hazardous environments.

2.4 THE PELEE ISLAND CRASH

On 17 January 2004, a Cessna 208 Caravan was destroyed when it crashed after take-off from Pelee Island Airport, Ontario, Canada (TSB, 2005b). On board were the pilot, his girlfriend, eight men and two dogs. All were killed. The aircraft had landed after a short 15-minute flight from Windsor, Ontario. It had picked up some icing in flight, and people on the ground commented on this to the pilot. After a short turnaround, the aircraft made ready to depart. By now, freezing rain was falling. The girlfriend was not ticketed, and no passenger manifest was produced. Contrary to regulations, the dogs were not carried in cages but roamed freely on the floor of the aircraft. The passengers, a hunting party, were not weighed before boarding. The aircraft was considerably overweight as well as carrying a burden of ice from the in-bound flight. The icing conditions prevailing when the pilot took off simply aggravated the problem. Observers noticed that the aircraft took most of the runway to get airborne. Shortly after take-off, the noise of a crash was heard.

This accident is of interest, in the context of this chapter, as much because of the circumstances surrounding the flight as of the actions of the pilot. Therefore, I need to explain the context of the flight that day. Pelee Island is the southernmost community in Ontario, Canada. Located at the western end of Lake Erie, the island is connected by ferry services to both Canada and the United States. The Owen Sound Transportation Company (OSTC) provided the main ferry service to the island. Once privately owned, OSTC was now a subsidiary of the Province of Ontario government. In winter, once Lake Erie freezes, communication links were maintained by a charter air service provided by Georgian Express.

Transport Canada's (TC) Commercial and Business Aviation Division (CBAD) is responsible for the oversight of commercial air operations that fall within its remit. When an operator applies for an approval to conduct a service, the CBAD is required to assess the safety and viability of the application. The functions of the CBAD are divided between certification – which deals with new applications and changes to existing operations – and operations, which deal with the ongoing oversight of air operator certificate (AOC) holders, of which Georgian Express was one. In order to fulfil this obligation, the operations subdivision appointed a principal operations inspector (POI) to work with Georgian Express.

The OSTC notified the CBAD of its intention to start the winter air service and was told, in a letter dated 11 December 2003, copied to the certification subdivision, that the request must come from the AOC holder, Georgian Express. As the OSTC did not hold an AOC in its own right, the operations subdivision of the CBAD was not copied in on the correspondence. The air link was conducted as though it were a scheduled service but under a subsidy arrangement and, as such, constituted a charter operation. Therefore, it did not require Georgian Express to apply for an amendment to its AOC. The department that had daily oversight of the company, then, was unaware of the plan.

The TC POI had a responsibility to conduct regular inspections and audits of air operators. A scheduled audit of Georgian Express had been started on 11 September 2001 and immediately suspended given the tragic events of that day. The audit was rescheduled for September 2004, based on the POI's ongoing assessment of the nature of Georgian Express' operations. However, special purpose inspections can be triggered by significant changes to a company's operations. The commencement of the Pelee flights would be just such a significant change but, as we have just seen, the operations subdivision had not been included in any communications, and the POI did not know about the new Pelee Island route.

Georgian Express flights were operated as pilot self-dispatch, meaning that operational control was delegated to the pilot in command. It was the company's responsibility to co-ordinate the services required for the operation, such as fuel, de-icing and passenger handling, and to make appropriate arrangements. The company's operations manager travelled on the first flight of the new service, on 13 December 2003, to check arrangements. As no de-icing was available at Pelee Island Airport, the company provided small, two-gallon containers of de-icing fluid, with an attached spray, to be carried on the aircraft for use at remote locations. According to procedures, a company flight plan was supposed to be prepared for each flight. The flight plan was used to record details of load, fuel, passengers and aircraft weight and balance. Passenger weights were calculated using a set of industry-standard figures. A copy of the completed flight plan was supposed to be left with the passenger agent before departure. No flight plan was found for the accident flight, nor were any de-icing fluid containers found in the wreckage.

The 208 Caravan is manufactured by the Cessna Aircraft Company. The design of an aircraft must satisfy international standards of airworthiness before it can be awarded a certificate of Release to Service. The characteristics of the aircraft under different conditions of flight are tested, and engineers are required to provide solutions to a variety of safety contingencies. The aircraft cannot be used until it receives a certificate from the airworthiness division of the appropriate authority, in this case, the US Federal Aviation Authority (FAA) as the state of manufacture. The effect of ice on aircraft performance is well known, as we saw in Chapter 1, and the regulatory authorities and manufacturers publish specific regulations, general guidance and seasonal reminders about flying in icing conditions. There are specific regulations that cover the removal of ice, particularly on the wings of aircraft, before take-off and take-off in conditions of freezing precipitation. The Cessna 208 Caravan is also fitted with devices to prevent the formation of ice on exposed surfaces critical to flight and to remove ice once it has started to form.

The weather conditions prevailing en route to and at Pelee Island were conducive to the formation of ice. The accretion of ice on an aircraft is an unfortunate application of the laws of physics. Therefore, under certain conditions, there is a known risk of ice formation. Ice is a hazard in that it adds to the weight of an aircraft and can affect performance. It can also disrupt the aerodynamic properties of the aircraft. Exposure to icing is a hazardous condition, but the risk is mitigated unless the rate and extent of ice formation exceed the capability of the aircraft's protection systems. The pilot of the Caravan at Pelee Island was aware of the icing conditions on that day, and he had needed to de-ice the aircraft on an earlier flight. On the outbound trip, the formation of ice did not seem to have impaired the aircraft's performance. Possibly, the pilot thought that the absence of any adverse effects would similarly apply to the return trip. His immediate experience may have shaped his construction of risk. But that is all conjecture. What we do know is that the combination of an overweight aircraft and the degraded aerodynamic properties of the ice-covered wing reduced the performance margins available to the pilot, thus rendering the aircraft unsafe. The crash of the aircraft in January 2004 resulted in the tragic loss of all on board. This brief synopsis of events, though, points towards a complex scenario being played out on that day.

2.5 LINEAR MODELS OF ACCIDENT CAUSATION

Lawsuits totalling more than US$63 million were brought as a result of the Pelee Island crash. In total, 11 defendants were cited, including the pilot's estate, the airline, the airport, Cessna as the aircraft manufacturer and the shipping company that chartered the aircraft. Of course, legal processes seek to attribute blame. Some person or body must be culpable. The primary concern of safety investigation and event analysis, though, must always be to generate understanding. However, the fact that outcomes usually flow from actions and that 'someone' must be called to account is the starting point for our discussion of safety simply because it seems to be the most natural way humans deal with failure. Models of accident causation that link cause and effect have a long tradition in accident investigations. Their popularity lies, in part, because of an apparently intuitive need for humans to want to create stories to make sense of the world.

Linear models – the 'accident chain' metaphor – try to establish a causal relationship between actions and outcomes. They represent the earliest organised attempts to understand failure. By working backwards from the adverse outcome, it is hoped that the factors that contributed to the accident will be identified. Remedial measures

FIGURE 2.1 Linear model of Pelee Island.

can then be put in place to reduce the probability of a repeat performance. Various classes of linear models have been proposed, including the cluster of activities under the umbrella of root cause analysis and a family of approaches derived from engineering that can be represented by failure mode analysis. Linear models attempt to show relationships and influences. They have also been used to generate taxonomies of error. By cataloguing and categorising error, it is hoped that future errors can be better identified, trapped or mitigated.

Figure 2.1 attempts to apply a linear model to a subset of the events of Pelee Island. The first thing to notice is that no single line can be traced through the accident: several chains of events converge. The model is not meant to be exhaustive and it only deals with two aspects: the way in which the aircraft loading was managed, and also the way in which the pilot dealt with the problem of ice on the aircraft. The figure shows how some features of the event had multiple effects. For example, the ice on the wings both degraded performance through reducing lift and also added weight to an already over-loaded aircraft. Some of the boxes contain statements of fact while others refer to actions or, rather, inactions on the part of the pilot. The representation does not capture the motive of individual actors. For example, two people on the ground mentioned the ice on the wings to the pilot, but he took no action – why not? The portable de-icing spray was not carried on this flight – why not? After all, the weather reports suggested that ice would be encountered, and the pilot had already had to de-ice the aircraft on a previous flight that day. He elected to carry an additional 1000 lbs of fuel, which added to the aircraft's total weight on take-off and aggravated the problem. This was a quantity far in excess of anything he would have required. Why? In order to understand how a situation becomes unsafe, we need to understand the motives of individuals and why actions made sense at the time. Of course, establishing motive will always be problematic, if not impossible, after a fatal accident. The arrows on the diagram indicate the notional direction of causality but, of course, there is no clear cause and effect relationship at work here except in the case of aircraft performance. Cause and effect are often judgements made after the event.

Linear models allow us to bring some order to the chaos and confusion that often surrounds an accident. Actions and outcomes are arranged in a temporal and spatial relationship. More complex varieties of analytical model suffer in that they identify causal factors from the perspective of the designers of the model and, so, are at risk of offering a skewed analysis. In effect, the analysis confirms the agenda of the investigator rather than establishing an objective version of events. For example, the Pelee schedule was designed as two sectors, out and back, with the appropriate planning and documentation completed for each sector. Pretty quickly, however, it seems that the pattern became a single sector with an intermediate stop, which possibly explains why the pilot did not complete paperwork for the return flight. From the company's perspective (and that of an outside investigator), the pilot failed to follow procedures, but from a line pilot's perspective, what we are seeing is how processes get adapted to make the job simpler.

Linear models are weak in terms of establishing the cognitive aspects of the performance. Although such approaches allow us to identify probable causal factors, they usually lack the specificity required to allow us to implement meaningful preventive measures. This point is important as the common response to a failure is to increase the level of oversight, change a procedure or introduce additional controls. This 'barrier' approach, as Hollnagel (Hollnagel et al 2006) points out, doesn't prevent outcomes, rather it changes behaviour: we find ways around the barrier. One final criticism of linear models is that the different methods will often result in different results when applied to the same event, and different analysts will come to different conclusions when using the same technique. They lack any internal consistency simply because they rely so much on individual judgement. The concept of an 'error chain' has been enshrined in safety folklore, but the metaphor, despite attempts to develop tools and techniques to bring structure to the analytical process, does not offer a full account of how unsafe situations develop (for a fuller description of the problems, see Dekker 2002). Perhaps the biggest fallacy of the linear model is the thought that 'breaking the chain' will save the day.

Linear models are not really helpful in identifying how competence might affect safety. You might argue that if everyone involved had just done their jobs properly, this event might never have happened. Alas, that cannot be said for certain. I once listened to Captain Dan Maurino, the then Coordinator, Human Factors and Flight Safety at the International Civil Aviation Organisation, talking about the fatal 1990 Avianca B-707 crash at Cove Neck, New York (NTSB, 1991). He said that he wished he could have spoken to the captain of the aircraft before it left Bogota and prevented him from getting airborne, but he knew that, such was the general condition of the airline, if this crew were saved then another would probably crash. Dan was reflecting on the fact that competence is demonstrated across an organisation, not just on the flight deck. With hindsight, we could say that the pilot of the Pelee Island Caravan should have weighed his passengers, de-iced before take-off and done the paperwork properly. We can also say that he knew all of this but, for some reason, chose not to do so on the day. The Avianca captain had poor English language skills and needed assistance in understanding the ATC controller's instructions. His aircraft's autopilot was not serviceable, and so, he was required to hand fly for the entire route, a physically and mentally demanding task. Both pilots were employed by airlines that, it could be argued, also bore responsibility for subsequent events. We need to look more broadly at an organisation's range of functions to identify the types of competence that support safe outcomes.

2.6 NORMAL ACCIDENT THEORY (NAT) AND HIGH RELIABILITY ORGANISATIONS

The concept of the 'normal accident' was developed in the 1980s, largely in response to concerns about safety in the nuclear power industry (Perrow, 1999). The complexity of power plants meant that the technology involved could sometimes behave in unpredicted ways. The main contention of NAT is that accidents in highly complex systems should be expected. On the other hand, high reliability organisations (HRO) theorists have tried to identify why critical industries and work contexts – like aircraft carrier deck operations – have remarkably fewer accidents than we might expect (Roberts, 1990). Perrow suggested that the interactive complexity of systems makes the task of the human operator more difficult when components fail. Tight coupling between subsystems makes them highly interdependent, and because of this, malfunctions will quickly propagate in unpredictable ways.

This idea of unpredictability is echoed in Reason's (1990) latent conditions and resident pathogen concepts. Reason was proposing that certain aspects of the design of a system or process are fallible although their effects may never be discovered until a particular, probably unique, set of conditions applies. Thus, potential sources of failure can exist undiscovered in a system for years. For example, on 31 December 1991, an MD-82 took off from Stockholm Arlanda airport bound for Warsaw (SHK, 1993). The aircraft had been standing overnight, and super-cooled fuel in the wing tanks, resting in contact with the upper surface of the wing, had resulted in ice formation. A ground technician failed to notice the condition of the wing and, when the aircraft took off, the ice broke away and went into the tail-mounted engines, damaging some of the compressor blades. This resulted in both engines starting to surge, disrupting the flow of air through the engines. The normal response to such a condition is to reduce power to restore a normal airflow. However, the aircraft had been fitted with a safety device called automatic thrust restoration (ATR). This system was originally intended for reduced thrust take-offs required under airport noise abatement procedures. In such a situation, should one engine be lost after take-off, the power on the other engine was automatically increased to the maximum. The airline in question, however, never used reduced thrust take-offs and, over time, the ATR system had dropped off the training programme. Although two disconnect buttons were located on the power levers, pilots were not told what they were for. As the pilots tried to reduce power to control the surging, the ATR automatically moved the thrust levers forward. The very system that was designed for safety in the event of an engine failure after take-off probably resulted in the destruction of both engines. During the subsequent accident investigation, the FAA, who approved the modification in the first place, said that they never envisaged a situation where both engines would fail after take-off.

In the Beech 1900 incident at Williamstown described earlier, the FO believed the engine to be on fire when he saw flames coming from the nacelle. Attempts to discharge the engine fire extinguisher failed because the bottle was electrically operated and the wiring to the extinguisher had been destroyed. The design of the system failed to accommodate the possibility of a serious fire in the engine nacelle that, in turn, would disable the system itself. On the one hand, we can identify a fallible design decision here but, at the same time, we can see the tight coupling of events in NAT terms. The anchoring of cables to fuel pipes in a location containing important control devices resulted in one malfunction disabling critical, unrelated, systems.

The accident at Pelee Island has some, limited, characteristics of a normal accident. The Cessna Caravan is a relatively simple, robust aircraft type. It has a high wing configuration and a fixed tricycle undercarriage. Although approved for flight into icing conditions, the aircraft, because of its design, has lots of exposed surfaces on which ice can form, and the manufacturer had provided protective devices in accordance with airworthiness requirements. Electrically heated panels on the propeller and pneumatically operated 'boots' on other critical structures provide the means to remove ice that might form in flight. The particular aerofoil section used in the design of the Caravan wing has been found to be susceptible to performance degradation once the ice begins to build up. In fact, global Cessna 208 accidents do show a marked seasonality, with more in the winter months. The most critical location for ice formation is at 12% chord but the de-ice boots only extend as far as 5% chord.

Ice removal by boot activation can be improved by accelerating the aircraft. However, ice increases drag, which can sometimes result in the aircraft having insufficient surplus power to allow it to accelerate in level flight. Similarly, a good rule for pilots once they encounter ice is to change altitude in order to get out of icing conditions. This can sometimes require climbing but ice adds to the weight of the aircraft and, again, performance limitations prevent the pilot from taking the best course of action. Some of the comments made by Caravan pilots in Chapter 1 alluded to these problems.

Aircraft design, in terms of aerodynamic properties and performance, and meteorological conditions exist in a tightly coupled state, and the manner in which these factors interact can be surprising, especially when we also consider the nature of normal work, as this example illustrates.

> On 29 April 1993, an Embraer 120 encountered icing conditions in the climb and subsequently stalled (NTSB, 1993). The captain had been asked by the flight attendant if it was possible to climb faster in order to reach the cruising altitude sooner: she wanted to start the cabin service but did not want to have to push the service cart along an inclined cabin floor. The captain used the 'pitch hold' mode on the autopilot to increase the angle of climb and, thereby, increase the rate of climb. The aircraft power was left at the 90% 'climb power' setting. The aircraft pitch angle in the climb was initially 3.2°. The pitch was increased, first, to 5.2° and then 6.4°. The rate of climb reduced to zero, the speed decayed and the stick shaker activated at 141 knots. The aircraft lost 12,000 ft in the subsequent stall.

The aircraft had probably started to accumulate ice once it had climbed into the freezing conditions. Ice starts to form at the stagnation point on the leading edge of the wing where the local airflow velocity is zero. When the captain increased the pitch angle, the stagnation point would have moved down the leading edge and the ice that had previously formed would now be in such a position to disrupt the airflow over the upper surface, leading to premature boundary layer separation and an increased stall speed. In this case, icing, aerodynamics and a desire to be helpful resulted in severe damage to an aircraft.

NAT generally applies to systems that can also be considered highly reliable. An HRO, as mentioned earlier, is one where the number of failures occurring is much less than the total number of accidents – or accident opportunities – that *could* occur. Amalberti (2001) identifies three bands of organisations based on safety:

1:1000 systems (1 catastrophic accident per 1000 trials): Activities such as mountaineering and bungee-jumping.
1:100,000 systems: Activities such as charter flights, helicopter operations, emergency medicine.
1:1,000,000 systems: Scheduled aviation, power generation.

An HRO would fall into the final category. Amalberti's analysis reveals some of the paradoxes present in the HRO concept. First, because accidents are extremely rare in this cluster of systems, the response to failure can often be irrational. Media-driven, short-term responses to accidents give the appearance of taking action while the real cause of the accident is often not addressed. Furthermore, the timescale needed for change to take effect often falls outside the attention span of media and the tenure of the decision-makers. Responses to infrequent failures usually take the form of increased regulation. This, in turn, makes the job of work more difficult and usually results in increased rates of rule violation simply to get the job done on a daily basis. In an examination of a military friendly fire tragedy in Northern Iraq in 1994, Snook

(2000) demonstrated that 'normal accidents' can easily occur in 'HROs'. Snook showed how organisations drift towards an unsafe condition despite efforts to control risk, and we will return to this idea later in this chapter.

Perrow's NAT focuses on technology and the unpredictable nature of combinations of systems. The HRO school does look at social systems and, indeed, uncertainty is introduced into systems largely through the variability of human operators (Marais, Dulac & Leveson, 2004). Perrow argued for greater redundancy to allow for failure, but redundant systems themselves fail. In the case of HRO theory, it is argued that a 'culture of reliability' contributes to safety. However, as both Leveson (2011) and Marais et al. (2004) observe, reliability and safety are different qualities. Accidents can occur as a result of unexpected interactions between normally functioning components in a system: reliable elements induce unsafe conditions. Ironically, in some situations, redundancy has been achieved by introducing more humans. Air taxi companies flying aircraft certified for single-pilot operation are often required, either by clients or their insurance companies, to use two pilots. Where the company has failed to implement effective two-pilot procedures, incidents have been caused because of uncertainty over roles and responsibilities.

Part of the problem we face is that much of our analysis of failure relies on hindsight (Dekker, 2002). We look back on events and see clearly where things went wrong. Hindsight couples cause and effect, even in situations where there is, in fact, only a weak causal relationship, if at all. By concentrating on the technology of systems, and not human operators in systems, NAT and HRO theory lack the ability to address the issue of why operators act in an apparently irrational manner. In order to achieve our work goals, we act in ways that make sense to us at the time. The manner in which we construct a local, bounded, rationality will be covered in more detail in Chapter 5.

2.7 ORGANISATIONAL ERROR

The fact that accidents can have an organisational genesis is recognised in Reason's model of accident causation (Reason, 1990). The model is based on the elements of a production system and is illustrated in Figure 2.2. At the top level, decisions-makers make strategic decisions about the structure and function of the organisation. Unfortunately, senior managers cannot anticipate all the ways in which things might go wrong and, so, some of their decisions will contain the seeds of a future failure. The design of the ATR system on the MD-83 described earlier is an example of a fallible decision, a plausible design solution to a problem that was subsequently found to exacerbate rather than moderate a situation under very specific circumstances. These are the latent errors and resident pathogens referred to earlier. Control of operations on a daily basis is delegated to line supervisors. In order to meet production targets, supervisors turn a blind eye to discrepant production activities. The third level in the model captures the underlying vulnerability of human operators, who are susceptible to stress and fatigue. In many ways, the level of psychological precursors is a boundary layer between management and production. The fourth level is that of production. While doing work, people take short cuts, modify procedures and commit errors. Finally, systems put in place measures to

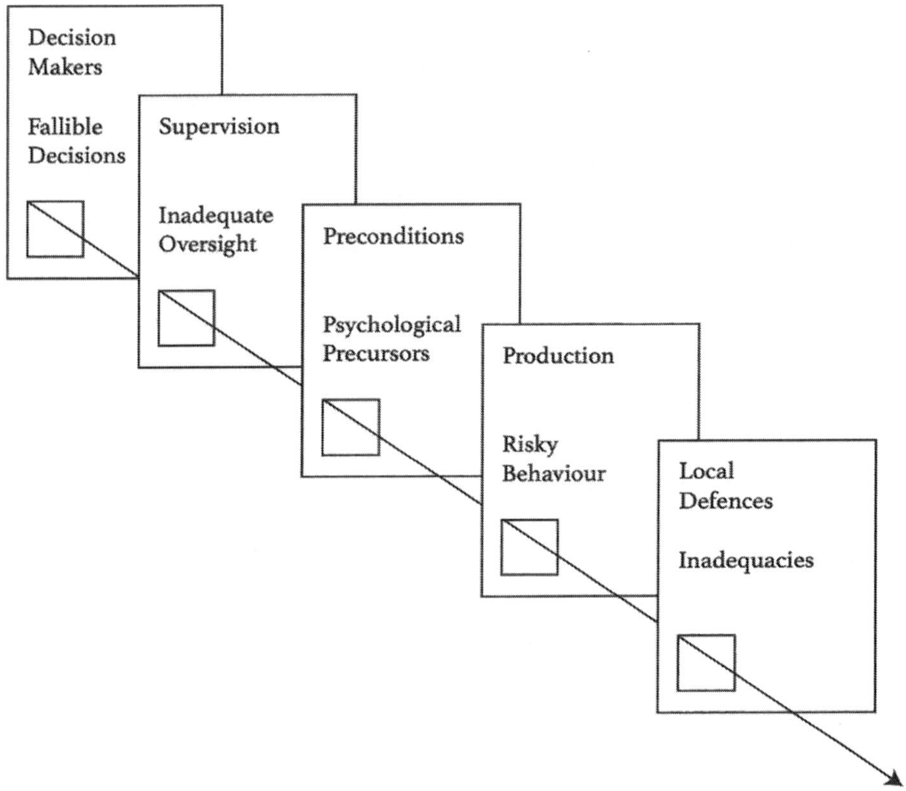

FIGURE 2.2 Reason's model.

keep the system safe. Local defences are, themselves, the product of human design activity and, so, are potentially flawed (think of the B-1900 engine fire extinguishers). The model represents the manner in which failings at one level may be trapped at the next level in the system. So, a flawed work process may be ameliorated by increased supervisory oversight. Operator failings will be captured by warning systems. However, it is conceivable that, as the figure illustrates, flaws at every level can align to allow a propagation path for failure.

The model is often described as a variation of the linear approach, and it is not hard to see why. Hollnagel, however, describes it as an 'epidemiological' model, and I have to agree with that. In simple terms, the model not only creates an agenda for investigating aspects of the organisation that can contribute to failure but also provides a framework for making comparisons between organisations that can then throw light on why things fail. For example, it has long been recognised that aviation in Alaska is riskier than that in the other States of the USA. Over the years, there have been various studies conducted by the NTSB, trying to understand why the accident rate is so high. At one point, as a commercial pilot, you were 20 times more likely to be killed in an accident in Alaska than if you were flying elsewhere. One study found that operators with a high turnover of maintenance managers had higher accident rates. Applying Reason's framework across a range of organisations could reveal epidemiological risk factors.

We can apply the model to the Pelee Island accident. First, we can start with the decision-makers. The division of responsibilities in TC had allowed Georgian Express to commence operations without any oversight from its POI. The initial trial flights to get the approval to run the service were based on each sector being a discrete entity with a complete dispatch process being conducted before each flight, but we have seen how that concept quickly seemed to have changed. Line supervision was delegated to the pilot who failed to complete some of the processes associated with the dispatch of the aircraft. Equally, the migration from two scheduled sectors to a single flight with an intermediate stop was not picked up by the management.

The pilot had had very little sleep the night before, having returned after a weekend away and getting up very early to start work; fatigue was a probable psychological precursor. Despite having his attention drawn to the ice on his aircraft, the pilot chose to take no action despite freezing rain falling at the time of departure. Here we see risky action that derives, possibly, from a desire to get the work done and maybe avoid getting stuck on Pelee Island. Georgian Express provided equipment for de-icing aircraft as a local defence, but this equipment was left behind on the day and Pelee Island Airport lacked alternative facilities. Pelee Airport had no scales for weighing passengers, another 'defence' of sorts, had they been present. Of course, even if these things had been available, there is nothing to say they would have been used. Like linear models, Reason's Model allows us to arrange factors in related clusters. However, dynamic aspects of accident propagation are not explained, largely because it is not a model, rather it is a metaphor for how production takes place in an organisation (Hollnagel, personal communication). The NAT and HRO concepts also tend to look at the structural properties of systems rather than the behaviour of individuals. From a competence perspective, we need to recognise that humans work within complex and unpredictable sets of relationships involving components such as technology, meteorological phenomena, commercial and social policy goals. Each individual's performance must carve a path through these, often competing, forces.

2.8 WHY HISTORICAL MODELS ARE PROBLEMATIC

The models we have looked at represent a challenge for developing a competence-based approach to CRM. I want to illustrate the problem by following one airline's experience of coping with operational performance issues. The starting point for the discussion is a tragic, fatal accident. On 12 January 1999, an F27 cargo aircraft carrying newspapers crashed after it stalled on the approach to Guernsey Airport, in the Channel Islands (AAIB, 2000). The immediate cause of the crash was the load shifting in flight. As a result, when the crew selected the landing flap, the aircraft's centre of gravity moved out of limits, and the aircraft stalled. The aircraft had been, first, incorrectly loaded and then, second, the load had not been properly secured, which is why it shifted in flight.

To fully understand the actual 'cause' of the accident, we need to go back a few years. The company originally operated between fixed bases with aircraft loading and administration being undertaken by company staff. The company then went through an expansion and took on a significant number of parcel contracts for the UK postal service. Parcels tend to fill the space available before they become weight limiting. Therefore, loading the aircraft was simple: you just packed in the parcels and threw a net over them. Weight and balance was never an issue. The next phase of growth saw the company moving into the ad hoc charter market. Cargo was collected from wherever it was available and ground handling was often undertaken by third parties who might never have dealt with an F27 before. This was the situation on the day of the accident but, crucially, the company's procedures for the oversight of loading aircraft had not kept up with the changes brought about by the growth and the opening up of new markets. These changes in the nature of the loads carried and the ground support available to deal with loading meant that the airline's risk profile had changed. In Chapter 1, we saw that an effective competence model should reflect the contingencies most likely to be encountered by the crew. In this case, commercial developments over time had changed the nature of the task and the company's processes – and crew competence – had become uncoupled.

The response to the accident was remarkably lacking in insight. Because the aircraft stalled when the crew selected the landing flap, the procedures were changed to prohibit the future use of full flap for landing. Nothing was done about loading procedures. Of course, the landing flap is designed to reduce both the touchdown speed and the stopping distance needed on the runway. On the morning of 1 July 2000 (AAIB, 2001), just over a year later, a company F27 was on final approach to Coventry Airport. The weather was mist and light rain with a cloud base of 400 ft. With a slight tailwind, the aircraft was high on the glide slope and 19 kt fast. The captain touched down 600 m along the runway with 1000 m remaining. The aircraft did not stop in the distance available. After that accident, the procedures were changed to restore full flap for landing. But the story did not end there. After the second accident, the company pilots were on their guard. One night, during an FO training flight, the captain said that they needed to be particularly careful about getting high and fast on the approach, his concern based on the folklore that had now developed in the airline. Neither of the pilots had been in the company when the fatal crash occurred but were, of course, aware of the event. The result was a heavy landing almost in the undershoot as the crew ended up slow and low on the approach path (personal communication).

This sequence shows how the classical 'barrier' model results in constraints being modified or new barriers being put in place. The response to one event, ironically, created the context for the next. As well as formal, procedural, responses to the investigation process being introduced, a social change occurred based on the company's expectations of tolerable crew performance. Failure begat failure.

2.9 CONCLUSION

This chapter has looked at events from the perspective of accident investigation models that have been in common use for many years. While they all have undoubtedly served a purpose or provided some insight, neither strategy really addresses the problem of changing workplace behaviour to build robustness into processes. By default, almost, linear models seem to result in more barriers to be circumvented. By placing ever-more constraints on operations, efficiency becomes difficult to achieve unless short cuts are taken. The NAT and HRO schools are too abstract to be of practical use, and Reason's model, too, lacks the specificity needed to generate inputs to training. In fairness, these various schools of thinking attempt to offer modes of explanation, not routes to change and, without a doubt, Reason's approach was a milestone because it dragged the focus away from the individual to the broader organisational context. A competence-based approach to CRM will need different tools of analysis and models of performance and, in the next chapter, I want to outline an approach based on more recent thinking about safety.

REFERENCES

AAIB. (2000). AAR 2/2000 Report in the Accident to Fokker F27-600 Friendship, G-CHNL near Guernsey Airport, Channel Islands, on 12 January 1999.

AAIB. (2001). AAIB Safety Bulletin AAR 4/2001 Channel Express, Coventry-Bagington Airport, UK.

Amalberti, R. (2001). Revisiting safety and human factors paradigms to meet the safety challenges of ultra complex and ultra safe systems'. In B. Wolpert & B. Fahlbruch (Eds.), *Challenges and Pitfalls in Safety Interventions*. Holland: Elsevier. pp. 265–276

ATSB. (2001). Occurrence Report No: 199904317 dated April 10, 2001.

Blockley, D.I. (1992). *Engineering Safety*. London: McGraw-Hill.

Cox, S. & Cox, T. (1996). *Safety, Systems and People*. Oxford: Butterworth-Heinemann.

Cullen, W.D. (1990). *The Public Inquiry into the Piper Alpha Disaster Vol 1 & 2*. HMSO.

Dekker, S. (2002). *The Field Guide to Human Error Investigations*. Aldershot: Ashgate.

Hollnagel, E., Woods, D.D., & Leveson, N. (2006). *Resilience engineering: Concepts and precepts*. Aldershot, England: Ashgate.

Hollnagel, E. (2009). *The ETTO Principle: Efficiency-thoroughness Trade-off. Why Things That Go Right Sometimes Go Wrong*. Aldershot: Ashgate.

Leveson, N. (2011). *Engineering a Safer World*. Cambridge, MA: MIT Press.

MacLeod, N. (2005). *Building Safe Systems in Aviation*. Aldershot: Ashgate.

Marais, K., Dulac, N., & Leveson, N. (2004). Beyond normal accidents and high reliability organisations: The need for an alternative approach to safety in complex systems. *Paper presented at the Engineering Systems Division Symposium*, March 29–31. Cambridge, MA: MIT.

NTSB. (1991). Report No: AAR-91/04 Avianca Boeing 707 Fuel Exhaustion Cove Neck, New York. January 25, 1990. Published April 30, 1991.

NTSB. (1993). Report No: FTW93-M-A143. In-flight Loss of Control. Continental Express Embraer 120. Pine Bluff, Arkansas. 29 April 1993.

Perrow, C. (1999). *Normal Accidents*. Princeton, NJ: Princeton University Press.

Reason, J. (1990). *Human Error*. Cambridge: Cambridge University Press.

Roberts, K.H. (1990). Some characteristics of high-reliability organisations. *Organization Science*, 1, 160–177.

Snook, S.A. (2000). *Friendly Fire*. Princeton, NJ: Princeton University Press.

SHK. (1993). Report C No: 1993:57. Air Traffic Accident on 27 December 1991 at Gottrora, AB County. Case l-124/91.

TSB. (2005a). Aviation Investigation Report No: A04O0336. Rejected Landing – Collision with Terrain. Oshawa Municipal Airport, Ontario, December 16, 2004. Released October 18, 2005.

TSB. (2005b). Aviation Investigation Report No: A04H0001. Loss of Control. Georgian Express Ltd. Cessna 208B Caravan C-FAGA Pelee Island, Ontario. 17 January 2004.

Tversky, A. & Kahneman, D. (1974). Judgment under uncertainty: Heuristics and biases. *Science*, 185(4157), 1124–1131.

3 A Systems Model of Aviation

3.1 INTRODUCTION

In Chapter 1, I suggested that the goal of CRM training might be to develop operational expertise but a problem we face is that we lack both a clear model of what airline pilot expertise looks like and also the tools for designing training in an operational context. In the previous chapter, I reviewed some safety concepts and considered how conventional approaches to accident analysis were weak at revealing underlying competence requirements. My intention was to see what light they might cast on performance in the workplace. The conclusion I drew was that a focus on preventing a repetition of a specific event offered little to help us understand the broader question of expertise. In this chapter, I want to take a different approach. We will start by looking at attempts to apply systems concepts to the issue of safety. Drawing on the key themes in a systems approach, I want to develop a model of aviation based on workplace performance. It is this framework that will shape the rest of the book as I attempt to derive a competence set that will help in the design of more valid training.

3.2 SYSTEMS THINKING AND SAFETY

The shift to systems thinking (Leveson et al., 2003; Rasmussen & Svedung, 2000) was primarily a response to the inadequacies of linear models and those safety initiatives based on creating barriers that supposedly trap failure. As we saw, linear models are predicated on cause and effect relationships derived from analysis after the event. Hollnagel (1999) observed that barriers do not prevent anything, rather they change behaviour. Adverse events will still happen. As well as moves to apply systems thinking to safety, complexity theory is also a potentially rich source of insights into how systems behave. At the time of writing, two main approaches to systems and safety are competing for popularity. Nancy Leveson, at MIT in the USA, has developed the systems theoretic accident model and processes (STAMP) (Leveson, 2011) while, in Europe, Erik Hollnagel advocates his functional resonance analysis method (FRAM) (Hollnagel, 2012). The two approaches share some common elements but their differences reflect the perspectives of their authors. STAMP is rooted in engineering. It maps the structural elements that interact in order to achieve a task and then looks at the safety constraints that have to be satisfied in order to operate in a safe manner. FRAM, on the other hand, is a psychological model. It sets out to describe what an operator is actually doing while completing a task (rather than what someone thinks they *should* be doing) and then maps the inputs needed to do the job and outputs that flow from activity. Because of its origins in engineering and software, STAMP is less adept at dealing with the performance of humans within the system being modelled. FRAM is underpinned by a set of concepts collectively known as 'resilience engineering' and which do offer some pointers as to where we need to look in order to understand how humans and systems interact. I want to draw on both STAMP and FRAM, first laying out some of the fundamental concepts here before offering a view that pulls the key ideas together.

In the discussion that follows, a 'system' comprises assemblages of humans and technology configured to achieve a performance goal. These represent the static components but the system exists to make decisions. Rasmussen and Svedung (2000) viewed organisations not so much as collections of interrelated functional components, the organisation chart approach, but rather as nested decision-making processes. To me this is important because it reflects the fact that systems exist to achieve a goal and every component in a system is either the output from a set of decisions or, itself, exists to make decisions in relation to its specific responsibilities. For example, a commercial airliner is more than just a piece of equipment. It is the embodiment of a set of decisions about how to make money by carrying a payload over a distance. The structure itself is a manifestation of a set of solutions to mathematical, aerodynamic and engineering problems. Its utilisation is the output from a set of commercial decisions. The Airbus A-380 was hailed as a remarkable feat of engineering and was very popular with passengers. The economics of operating the aircraft were such that there were few markets it could profitably serve. As the embodiment of commercial decision-making, it was flawed. Very few carriers could adapt their networks to make the aircraft viable. In this view, then, failure is no more than suboptimal decision-making.

Leveson's STAMP model proposes four characteristics of a system. First, functional units are arranged in a hierarchy with higher-order elements exercising control over lower-order elements. Position in the hierarchy embodies both authority and legitimacy but, on examination, we see that position is simply a reflection of the order of magnitude of the decisions made by the functional component. Second, control is exercised by higher-order functional units over those below it in the hierarchy. Control is exercised through artefacts that are the output from decisions and have different levels of force. Some are rules and regulations that carry the weight of law. Breaches of policies and procedures can often attract actual punishment or admonishment. Breaches of social contracts can attract ostracism. These controls constrain the freedom of operation of the lower-level units in order to keep activities within expected bounds. Third, systems function through communication. This includes the signals that serve to direct and validate actions and verify system status and progress. It includes rules and regulations, policies and procedures that are issued to guide the actions of others. Weather charts and reports, flight plans, notices to airmen are forms of communication that support a specific flight. Conversation facilitates collaboration. Different types of communication have different levels of force in terms of requirement for compliance and also different levels of validity. Communication can differ in terms of specificity i.e. some forms are generic while others are specific to a moment. Communication includes those signals that relate to actions taken and indicate the probability of success or failure. Finally, systems have emergent properties that cannot be explained by simple descriptions of how the elements at each level work. For example, safety is not an attribute of a functional component, it is an emergent property of the way the component works. It is important to remember that these characteristics are functions of processes rather than structural properties. They describe activity in a system rather than physical elements.

Systems have to contend with uncertainty. Bookstaber (2017), in his book on the financial crash of 2008, makes the point that complex systems demonstrate non-ergodicity: put simply, nothing happens the same way twice. The implication of this is that, despite the fact that most of commercial aviation is heavily proceduralised and takes place within defined spatial arrangements, on every flight pilots have to create new solutions to achieve the same goal. Every flight is a subtle blend of 'the same as before' and 'not quite the same'. Furthermore, the author talks of 'radical uncertainty'. Because of emergence and the issue of non-ergodicity, events can often unfold in manner that cannot be anticipated or easily explained. Competence can therefore be seen as a set of behaviours directed at accomplishing goals or solving problems in a dynamic, uncertain workplace.

The importance of taking a behavioural perspective is reinforced by Hollnagel's view that success and failure are equally probable outcomes from the actions of individuals. Therefore, we need to have a way of looking at how systems create contexts for individual behaviour. This concept is an important contribution to safety thinking. It forces us to move away from retrospective judgements of performance to focus, instead, on the fitness of the goal-directed behaviour of individuals. In order to do work, we are all engaged in making judgements about the effort needed and the process to be applied to get a job done. We make trade-offs between being very thorough and being just effective enough – or efficient – to achieve the goal. Hollnagel (2009) calls this the efficiency/thoroughness trade-off principle. These trade-offs affect the probability of a successful outcome. In all areas of activity, systems can cope with variations in performance up to a degree: there is buffering capacity. However, there comes a point where a threshold is reached and continued operations are impossible: we have reached a system boundary or limit of viability.

From an individual's perspective, what resilience engineering tries to identify is how we detect the approach of a boundary and how we manage performance in those boundary layer states such that safe operations are restored. Both the identification of warning signals and the activity needed to manage marginal performance are forms of competence. There are two key problems that emerge from the discussion so far and which were touched on in previous chapters. First, procedures and technology are designed to accommodate a core set of anticipated conditions but workers still need to be able to maintain control while dealing with unplanned contingencies. Second, problems inherent in design will emerge at a point removed in space and time. Systems demonstrate cross-scale effects. Therefore, the relevant knowledge necessary to understand the specific problem might not be currently available. Both are serious challenges when considering what competence might look like. With that in mind, I now want to outline an organising framework that describes aviation from a systems perspective.

3.3 THE STRUCTURE OF THE AVIATION SYSTEM – AVIATION AS HIERARCHICAL DECISION-MAKING

Although I have proposed that a system comprises physical entities – people, organisations, technology – assembled to deliver a performance, in this section I want to map out how that system behaves. The system is a framework for making decisions and the behaviour of that system reflects the manner in which those decisions are enacted. Leveson's hierarchy describes the different types of decision that are required for aviation to work. Because different actors reserve to themselves the right to make decisions in specific areas, the system is rendered tangible. For example, whereas an individual makes decisions about job-related demands, a company has the right to make decisions about who to employ and what aircraft to buy, for example. I have identified four levels of decision-making in my model but we also need to recognise that they are all embedded in a broader environment. The environment comprises those factors that shape activity but are not, in return, shaped *by* that activity. Environmental factors include the climate and weather, economic cycles, geopolitics. The four levels are as follows:

Level 1: Production. This is the lowest level and represents the individual in the workspace. This is where skills, tools and processes come together to generate work. Fundamentally, Level 1 describes how an individual engages with the world in order to achieve a task-related goal. I refer to it as the level of production and decisions are typically those that result in individual goal-directed action.

Level 2: Collaboration. At this level, we are recognising that work is distributed across space and time but, more important, it also involves many players. Level 2 is concerned with all those actors collaborating as part of a task. They may be on-board the aircraft or off-board. The off-board crew forms and disperses according to needs dictated by the phase of flight and could be considered a virtual team that support the aircraft throughout its trajectory. Decisions at this level are typically those associated with the organisation and conduct of work, with agreeing solutions and the allocation of responsibilities though not necessarily the actual 'doing'. Level 2 exerts control over Level 1 typically through social rules, formal job descriptions and group authority structures.

Level 3: Exploitation. At this level, we are concerned with both the profit-making and not-for-profit entities represented by airlines, maintenance organisations, ground handling companies, air navigation service providers, equipment manufacturers, etc. The role of the commercial entity is to generate returns on investment through the use of aircraft, crew and other resources. In the case of not-for-profit organisations, capability has to be provided within a budget. Level 3 is where decisions are made about the organisation of production. Control is exercised through commercial and employment contracts, codes of conduct, policies and procedures.

Level 4: Facilitation. The goal of decision-making at this level is to facilitate activity through the provision of tools of production, capital assets, regulations for the conduct of work, etc. Regulatory authorities define the manner in which companies can deliver services. Control over the lower levels is through granting approvals and permissions: we are concerned with constraint.

Of course, a 'system' is a disembodied entity: it does not 'make' decisions. Individuals make decisions. The key point is that, at each level, individuals make those decisions to achieve a goal that falls within the area of responsibility of that level. Remembering that this hierarchy simply describes a nested structure as proposed by Rasmussen and Svedung, applied to aviation, what is of more importance is how the system functions.

3.4 BEHAVIOUR WITHIN A SYSTEM

Each level in the system is focussed on a different set of goals and, as we move up the hierarchy, decisions are further removed from the point of action. The configuration of the elements of the system shapes behaviour. Furthermore, systems are not static structures fixed in space and time: they are dynamic, fluid entities constructed to solve problems of production. As we have seen, the structural aspects of a system comprise the component parts, the tools and processes and the formal linkages between those parts. These structural aspects provide the framework to support operations and are important but the observed outcomes flow from individuals trying to achieve goals while working within the constraints of the system. For example, the lack of a set of scales to weigh passengers at Pelee Island Airport did not cause the aircraft to crash. The need to complete a weight and balance calculation as part of the aircraft dispatch process represents a solution to satisfy a decision made by the regulator about the safe conduct of operations. On the day, the pilot created an alternative solution to an operational problem. The ability of a system to cope with the inherent variability in normal operations reflects its level of resilience. Hollnagel has defined resilience as:

> The intrinsic ability of a system to adjust its functioning prior to, during or following changes and disturbances so that it can sustain required operations under both expected and unexpected conditions.
>
> **(Hollnagel et al., 2011)**

Resilience, then, is about the recognition of potential disturbances, developing adequate responses and restoring performance predictability. It is about retaining and regaining control of events. Predictability is an essential property of a system from an organisation's perspective while unpredictable outcomes jeopardise viability. Change and disturbance need not be catastrophic to bring about a significant breakdown.

Action takes place in a bounded space in that activities in dynamic processes typically have a point, under normal circumstances, by which they must be completed. The concept of boundaries, though, is problematic. Woltjer (2008) comments that 'no one knows where exactly the absolute boundaries of safe intervals lie'. Mendonça (2008) suggests that systems will have multiple boundaries and that measures of margin might include limits of performance (effort) and the borders between organisations (entry/exit points). Broadly speaking, a boundary reflects the freedom of action a system possesses. A boundary may be defined in absolute terms, such as the speed at which an aircraft will stall, or in relative terms, such as society's willingness to tolerate certain types of behaviour. A boundary acts as a significant constraint on performance and represents the point at which operations are no longer viable. If we follow Mendonça's argument, however, we can also identify interfaces in a system that are transactional boundaries between operational entities. Interfaces exist between collaborating individuals and between different work teams. For example, it is well-understood in aviation that the flight deck door is a significant, if metaphorical, interface between the flight deck and the cabin crew. Boundaries are edge states that demand attention and a failure to act effectively in relation to a boundary or an interface will increase the risk of failure. The management of boundaries and interfaces requires effort.

Woods (2006) proposes two properties of system performance in relation to boundaries: margin and tolerance. Margin describes 'how closely or how precarious the system is currently operating to a performance boundary' while tolerance describes 'how the system behaves near a boundary'. Margins, then, represent the zone approaching the point at which a system can transition from a desired to an undesired state because task demands overwhelm the control structures. Some authors talk about margins being 'eaten up' or experiencing 'slow erosion'. I would argue that the margin is the point at which a need to intervene to achieve the desired goal can be identified. If we think of declarative knowledge as constraint sets, then the point at which the probability of violating a constraint is discernible marks the distal edge of the margin. Margins are zones in which discrepant signals emerge.

Tolerance has been interpreted as the nature of failure at a margin: is the system brittle or is it capable of graceful degradation (Woltjer, 2008)? I want to add a third state, which is that of inertia. Galbraith (1954), in his study of the Great Depression, talks of 'the bezzle'. In all financial systems there is an element of corruption, the true extent of which is usually only revealed when the system comes under stress. In a financial crash, every penny is needed and only then do you realise that some have gone missing or have been embezzled. In better times, while graft may be suspected, there is little incentive to invest the additional effort needed for its eradication. An inert failure, then, is a mode in which the system still functions, albeit in a recognisable suboptimal state, but it usually takes an additional shock to bring about failure.

Mendonça proposes that tolerance is not simply the nature of performance but how that performance is achieved. He suggests that measurement will need to capture process-level descriptions of behaviour. For example, Neerincx (2003) describes a three-dimensional model of cognitive task load that includes information processing, task-set switching and percentage of time occupied by a task. If a system's margin represents the time and space available in which to complete an actual task then tolerance, in this formulation, would be manifested in the ways in which operators manage task switching, the manner in which they track the time available and their ability to process and manipulate information to maintain control. I prefer to reserve tolerance for the characteristic of failure and, instead, describe performance in the margin in terms of the efficacy of the operator's action in anticipating change, sustaining the systems status and restoring stability.

What emerges from this discussion is that competence in normal operations must cope, primarily, with situations where signals might be tentative or obscure but small changes can have a significant negative impact. Behaviours associated with information seeking and signal interpretation, around efficient load shedding and task modification are needed to contain operations and prevent a potential breach of a boundary condition.

Before we apply this formulation of a system to some examples, I want to just summarise the key features. A system comprises people and technology assembled to achieve a goal. Action flows from decisions about how to configure technology and organise work to achieve the desired goal. A goal will have an associated set of constraints that must be satisfied if the goal is to be successfully accomplished. Action takes place in a bounded volume of space and time and a failure to act effectively in relation to these boundaries will increase the risk of a degraded outcome or, even, failure. Boundaries have associated margins in which signals relating to the status of the task become apparent. Under normal circumstances, these signals simply relate to the approaching state change as we shift from one goal to the next. Occasionally, signals relate to a need to intervene to restore system status. The actions of agents involved in managing the system can be assessed in terms of their efficacy in relation to accomplishing the goal. Failure at the boundary can take one of three forms: graceful, brittle or inert. Having mapped out the key features, I want to go back to the previous discussion of the Pelee Island accident and look at the elements through the lens of a systems model.

3.5 PELEE ISLAND AS A SYSTEM

To start with, I want to arrange the key actors in this scenario using the hierarchical model (Figure 3.1). Starting with the environment, the most obvious representation of Level 5 in this accident is a climate that resulted in Lake Ontario freezing in winter and, thus, requiring alternative provisions for travel. The local weather at the time is also an environmental factor. It directly impacted activity at Level 1 through its effects on aircraft operations and the pilot's subsequent need to respond appropriately. It also triggered behaviour at Level 2 through the need to undertake de-icing, a process often requiring outside support. Weather has an effect at Level 3 because of the increased probability of rescheduling flights in adverse weather with subsequent knock-on effects for rostering, crew duty times, aircraft positioning, etc. The distribution of the population can be considered an environmental factor in that it created a demand for year-round communication links for isolated communities. It was this policy that resulted in the need to substitute an air link in winter when ice prevented ferry sailings.

Transport Canada represents Level 4 in the systems model. Transport Canada has a responsibility to ensure the safe and efficient conduct of aviation within its jurisdiction. To achieve this it has processes in place to ensure that air routes are viable and that air carriers' work within the rules. The primary vehicle for exercising control is the Canadian Aviation Regulations. This framework of rules, guidelines and

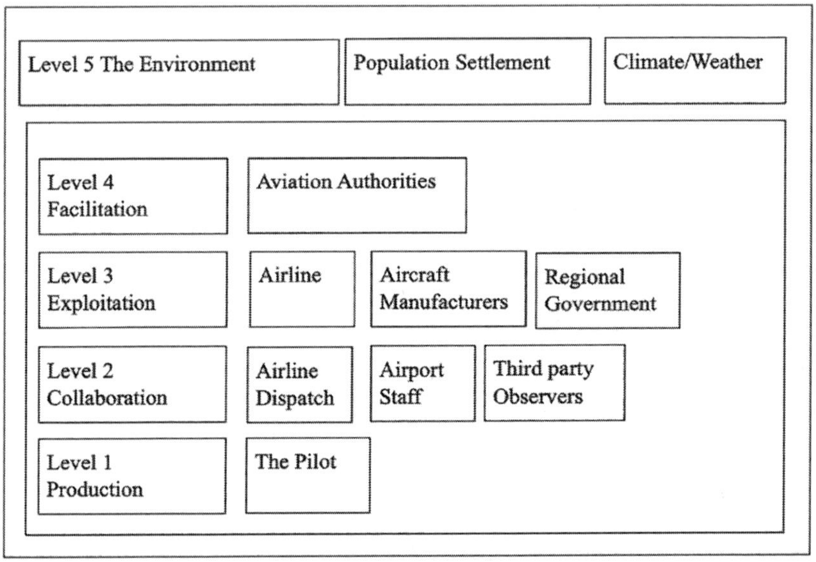

FIGURE 3.1 A hierarchical model of aviation.

standards represents the constraints that must be met in order to hold an approval to operate. The regulations are amplified through the air operator's certificate (AOC) manual that specifies the steps needed to acquire and maintain an AOC. One requirement of an AOC holder is to develop a set of manuals containing information for crews on the operation of aircraft and the conduct of flight. These manuals are open to inspection by the regulator.

Oversight of air operators is conducted through a network of operations inspectors and the work of the inspectors is described in the operations inspector manual. Inspectors conduct scheduled and no-notice checks and audits and have the power to demand changes to company processes. Transport Canada has the authority to impose various forms of punishment on offenders ranging from fines to the cancellation of an AOC, thereby causing an air operator to cease functioning. This body of rules and regulations represents a historical trace of the decision-making process at the level of the regulator in relation to risk management.

The division of responsibilities within Transport Canada that resulted in a significant change to an AOC not being communicated to the Georgian Express Principal Operations Inspector (POI) flowed from a decision about how to organise resources within Transport Canada and the subsequent allocation of responsibilities. The ambiguous nature of the Pelee Island route – a subsidised charter on behalf of the Government of Ontario being operated in the guise of a scheduled service – further clouded the question of responsibility for oversight. These confusions resulted in a breakdown in control.

The Government of Ontario, Georgian Express and the Cessna Aircraft Corporation represent Level 3 in my framework. The regional government allocates resources in relation to policies associated with supporting communication between remote communities. The role of the company is to organise and provision, to initiate, sustain, modify and to terminate production activity as required. This is the domain of management decision-making. It put in place a process for documenting flights using load sheets and passenger manifests. It made available equipment for de-icing the aircraft at remote locations. The company exercised control over line operations through procedures. Cessna builds aircraft, which are tools of production. The aircraft, itself, represents the output from a set of decisions taken in relation to meeting a demand for a particular class of aircraft suited to a specific set of conditions. Control is exercised through operating procedures and performance limitations. The township of Pelee operates the airport and also represents a Level 3 actor. Because of its proximity to the USA it is considered a port of entry into Canada and has immigration facilities but being a small regional airport, the size of aircraft it can handle is limited and it has no air traffic control tower. The range of traffic management services it offers is limited. The sophistication of services offered by the airport, then, reflects decisions made about the need and available finance. The airport staff had no control over events that day but the configuration of the asset – the airport – shaped the pilot's options in dealing with the aircraft's performance issues.

Level 2 in my system model captures collaboration. In a formal sense, although the Cessna was operated by a single crew member, the pilot interacted with the company dispatchers and also with air traffic control en route to Pelee Island. We can also see elements of informal collaboration. Observers on the ground commented on the presence of ice to the pilot. From the accident report, there is little evidence to illustrate control being exercised at this level other than this interaction between the pilot and the person on the ground. It is conjecture but, for the purposes of this discussion, this event represents communication across a notional interface in relation to an operational issue. The presence of ice was significant enough to trigger the comment. It is possible that the differing status of the two actors – pilot v non-pilot – captures the power dynamic and, thus, who exercised control.

Level 1 has, historically, been the main focus of accident investigation. It was here that the pilot applied his skills of operating the Cessna Caravan within the procedural context of flight management. He made decisions about loading the aircraft, preparing it for flight and manipulating the device to get airborne. Control was exercised through inputs to the aircraft. Feedback was available through the performance of the aircraft.

Systems are configured with sufficient controls in place to afford predictability. Feedback loops generate signals that confirm that the system is behaving the way stakeholders expect. However, operations on a daily basis are subject to perturbations and deviations and actual performance can often be at variance with what was anticipated. Systems can accommodate departures from intended action to the extent that the operation has sufficient buffering capacity. Accidents and incidents occur when activities exceed the bounds of control.

3.6 EXPLORING STRUCTURAL ELEMENTS OF A RESILIENT SYSTEM

The previous section elaborated on the idea of systems as enacted decisions. The system as envisaged in my hierarchical model is an abstract framework that describes the space where decisions are made but the purpose of the system is to achieve an outcome. It is a dynamic entity, not a static structure. The next case study illustrates the space in which action occurs. I want to explore boundaries, margins and tolerance. The event itself is relatively straightforward, albeit with a tragic outcome. It is another example of the banality of failure. It is the circumstances that allowed the event to happen that are of interest to us.

> On 23 December 2003 a Learjet 24B took off from San Bernadino County Airport bound for Hailey, Idaho (NTSB, 2006). Climbing through 26,000 feet the FO requested a return to San Bernadino. When asked if he wanted to declare an emergency he declined. The controller cleared the aircraft to 24,000 feet. The initial descent was normal but the aircraft quickly pitched nose down, losing height at 10,000 fpm. Ninety seconds later the aircraft crashed into the high desert. Such was the impact force that little of the aircraft was recoverable. Tissue samples from the captain were collected but no identifiable remains of the FO were found. The weather was fine, no problems were reported with the aircraft and no other aircraft were in the vicinity. Apart from two recent incidents where the left engine flamed out in flight, we have no idea of what caused this accident.

It is a sad, tragic irony that the complete lack of evidence about the performance of the crew allows us to fully focus on the system the crew were working within. The context of the operation is summarised below and illustrated in Figure 3.2:

- The aircraft was registered to and operated by Pavair Inc. of Santa Monica, California.
- Pavair is a Part 91 operator and, therefore, could not conduct commercial flights.

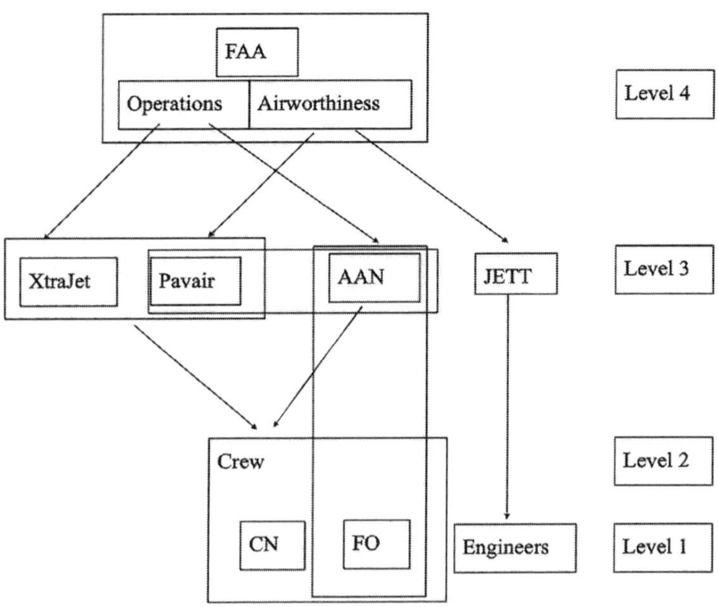

FIGURE 3.2 The Learjet 'System'.

- The flight was dispatched to collect a 'loyal paying customer'.
- Pavair and XtraJet were owned by the same person.
- The captain occasionally worked for the owner of XtraJet Inc. but was not an employee of that company.
- The captain and First Officer (FO) were trained by American air network (AAN).
- AAN is a Part 135 certificate holder that arranges lease agreements for aircraft and crews and charges a management fee to operate aircraft under its certificate.
- The FO was listed as an employee under AAN's certificate.
- The aircraft was listed under AANs certificate but was removed on 17 November 2003 because Pavair had provided inadequate maintenance records.
- Maintenance was provided by jet executive transport technologies (JETT).
- The following maintenance records for the aircraft were not available:
  - Airframe logbook 1969 – October 1973
  - Left Engine Logbook 1981 – December 1998
  - Maintenance logbooks from April 1999 to accident date
- Flight logs, manifests and discrepancy logs were provided by XtraJet for the period 27 February–17 April 2003.
- JETT provided a major repair and alteration form (FAA Form 337) that showed that all the installed communications, transponder and navigation equipment had been removed and replaced over the period July to mid-October 2003. The form was not signed but the work had been inspected by XtraJets FAA-approved inspector. The Form 337 had not been submitted to the FAA as required by regulations.
- Because of two very recent left engine flame-out incidents an FAA airworthiness inspector requested the aircraft's maintenance records. When Pavair failed to provide the information, the aircraft was placed on a condition notice on 12 December 2003 preventing further flights. On 22 December 2003, a JETT engineer called the inspector who was not in the office. The engineer informed another inspector, on behalf of Pavair, that the documentation had been provided and requested the lifting of the condition notice, which was done verbally. The documents were never received.

The goal of the system, on this day, was to undertake a flight on behalf of a customer. In order to be considered successful, two constraint sets needed to be satisfied. The first related to the condition of the aircraft and the second related to the nature of the dispatch of the aircraft. The boundary was created by the decision to undertake the job for the 'loyal paying customer', which set the point by which the constraints needed to be satisfied. Although the aircraft was technically owned by one company, it was managed on a daily basis by another and maintenance was provided by a third. This complicated arrangement is commonplace and allows private owners of aircraft to generate revenue. It also adds to complexity in that the behaviour of the system cannot, necessarily, be explained by the functioning of the component parts. This configuration illustrates the concept of interfaces: internal boundaries, in effect.

Interfaces in this event existed between the FAA and both the aircraft owner and the maintenance provider. There is also an interface between Pavair and JETT. Each party engaged in transactions across their respective interfaces. Transactions included work requests, completion sign-offs, form submissions, provision of requested information, checking and verification of documents, communication of decisions, issuing approvals. These various processes were effortful in that they required time and activity to satisfy and, then, to validate that they met requirements.

Dispatching aircraft under Type 91 or 135 approvals requires adherence to different sets of rules: one is simpler than the other. One set applies to private flying while the other covers commercial operations. It happens that a business jet might be dispatched under Type 91 regulations for an empty, positioning sector and then Type 135 rules are followed for the revenue sector. In this case, it appears that Pavair was operating in its own right and, thus, because there was no passenger on board when the aircraft departed, the first sector was technically legal. We have no way of making a judgement about the second sector with the 'loyal customer' on board. The first set of goal constraints were initially met.

The second set of constraints that needed to be satisfied related to the condition of the aircraft. The FAA provided the regulatory framework for both airworthiness and also operations. It exercised control by establishing formal requirements for work processes, for documentation, by setting standards of evidence to satisfy requirements, by authorising agents to act on its behalf and, ultimately, by granting or withdrawing permissions and approvals. Logbooks and official forms provided the feedback the FAA could use to renew permission to fly in addition to formal inspections. The FAA had requested evidence in relation to the aircraft, which did not arrive.

The system boundary was the last point at which the constraints we have identified could be satisfied before the system moves from a valid state to one that is invalid. A valid state is one that can be considered safe or viable. On this day, the boundary was established once the decision had been made to collect the passenger. The margin is the zone in which behaviour relative to the specific boundary is manifested. There were two signals of system status: the technical condition of the aircraft and the probability of compliance with the dispatch regulations for the second sector. We have no evidence of how Pavair intended to resolve the latter condition. We do know how the first condition was managed. A request was made by a third party for permission to operate the aircraft.

Buffering describes the variability of performance the system can accommodate. In effect, it captures the range of alternative behaviours that might be seen to depart from a notional ideal template and, yet, still allow the task to be done. It seems that significant periods of the aircraft's maintenance history were undocumented. Information was held in different locations and, yet, the aircraft continued in use. If we look at the action taken to release the aircraft, we see surrogates interacting. The management of the part-processed FAA Form 337 is a further example. Despite all of this behaviour, while chaotic, the system remained intact. In lieu of actual sight of the requested documents, permission was granted on the basis of an understanding that the information had been sent. There is no evidence to suggest that either party knew that statement to be true. My analysis makes no suggestion that either the FAA inspector or the JETT engineer were culpable. They were simply acting to achieve an outcome. The act that breached the boundary was the release of the aircraft on the basis of a promise. Of course, this did not cause the aircraft to crash but even if the crew had been successful in fulfilling the promise to the loyal customer, the system would still have failed. This suggests that systems can possess significant buffering capacity. This example raises questions about the legitimacy of actors. Although the aircraft was owned by Pavair, communication was left to an engineer from JETT. Equally, the approval was granted not by the inspector with responsibility for the case but another, who happened to be in the office and took the call.

Tolerance describes the robustness of the system when operating close to the boundary. The one element of this story that we have not been able to examine is the behaviour of the flight crew. All we can say is that, in the space of a few minutes, the flight went from being within the control of the crew – the FO had declined to declare an emergency when asked – to catastrophic. In this case tolerance was brittle. If we look at the management of the aircraft, we cannot say that degradation was graceful. The system did not degrade to a lesser but still safe state. This is an example of what I consider an inert state. Inadequacies in the management of the aircraft's airworthiness status were apparent and action was demanded but no meaningful progress was being made.

Efficacy describes the probability of an intervention achieving the goal within the margin, before the boundary is breached. The only evidence we have to assess efficacy in this example is the use of an informal telephone call and a subsequent verbal approval based on a promise that action had been taken. The formal process had been usurped by, probably, well-intentioned actors. The action was expedient but not efficacious in that the problem was not solved.

This first exploration of the systems concept illustrates a number of key points. First, actors in a system engage with one another to achieve a goal. The goal must not be confused with an immediate task. The goal was to undertake a legal flight. The tasks were those processes associated with achieving the goal. Although the FAA was the controlling authority, the manner in which control was exercised allowed sufficient variability for the process to fail. The boundary and its associated margin, in this case, is a negotiated space in which work is accomplished. It is not an absolute. Any intervention that changes the available margin, such as the intention to operate the flight, modifies the risk attached to the situation. Buffering describes a broad range of activity. It is important to understand that we are not making a judgement about the correctness of the various actions taken. Buffering recognises that normal work comprises a multiplicity of solutions and describes a system's ability to cope with non-ergodicity and radical uncertainty. Efficacy describes the value of work, by which I mean given the current conditions, will the actions of individuals satisfy the applicable constraints before the boundary is reached? Efficacy is a measure of fitness, not 'rightness'.

This example also illustrates how organisational complexity increases the opportunity for safeguards to be circumvented, for mechanisms of control to become confused and for communication to become fragile. The distributed nature of ownership, management and responsibilities in this case – who was actually accountable for what? – increased the probability of failure. The system functioned in absence of the formal, expected protocols and artefacts.

3.7 SYSTEMS AND SCALE EFFECTS

The Learjet example we have just explored ultimately hinged around behaviour in the near-term: a phone call used to expedite an administrative process. Systems, though, work across multiple time scales. In the last chapter we saw that fallible decisions occur at the level of strategic management and can give rise to latent errors. We also saw that complexity in technology results in tight-coupling and the associated unpredictability in fault propagation. In a systems context, the relationship between aircraft manufacturers and airworthiness authorities represents an interface condition. The airworthiness requirements – controls – are published and the manufacturer then tries to satisfy those requirements in a cost-effective manner. The robustness of the solution might not be immediately apparent, as this next example illustrates.

On 5 December 2001 a Saab 340B was climbing through FL 190 when the two electronic flight information system (EFIS) screens on the right-hand (FOs) side failed (ATSB, 2003). The aircraft had been dispatched, in accordance with the operator's minimum equipment list (MEL), with the left-hand (captain's) upper screen unserviceable. In addition, the right-hand starter generator was annotated as requiring removal for overhaul. A 10% extension had been applied to the life of the generator to allow the flight to be accomplished. The crew applied the EFIS failure/disturbance checklist and tried to restore the screens using the EFIS drive transfer system. The aircraft then experienced the following cascading progression of system failures:

The right engine ice protection annunciator illuminated (failure)
Cabin pressure annunciator illuminated (spurious)
The flight attendant call bell activated (spurious)
The stall warning clacker activated (spurious)
The right stall fail annunciator illuminated (failure)
The rudder limit annunciator illuminated (failure)
The global positioning system (GPS) failed
VHF 2 radio and intercom inoperative (failure)
The FO's communications system failed
The radio magnetic indicators failed
The automatic direction finder 2 failed
The right DC generator out light failed
The right engine instrumentation failed
The right hot battery warning illuminated (correct indication)

It is likely that other services powered from the right-hand electrical system DC bus were not available but the crew did not know of their loss. Once the voltage dropped below 18 V, Inverter 2 failed. As a result, the flight data recorder was inoperative for a period of 18 minutes until the crew selected Inverter 1. The ground proximity warning system (GPWS) was unavailable, as were the right fuel gauge and fuel shut off valve. No heading reference information was available for the 18-minute period. Once the generator out light came on, the crew used the appropriate electrical failure drill to restore power. They landed safely.

The immediate cause of the event was quite straightforward. Having started the engines, the crew completed the after-start checklist. This included a check of generator output and the indications were normal. At some point soon after the check was completed, the wear on the generator brushes reached a point at which the right-hand generator was no longer fully functioning. The generator control units (GCU) of this production standard Saab 340Bs were not capable of detecting a low voltage condition and so no direct warning was available. The crew could only have discovered the deteriorating condition of the generator if they had chosen to randomly check the output. There were no further mandated generator checks in the normal procedures. Once the output from the generator began to drop, power was supplemented by the battery. Once the battery became the sole supply of electrical power – that is, there was now no electrical power coming from the right-hand generator – it started to drain. A low voltage light operated when the output from the battery dropped below that required to operate the GCU. Throughout the episode, the bus tie relay failed to connect, which would have connected the left- and right-side electrical distribution systems, because, with the switch set to AUTO, the system only functioned when the generator out light illuminated, which did not happen for some 18 minutes.

I want to examine this scenario in three parts. First, I want to look at aspects of the original design of the aircraft. Next, I want to look at the maintenance event that precipitated the failure and then, finally, I want to look at how the crew dealt with the situation. Each of these elements represents a slice in time. The Saab 340 first flew in January 1983 and the second-generation 340B entered service in 1989. Production ceased in 1999. By the time of the event, the Saab 340B was already a relatively mature product. I want to explore how latency – Reason's latent error – and cross-scale interactions shape the behaviour of systems.

Establishing the fundamental airworthiness of the aircraft design is a superior goal with many subordinate goals that must be met. Saab, the manufacturer, needed to demonstrate that the decisions its engineers had taken in relation to the relevant airworthiness requirements were valid. Originally a manufacturer of military aircraft, the 340 was its first venture into the commercial market, having identified a need for a 30 seat regional airliner. The boundary of the system might be considered the point at which the effort needed to gain the type certificate did not exceed the available design resources or reduce the potential future returns, rendering the project no longer economically feasible. The UK Government's attempts to modify the Nimrod maritime patrol aircraft, itself a development of the 1950s Comet commercial airliner, to provide an airborne early warning capability is an example of development costs resulting to a system boundary being breached (Connor, 1986).

The aircraft was approved by the Swedish Luftfartsverket, having satisfied itself that the design was compliant with the European and United States requirements. In this specific instance, in the absence of a GCU able to detect an under-voltage condition, the manufacturer relied on the failure of the EFIS screens as an indirect indication of a problem rather than develop a direct warning. This was a design compromise but one that was considered sufficient to meet the requirement. Once the aircraft entered service, the risk of an undetected low voltage condition had become apparent through operational experience and redesigned starter generator brushes were introduced in 1995 in response, together with a new checklist. A modified

GCU, with a low voltage detection capability, was also produced but its installation was optional. The Luftfartsverket was satisfied that the new generator brushes and checklist afforded adequate protection but the investigation report could not, subsequently, establish if the original design of the electrical system actually met airworthiness requirements. The aircraft involved in this incident was number 328 in the production sequence and the new GCU with low voltage detection was fitted to all aircraft as standard after serial number 367. The margin, in this case, extended over many years, from the point at which a design solution had crystallised to the point at which it was finally confirmed as being adequate. Signals relating to the adequacy of the design compromise did emerge through operational experience and these were known to the actors involved.

Buffering, as we have seen, represents the range of behaviours the system can accommodate. In this case, we see the original plan to use the EFIS screens as a surrogate for a dedicated low voltage warning, modifications to generator brushes, checklist revisions and, finally, a redesigned GCU. There might also have been others considered by the manufacturer at the design phase. If tolerance describes the nature of failure then given that the investigation could not decide if the electrical system met the required standards, it seems that we have another example of an inert state.

The significance of this element of the event is that it is suggestive of Reason's 'latent conditions'. A decision that satisfied one set of criteria created the opportunity for failure elsewhere but that risk only became apparent once other factors came into play. The solution to protection against a low voltage condition also illustrates how cross-scale interactions shape system behaviour. Design decisions taken probably 20 years prior to the event created the context for the actions of both the maintenance engineers and the pilots. It is maintenance that we will look at next.

The awarding of a release to service represents an actor at Level 4 – the authority – granting permission to one at Level 3 – the manufacturer. Once in service, as we saw with the Learjet, there is a need for ongoing maintenance to maintain the specific airframe in an airworthy condition. The operator used a third party to provide maintenance – just as in the Learjet example – but, under the terms of its AOC, it was required to have its own approved maintenance manager, together with a formal management process. Specific communication relating to the maintenance of the generator was contained in the manufacturer's maintenance review board (MRB) report. The MRB is an approved document that describes a minimum number of tasks necessary to keep the aircraft in an airworthy condition. The manufacturer also provided detailed maintenance task descriptions as individual Job Cards. The MRB and the Job Cards are communication artefacts that act as forms of control. The MRB specified that the aircraft generators were to be removed for overhaul every 1200 in-service hours. The life of the generator could be extended to 1600 hours if the brushes were replaced at 800 hours. The Job Card relating to this task stated that the brushes need only be replaced at 800 hours if they exceeded a certain level of wear. We have a discrepancy here. One document stated that brushes must be replaced to extend the service life while the other document stated that brushes only need replacing if they exceed a certain limit of wear. The manufacturer's guidance was that the MRB took priority over Job Card information. The operator reported that it had been led to believe that the MRB incorporated maintenance manual procedures and, so, the Job Card was an adequate reference.

The left-hand generator had been maintained in accordance with the MRB. The right-hand generator had been inspected at about 800 hours and, although a note to the effect that the brushes were 40% worn was recorded in the aircraft maintenance log, the brushes were not replaced. On the day of the flight, it was noted that the generator was due for servicing but the maintenance log had been annotated to the effect that a 10% extension to the generator life was authorised. No company maintenance procedures allow for such action. The generator had been installed for 1601.9 hours when the EFIS screens failed and, so was already at its maximum installed life when the aircraft departed.

The maintenance 'system' we see here has similarities to the Learjet example we saw earlier. The system comprises those components brought together to provide maintenance support. Control was exercised by mandating a system of maintenance oversight and by documenting the processes by which maintenance tasks were to be conducted. Servicing the generator was a subset of the broader process of conducting periodic maintenance, contributing to achieving the goal of sustaining aircraft serviceability. The boundary was the point at which the aircraft maintenance state must be declared valid prior to further use. If a task was not completed or an item was beyond its notional life then that state would be false. The margin was the notional space between the first task demanded by the MRB and full compliance. In effect, the boundary and its margin were forming and reforming as demanded by the cyclical nature of maintenance.

Buffering is represented by the range and scale of maintenance actions associated with, in this case, the generator. Maintenance tasks can be done correctly, part finished, done using the wrong tools or parts or completed using inappropriate techniques. For example, in a study of Piper PA-31-350 Navajo Chieftain accidents, 35% were the result of maintenance, primarily the use of the wrong part or the wrong maintenance technique (over-torquing retaining bolts or misaligning turbocharger section pipes being commonplace). One aspect of buffering is the fact that inconsistencies can be codified into processes. The discrepancies between the MRB document and the Job Card allowed for interpretations that increased risk. At a practical level, buffering could accommodate the fact that one generator could be maintained using the MRB process while the other followed the Job Card. In the latter case, if the brushes were not replaced at 800 hours, the generator would need to be replaced after 1200 hours. Buffering also needs to capture non-standard activities. It reflects the variability that is a part of normal operations. We saw that, after the 800-hour inspection, the extent of the wear on the brushes was simply annotated in the log, which brings us to the extension of the generator's life. The report did not explain what was meant by the notional '10% life extension' nor when it was entered into the logbook. It was probably a local fix to allow the aircraft to remain in service for a few more days but with no context we only have recourse to conjecture. With hindsight, it appears that an ad hoc act of creativity finally exceeded the system's buffering capacity. In this instance we see a variety of behaviours, all but one being accommodated by the system. Buffering needs to cope with latent errors (the discrepancy between the MRB and the Job Card) and local activity. The systems tolerance, in this case, was brittle. Breaching the boundary resulted in a significant failure.

Electrical generator brush wear is widely understood and periodic inspection and replacement is the standard control process. These are mitigations as the issue is a constant requirement and simply a function of the design of generators. Saab selected a generator produced by Goodrich to satisfy the requirements of the Swedish Authority. In so doing, it created a need on the part of its customers to enact solutions to ongoing maintenance. These cross-scale interactions, events happening removed in space and time and at different levels in the hierarchy, are additional sources of failure.

Turning, now, to the performance of the crew, although the aircraft remained controllable and standby flight instruments were available throughout, the tight-coupling inherent in the technology and the subsequent cascading failures – some spurious and others real – presented a challenge to the crew, who were already having to cope with one EFIS out of commission. For a period of around 18 minutes, the exact status of the aircraft was not understood by the crew. The weather conditions were relatively benign and so the loss of the GPS, GPWS and heading reference was not as significant as it might have been. The crew on the day comprised a captain conducting supervised line flying with an FO on his second operational flight. The captain was very experienced, in terms of total hours, and had been employed as a supervisory pilot, involved in the training of FOs, since March 1999. Although also an experienced pilot, the FO had completed the 5-hour Saab 340 type conversion course on 1–2 December 2001, just 3 days before the event. There was, therefore, a significant imbalance between the two crew in terms of on-type experience. Ground training did address a range of malfunctions pilots could encounter and both crew commented that they were aware of some aspects of a low voltage condition, such as the difficulty of detecting the condition prior to the EFIS screen blanking and the possible cascading sequences of failures.

The aircraft had been dispatched with one EFIS screen inoperative, a condition permitted under the terms of the MEL. When the next two EFIS screens failed the crew believed that this was a related problem, which was why their initial response was to action the EFIS checklist. The captain also reported that he had been reading about the EFIS system in the weeks prior to the flight, in particular how the EFIS drive transfer system worked. It was his opinion that the MEL item coupled with his recent reading led him to conclude that the problem lay with the drive transfer system. He handed control of the aircraft to the FO while he dealt with the 'EFIS failure/disturbance' checklist. The first step on the checklist was to check the generator voltage. This step was missed. Several senior pilots in the company later said that Saab checklists were poorly designed and it was not uncommon for crew to miss steps in checklists. Having completed the checklist, the EFIS screens were not restored. By now, the pilots were wearing oxygen masks and in a precautionary descent because of a spurious cabin pressurisation warning.

The goal of the crew was to maintain control of the aircraft and manoeuvre it to a safe landing place. To be successful, they had to prevent any further degradation and to restore the aircraft to its highest level of technical capability. The system boundary was the point at which further degradation would place the aircraft in jeopardy. This would mean that either aircraft controllability was impaired or the opportunity to make a safe landing was reduced.

The margin associated with the boundary was marked by the point at which the generator brushes became degraded and output was reduced. Although prior to the failure, there was no formal requirement to check generator output, a signal relating to system status was, nonetheless, accessible. The margin became tangible once the EFIS screens blanked.

Buffering, in this case, was reflected in the actions of the crew. From an organisational perspective, the blanking of an EFIS screen was considered sufficient to trigger the crew to execute the low voltage checklist. In reality, crews could implement a correct or an incorrect checklist and the chosen checklist could be actioned accurately or inaccurately. In this case, the crew implemented the wrong checklist incorrectly. What we see here is that the system was able to absorb irrelevant activity with no further degradation until a more unambiguous cue directed the crew to more constructive behaviours.

Fundamental to system functioning is the efficacy of the crew's behaviour. At this point we need to consider the other boundaries in the system, the interfaces. These are the notional internal partitions within the system that need to be negotiated. Interfaces will shape efficacy. In this case, the cockpit controls and displays represented an interface through which the pilots were trying to make sense of the situation and restore control. As we saw, the blanking of the two EFIS screens was misdiagnosed, in part because it occurred in association with an already disabled screen. This was followed by a sequence of warnings and systems failures, some real but many of which were spurious. This added to the challenge the crew faced. I have suggested that efficacy describes the probability of actions satisfying the constraints of the active goal and is shaped by Neerincx's cognitive task load model. The initial diagnosis by the crew was shaped by the existing malfunction, the pre-existing disabled EFIS screen, but was also influenced by the captain's recent study of the EFIS drive transfer system. We cannot discount the inexperience of the FO, a potential source of corroboration or alternative interpretation. Their response was to implement the most appropriate checklist, which was the one that matched their construction of the apparent problem. It takes time to complete a checklist. However, once started, the crew were presented with a series of unrelated indications, some real and some false, all of which were competing for the attention of the crew and possibly explains the fact that the first item on the checklist being actioned was missed. They would have been faced with the problem of trying to fit the new data into their active schema, that of an EFIS problem. The alternative was to break out into an alternative mode of action but information processing consumes cognitive resource and task switching is effortful. This will be discussed in more detail in Chapter 5.

The incident report notes that the crew donned their oxygen masks and initiated an emergency descent. This was, presumably, in response to the spurious illumination of the cabin pressure indication. This warning was powered by right-hand electrical busbar. The actual pressurisation system was controlled by the functioning left-hand busbar. It is not clear from the report if the aircraft had actually lost pressurisation but the significance of the warning was sufficient to trigger the crews' response. An emergency descent would have added to workload and further inhibited information processing. The problem was resolved by a new cue: the generator out light. It seems that the efficacy of the crews' actions in relation to the specific problem was low but the fact that the aircraft remained controllable suggest that the mode of failure, the tolerance, was graceful rather than brittle.

This example illustrates the effects of cross-scale interactions in systems. Decisions about the design of the aircraft, in this case the selection of a source of electrical power and, importantly, the provision of a warning of reduced power output, had two significant, interrelated, consequences. First, the design of routine maintenance schedules afforded opportunities for unintended activity. Second, the solution to the need to warn of a reduced output, when coupled with incorrect maintenance, created an opportunity for the crew of an aircraft to follow the wrong checklist. Systems interact across scales with events removed in time, and undertaken in completely different contexts, interacting in unintended ways. The arrangement of components in systems affords opportunities for control of processes to be degraded or even lost. Developing and sustaining complex systems is problematical. The way in which a system is configured represents a solution to a specific problem achieved within specific resource availability. The way in which the structure of the system influences its behaviour is part of the problem. The Learjet and Saab 340 examples illustrate a property of all systems, which is to migrate from a designed, intended state to an extant unintended operational condition. This process is known as drift.

3.8 SYSTEMS, DRIFT AND THE DISTORTION OF BUFFERING

Complex systems exhibit self-organised criticality (Bak, Tang & Wiesenfeld, 1987), the classic illustration of which is the sand table demonstration. Imagine a table where we continually add single grains of sand to a pile in the centre. The pile will build and maintain a specific shape. Occasionally, the addition of the next single grain will trigger slides of grains down the sides of the pile until an equilibrium is restored. The pile has sufficient integrity to maintain its shape but the addition of just a single additional grain will render the pile temporarily unstable until the new stable state is achieved. The idea is that systems migrate to states of sufficient stability but inherent criticality. This dynamic, fluid nature of systems behaviour is reflected in the concept of 'drift'. Dekker (2004) points out that most commercial enterprises function in conditions of resource scarcity and competition. In order to generate a sufficient return on investment in the face of competition for customers, organisations have to survive in highly marginal production environments. System resilience can be degraded as modes of production become risky and safety begins to drift. To illustrate what we mean by drift consider this example from the Pelee Island accident. When the charter was initiated it was expected that each sector, the outbound leg and the return, would be treated as two distinct flights. As such, each sector would be subject to the same standards of flight planning and pre-flight preparation. From the evidence of minimal turn-round times and load sheet production, it seems that crews viewed the task not as two separate sectors but as a single flight with an en route stop. This conceptualisation of the flight process allowed important safeguards for the return flight to be breached. In a similar manner, on 6 December 1999 a Piper PA31-350 Navajo Chieftain crashed after take-off from Johannesburg in South Africa (SACAA, 1999). The aircraft, engaged in a weekly charter transporting computer programmers to Namibia, was overloaded. Of nine flights flown as part of the contract, an examination of load sheets revealed that eight had been overweight for take-off. When the operator bid for the contract, the plan was to make the flight in two stages, stopping to refuel en route. After the initial proving flight, all subsequent flights were made with a full fuel load and continued direct without stopping. When fully loaded with passengers and their luggage for a week, the aircraft now had inadequate performance margins should an engine fail after take-off, as happened on this day. Drift, then, describes the manner in which working practices become aberrant and activity now takes place in ways that erode margins.

There is a grey area between drift and the concept of violation. The application of the '10% extension' in the Saab 340 event described earlier might be considered a violation. Violations will be considered in more detail when we look at human error but, in the context of this chapter, I will use 'drift' to mean the gradual modification of work processes that have a higher level of risk but, importantly, that change in risk exposure is not detected by the organisation.

I want to illustrate three modes of drift at the level of the system. The first we can call 'rational change', the second 'normalisation of deviance' and the third, 'asynchronicity'. Dekker (2004; NTSB, 2002) illustrates the process of rational change in relation to the failure of the Alaskan Airlines MD83 elevator trim jackscrew that resulted in the aircraft losing control over the Pacific Ocean, off the coast of California. The original aircraft design called for maintenance at set intervals. The maintenance task was codified in a set of schedules of work to be completed at set intervals. In Table 3.1, we show how the interval between maintenance events inexorably increased over time. The figures in parenthesis represent the actual time, in flying hours, between servicing events.

The table shows how, over a prolonged period, the interval between lubrications grew almost by an order of magnitude. Within a year of the original aircraft type, the DC-9, entering service jackscrew thread wear in excess of the predicted value was being reported and so a safety constraint was put in place. The extent of the wear was to be tested at intervals. Table 3.2 shows how the interval between safety checks experienced the same rational change process.

Rational change occurs when decisions are made based on factors that fail to capture the risks present in a system. In the case of aircraft maintenance, checking regimes are based on predicted values for wear and component failure. Maintenance aims to replace

TABLE 3.1
Jackscrew Lubrication

Aircraft enters service: Lubrication planned every 300–350 hours
1985: Every alternate B Check (700 hours)
1987: B Check interval increased to 500 hours (1000 hours)
1988: B Checks discontinued and tasks spread across A and C Checks. Jackscrew lubrication every 8th A Check (1000 hours)
1991: A Check interval increased to 150 hours (1200 hours)
1994: A Check interval increased to 200 hours (1600 hours)
1996: Screw jack lubrication removed from A Check. Task card raised to complete lubrication every 8 months (2250 hours at company rate of flying).

TABLE 3.2
Jackscrew End Play Check Intervals

End play check every C Check (2500 hours) or 3600 hours
1985: End play check every alternate C Check (5000 hours)
1988: Interval between C Checks extended to 13 months (6400 hours)
1996: Interval extended to 15 months (9550 hours)

components before they fail or to control the rate of degradation. To achieve this goal, managers of maintenance need reliable information about mean times between failures and rates of wear. Given limited resources, one way of improving efficiency is to increase intervals between maintenance events such that interventions are more closely aligned with need. Decision-making is often based on the absence of failure events. So, as the servicing intervals increased and no adverse effects were reported, the rational approach assumes that the new interval is no more risky than the old schedule. By locking the safety check into the same interval-driven checking process, the constraint in the system is, similarly, becoming more risky. Rational change is predicated on the fact that modifications to processes do not give rise to anomalies. However, systems can approach boundary conditions. The last jackscrew end play check made before the accident occurred was done in 1997. When the check was instituted, acceptable tolerance limits were defined for the measured play. These were set at between 0.003 and 0.040 inch. The final check on the accident aircraft measured the play to be 0.040 inch. After much discussion, it was decided that this was acceptable in that it did not *exceed* the limit. It took 6 days of discussion before the aircraft was released and a further 3 years before the trim system finally failed. The activity that underpinned the decision to release the aircraft characterises the second drift process, the normalisation of deviance.

In her study of the loss of the Challenger Space Shuttle in 1986, Vaughn (1996) describes the process by which exceptional events become rationalised. In the case of the space shuttle, tolerance limits were exceeded but with no subsequent failure. As a result, the tolerance limits were increased. Events that were deemed unacceptable at the design stage became acceptable once data was gathered to show that excursions from expected performance had no apparent ill effect. In the case of rational change, no evidence existed to indicate that the system was unsafe. With the normalisation of deviance, anomalies are a feature of the operational environment but data is used to minimise the significance of those anomalies. One of the ways anomalies become 'institutionalised' is thorough the embedding of warning signals in other information. Vaughn reports that danger signs can be interpreted as 'weak' or 'mixed' so that their significance is lost. Within hierarchical systems, information is transmitted upwards in abbreviated, generic forms and unfavourable information is lost. What people know and understand depends upon their position in hierarchy. Finally, deviation is normalised when oversight, such as is offered by safety departments or quality management systems, lacks independence. Where the responsibility for oversight is physically removed from the site of operations, the level of awareness is degraded. Where oversight is co-located, it becomes absorbed into the construction of understanding and is, therefore, complicit in the organisational view of safety.

The final source of drift is asynchronicity, in which changes in one part of the system are not made with reference to the rest of the system. Examples of asynchronicity include a change in procedure without reference to all agencies involved or, say, a change in a component supplier. One factor in the Pelee Island accident was the fact that the standard passenger weights used for load planning now no longer reflect the average weights of actual passengers carried. Society at large has grown heavier faster than regulations have been updated. Societal change has resulted in the planned and the actual loads becoming uncoupled. In this case, asynchronicity is a function of a change at Level 5 not being reflected in guidance issued by lower levels. The discrepancy between the MRB and the Job Card in the Saab 340 example suggests asynchronicity.

In a systems context, drift distorts the efficacy of interventions by arbitrarily restricting buffering: permissible solutions are engineered out of the range of possible responses to a problem. As a result, tolerance is modified as modes of failure shift towards brittleness. Drift is another illustration of cross-scale interaction in a system.

3.9 CONCLUSION

Woods' competence envelope mentioned in Chapter 1 – the range of potential demands and the associated performance controls – is a bounded space, those boundaries being properties of the model that map onto the uncertainties and perturbations it was designed to cope with.

In the previous chapter, I suggested that historical models of accident causation provided little guidance for developing a competence-based approach to CRM. In this chapter, I have taken more recent thinking about systems and safety and used them as a springboard. Drawing on the key concepts, I have developed a framework that elaborates on Wood's model of competence.

Airline pilots function in a space shaped by decisions about technology and rules. That space is not fixed but, rather, is dynamic and so pilots need to be able to recognise change, identify any implications and create workable solutions to ensure adequate control of the task. Control is the key outcome of system behaviour from a safety perspective. On a daily basis, aircraft operations are subject to variability and, on occasions, the solutions generated by individuals and crew are marginal. The knowledge structures I described as 'expertise' in Chapter 1 must allow for adaptive performance while remaining sensitive to the cues associated with increased risk of loss of control. Cross-scale interactions represent a considerable problem because they build opacity into systems.

By taking a systems perspective we start to get closer to identifying the behaviours required of individuals to keep the operation safe. For a start, rigid adherence to procedures is unlikely to be a defence if the procedures themselves are flawed. However, any departure from a prescribed process must consider changes to the current exposure to risk. How do we train crew to assess future risk? This involves assessing the efficacy of planned actions. Coping with cognitive load is a key aspect. Information processing, time management and the implications of task switching need to be addressed. Fundamentally, the centrality of decision-making needs to be understood. Alas, decisions do not always leave a coherent trace while decision-makers act in isolation and, often, in ignorance of the needs of others.

These first three chapters set the scene for a discussion of competence as a set of behaviours directed at maintaining control in dynamic systems. The rest of the book explores how processes at the different levels in the hierarchy shape individual performance.

REFERENCES

ATSB. (2003). Air Safety Investigation Report No: 200105715 Saab 340B, VH-XDZ 93 km North East of Trepell, QLD. 5 December 2001.

Bak, P., Tang, C., & Wiesenfeld, K. (1987). Self-organised criticality: An explanation of 1/f noise. *Physical Review Letters*, 59(4), 381–384.

Bookstaber, R. (2017). *The End of Theory: Financial Crises, the Failure of Economics, and the Sweep of Human Interaction*. Princeton, NJ: Princeton University Press.

Connor, S. (1986). Nimrod reaches to the end of the runway. *New Scientist*, 31 July 1986.

Dekker, S.W.A. (2004). Why we need new accident models. *Human Factors and Aerospace*, 4(1), 1–18.

Galbraith, J.K. (1954). *The Great Crash, 1929*. Boston, MA: Houghton Mifflin.

Hollnagel, E. (1999). Accident analysis and Barrier Functions, Project TRAIN, IFE.

Hollnagel, E. (2009). *The ETTO Principle: Why Things That Go Right Sometimes Go Wrong*. Farnham, UK: Ashgate.

Hollnagel, E. (2012). *FRAM - The Functional Resonance Analysis Method: Modelling Complex Socio-technical Systems*. Farnham, UK: Ashgate.

Hollnagel, E., Paries, J., Woods, D., & Wreathall, J. (2011). *Resilience Engineering Perspectives Volume 3: Resilience Engineering in Practice*. Aldershot: Ashgate.

Leveson, N. (2011). *Engineering a Safer World*. Cambridge, MA: MIT Press.

Leveson, N., Daouk, M., Dulac, N., & Marais, K. (2003). A systems theoretic approach to safety engineering.

Mendonça, D. (2008). Measures of resilient performance. In Hollnagel, E., Nemeth, C.P., & Dekker, S. (Eds.), *Resilience Engineering Perspectives, Volume 1: Remaining Sensitive to the Possibility of Failure* (pp. 29–47.). Aldershot, UK: Ashgate.

Neerincx, M. A. (2003). Cognitive task load analysis: Allocating tasks and designing support. In Hollnagel, E. (Ed.), *Human Factors and Ergonomics. Handbook of Cognitive Task Design* (pp. 283–305). Mahwah, NJ: Lawrence Erlbaum Associates Publishers.

NTSB. (2002). Aircraft Accident Report Number AAR-02-01: Loss of Control and Impact with Pacific Ocean, Alaska Airlines Flight 261, McDonnell Douglas MD-83, N963AS, about 2.7 Miles North of Anacapa Island, CA, 31 January 2000. Adopted on 12/30/2002 NTSB. (2006). Aircraft Accident Report LAX04FA075.

NTSB. (2006). Accident Number LAX04FA075 Learjet 24B. 23 December 2003.

Rasmussen, J. & Svedung, I. (2000). *Proactive Risk Management in a Dynamic Society*. Karlstad, Sweden: Swedish Rescue Services Agency.

SACAA. (1999). Report No: J10/2/7132. Piper Navajo Chieftain loss of power after take off. Johannesburg Airport. 6 December 1999.

Vaughn, D. (1996). *The Challenger Launch Decision*. Chicago University Press. Chicago.

Woltjer, R. (2008). Resilience assessment based on models of functional resonance. In Hollnagel, E., Pieri, F., & Rigaud, E. (Eds.), *Proceedings of the 3rd Symposium on Resilience Engineering* (pp. 299–302). Paris: Mines-ParisTech.

Woods, D. (2006). Essential characteristics of resilience. In Hollnagel, E., Woods, D., & Leveson, N. (Eds.), *Resilience Engineering*. Aldershot(pp. 21–34). Aldershot, UK: Ashgate.

4 On Being Human – Frailties, Vulnerabilities and Their Effect on Performance

4.1 INTRODUCTION

In the last chapter, I proposed that a system is a configuration of assets directed at achieving a goal. We know that technology can fail, and it is clear that rules, regulations and procedures are not infallible either. In this chapter, I want to look at some general factors that affect the ability of humans to function within a system. We start by looking at a cluster of factors that, under the generic label of 'Personality', shape us as individuals. We then go on to look at two processes that inhibit performance. Usually labelled 'stress' and 'fatigue', we will see that, in fact, there are several factors at work here, not all of which are well understood. The outward manifestation of competence in the workplace will be mediated by the processes described in this chapter. There are, though, two sides to this particular story. On the one hand, there is the way the individual experiences and copes with the demands of the world. On the other hand, we need to consider how the workplace, itself, contributes to the problem. I will deal with aspects of the organisation in Chapter 9 but, for now, we will focus on the individual. In the discussion on fatigue, in particular, I will be drawing on findings from surveys I have conducted with line pilots and data collected using the line operations safety audit (LOSA) process (Klinect, 2005). LOSA, as the name implies, is an audit organised using a framework known as threat and error management. Trained observers identify elements in the operational environment that require additional attention on the part of the crew. Known as 'threats', these aspects bring complication to the planned task. The response of the crew to the threat is recorded by the observer. The next step is to record any errors made by the crew, how those errors are managed and, finally, whether the outcome of the performance is benign or result in the erosion of safety margins. 'Error management', in LOSA terms, starts to look a bit like efficacy in that it throws some light on how crew respond once the performance starts to degrade. Jim Klinect, whose PhD research underpinned the development of LOSA, describes it as 'an anthropologically-rich narrative' (personal communication) and that is very much its strength. I start the chapter by looking at the paradoxical concept of supposed individual differences in an apparently predictable framework: personality.

4.2 'PERSONALITY' – HOW EVOLUTION SHAPES BEHAVIOUR

Any discussion of human interaction in teams needs to begin with an examination of how evolutionary pressure on a social animal has resulted in consistent repertoires of behaviour. We recognise that humans apparently exercise agency – they can choose how to behave – but it is also a fact that our actions have a degree of predictability. To participate in a group, I have to be able to anticipate, to some degree, how others might respond to my actions. I can then decide whether I need to modify my behaviour based on that calculation. Evolutionary pressure has been at work on our predecessors for millions of years but Homo sapiens emerged just 300,000 years ago. We think that language emerged between 100,000 and 30,000 years ago. With the demise of the Neanderthals about 40,000 years ago, we became the dominant hominid. In evolutionary terms, it is likely that there has been little fundamental change to our genome in that time. Modern humans may be little more than smart apes in contemporary clothes but the forces that shaped us have been at work for millions of years.

When our predecessors adopted a terrestrial, as opposed to arboreal, lifestyle, they encountered a serious challenge. Weak, quite defenceless and diurnal, early hominids needed to form groups during the hours of darkness for protection. Sleeping on the ground left you vulnerable to nocturnal predators and being part of a group offered an element of the defence. Fish form shoals, birds flock together and savannah mammals congregate in herds for the same reason. Many species, though, when forced into proximity show signs of stress, get aggressive and even fight. Early hominids were no different. Competition for access to sexual partners and resources, such as food, created hierarchical, status-driven structures maintained by aggressive acts. To maintain social cohesion there was a need to defuse tension. Grooming, laughing, communal eating of cooked food (once we had harnessed fire) and communication all progressively allowed for ever-larger groups of hominids to come together, to collaborate, to hunt large animals or to fight against competing groups (Dunbar, 2014). Even religion can be seen, in its earliest Shamanistic forms, as a communal rebooting of the social system. The evolution of the human eye illustrates social selection in an evolutionary context. Over several millennia, the eye has been shaped to support social living. We have one of the largest ratios of the white sclera to the coloured iris of any animal, and as a result, we are able to derive information about the thinking of others based on our ability to detect the direction of their gaze. Knowledge of where you have directed your attention allows me to make a fairly reliable guess about what is on your mind (Graziano, 2013). Performance in a social group relies on predictability supported by anticipatory priming – our actions are formed by an expectation of what constitutes an adequate response. Therefore, having some idea of what you might be thinking about at the time makes that process more reliable.

The fact that people display fairly recognisable, consistent patterns of behaviour has been long understood. Unfortunately, how best to describe those patterns and how to even get close to a causal explanation has met with serious difficulties. Most people do not act in a random fashion. Behaviour is directed: people act in relation to their immediate environment. What we need to understand is *how* behaviour is 'directed' and for what purpose. The construct we call 'personality' is probably the oldest model used to explain this conundrum of consistent trends overlain with individual differences. Personality describes persistent behavioural traits but the rise of social psychology as a discipline gave rise to research into 'attitudes'. Attitudes are seen as 'proto-behaviours': they are 'tendencies' to act in a particular way. They are malleable and can be changed whereas 'personality traits' are considered to be more enduring.

More recent thinking takes an evolutionary approach. Such models assume that behaviour is, primarily, intended to increase my evolutionary fitness, which is my likelihood of passing on my genes. To that end, the behaviour is adaptive in that the way I generally act, how I act within the social group or in encounters with outsiders, will all have an effect on my probability of survival. Of course, by positioning modern behaviour in an evolutionary timescale forces us to confront the fact that how we act today can only be understood in relation to how people needed to act in order to survive millions of years ago. As we will see with the stress mechanism, some aspects of our evolutionary development have left us maladapted to the modern world, but generally speaking, evolutionary selective pressure can still offer a useful framework for understanding modern behaviour.

Psychobiological models of personality (Zuckerman, 2012) contend that the fundamental template for behaviour is coded in genes and manifested through gene expression. Gene expression simply means that the demonstration of a property is not binary. For example, variation in the height of humans can partly be explained through diet and environment but, also, reflects the way in which the gene for growth is triggered in individuals. As a result, height can be expressed as a normal distribution that reflects the interaction between genetic and environmental factors. Similarly, differences in behaviour probably also reflect the normal distribution of a particular gene complex. A gene-based model of personality implies that behavioural differences can be recognised in physiological differences, and this is proving to be increasingly true. That said, as in the case of height, environmental factors will also have an influence on the manner in which behaviour develops. Schools of thought that assign differences between races and genders to some fundamental genetic force abuse the science with malicious intent. Table 4.1 presents a five-factor model of personality with examples of consistent behavioural expressions.

TABLE 4.1
Personality and Behaviour (Opposite Expressions Implied)

Dimensions	Behaviours
Extraversion	Dominance, assertive, sociable, adventurous, forthright and outspoken. Linked to the functioning of the ascending reticular system in the CNS
Neuroticism	Anxiety, depression, guilt and low self-esteem. Only dimension linked to a psychiatric disorder. Linked to seratonin
Psychoticism	Impulsivity, sensation/novelty-seeking
Conscientiousness	Linked to socialisation, cautiousness, self-control v impulsivity, sensation-seeking and antisocial behaviours
Aggression/Agreeableness	Behavioural (fighting), emotional (anger) and attitudinal (hostility) emotional regulation

A psychobiological approach to personality reflects the fact that generic, persistent behavioural clusters are the result of adaptive responses that have, in evolutionary terms, enhanced fitness. These behaviours have proven useful in promoting gene survival. If you look at the dimensions of a personality, it seems that we can identify two broad dimensions: the degree to which we approach the world (exploration) as opposed to our desire to avoid the world (avoidance). Being approachable to strangers, seeking sensation or novelty, and coping with strangeness are all exploratory behaviours that increase the probability of discovering new resources or finding a mate. However, in the extreme, they expose us to risk. High levels of exploratory behaviour will probably be selected out of the gene pool simply because, millions of years ago, death was the likely outcome. On the other hand, excessively cautious behaviour reduced the probability of discovering new food sources, encountering prospective mates, etc. Again, excessively cautious behaviour will be selected out – you live a lonely life and eventually starve to death. We are left with the current repertoire of traits that allow us to engage with the world in a way that gives us a chance of survival, has a range of manifestations that allows us to adapt to circumstances but also afford sufficient predictability to support communal living.

Studies of pilot personality are remarkably rare, and any comparisons between reports are hampered by the use of different measures. Initially, studies were of military pilots, and the focus was on a better prediction of success in training. One recent study of commercial pilots (Mesarosova et al., 2017) found that, on a standard measure, there was a sufficient difference between a group of 591 airline pilots and norms constructed using the general United Kingdom (UK) population to suggest that there was a distinctive pilot personality type. This should not be a surprise given the selection involved in becoming a pilot in the first place. One consistent finding across this and other studies is that pilots tend to score lower on neuroticism than the general population. Mesarosova et al. also found in their study that pilots scored lower on extraversion and openness to experience, which was not consistent with other studies in which military pilots tended to score higher on extraversion. However, of specific interest in the context of this book is evidence of how personality might shape behaviour in an operational context. Behrend et al. (2017) looked at the trait of impulsivity and pilot decision-making. Pilots low on impulsivity were more likely to comply with procedures, but interestingly, high-impulsivity non-compliers reported that they sometimes made decisions outside of procedures in order to avoid worst-case scenarios. Impulsivity did not seem to be affected by training, but more experienced pilots expressed higher levels of indecisiveness which the authors attributed to the fact that the more experienced pilots 'recognise the uncertain character of the decisive situation and its complexity by delaying their decision'. Finally, in a study of German military helicopter pilots, it was found that those who score higher on extraversion were more likely to continue flight into worsening conditions than those of a more introvert disposition. It seems, then, that aspects of personality may shape the efficacy of performance in a system's context.

Behaviour is also shaped by the attitudes we possess. Personality traits are generic and relatively stable in disposition, whereas an attitude is a tendency to behave in a particular way in relation to some object in our world. Attitudes are short term and malleable: they are modifiers of behaviour. They possess three component parts: a cognitive component, a conative component and, finally, a behavioural response. The first describes what we know about the object; the second is what we feel; finally, the third is what we actually do. Attitudes can often be changed by providing information, which is to address the cognitive component directly. But, just as 'personality' is not a cast-iron guarantee of behaviour, attitudes are not programmed responses. For example, I know that smoking is a health risk, and I do not like smoking or being close to smokers, but I have tried cigarettes in the past, just to see what the fascination is. In a more rigorous experimental example, women who were provided with information about the health benefits of breastfeeding and who stated that they believed breastfeeding was the best method of weaning infants nonetheless still chose to bottle feed their own babies. In Chapter 6, we will look at pilots committing violations. In a sample of LOSA data, I found that some captains chose to continue using procedures they were familiar with when operating a new aircraft type (in this case, A-330 procedures were applied to the A-350 because the pilots were more familiar with the older processes). Those same captains would probably stress the importance of following standard operating procedures. This is an example of an attitude in an aviation context. Interestingly, it is typically captains who demonstrate this behaviour, not first officers (FOs). In a social context, captains have the status that affords them the opportunity to act unilaterally.

The topic of interest here is the idea of procedural compliance. All pilots know that procedural compliance is a requirement, often mandated in regulations, but pilots occasionally believe that 'the old way was better' or 'we could do this a simpler way'. The outcome is labelled as 'non-compliance' or a violation, but it is actually a manifestation of the attitude held by the pilots towards 'the rules'. It is also an example of 'drift' as discussed in the previous chapter.

Closely related to attitude is the concept of motivation. Motivation is described as those behaviours directed at achieving desired, or avoiding undesired, goals and can be either intrinsic or extrinsic. Intrinsic motivation is drawn from within the task or topic whereas, in the case of extrinsic motivation, the action is driven by external factors. For example, I once interviewed a class of sponsored students at a flight school about their reasons for becoming airline pilots. Some said that they loved flying and always wanted to become a pilot. Their motivation was intrinsic. Others made comments such as 'my father and grandfather were airline pilots' or 'people look up to you if you are an airline pilot'. Their motivation was extrinsic. Studies of performance in education have found that students who possess an intrinsic interest in their subject perform better in exams than students with extrinsic motivation: the subject was chosen because it had to be done, not because the students wanted to do it. Like personality, motivation has clear psychological and physiological mechanisms. Risk aversion is a well-established trait and represents motivation at work. Later in this chapter, we will look at disruption in the workplace. Pilots who find disruptions, or variability, to be an enjoyable part of the job are probably intrinsically motivated by the challenges represented by a demanding work schedule.

Behaviour is also shaped by our level of self-esteem. Broadly speaking, self-esteem captures an individual's belief in their self-worth. It derives from beliefs about ourselves as people, our relationships with others and our membership of larger social groups. The implication is that action will be moderated in relation to the potential impact on self-worth. For example, a common reason given by junior or inexperienced pilots and cabin crew (and nurses in a hospital setting) for not voicing concern at work is that they do not want to seem ignorant in the eyes of more experienced peers. Self-esteem is linked to subjective well-being and, therefore, is relevant to the discussion later about, among other things, work/life balance.

One final attribute we need to consider is self-awareness. Possibly, a dimension of metacognition, our ability to judge the impact of our performance on others and our willingness to reflect on the causes of unsuccessful interactions will influence our effectiveness. Self-awareness is an internal feedback loop that allows us to assess and modify our behaviour.

In the next section, we will look in more detail at the stress mechanism, and we saw in Table 4.1 that the personality trait of neuroticism is associated with aspects of stress. Those more disposed towards neurotic behaviour are more susceptible to stress. Furthermore, fear, a response to an extreme stressor, can be viewed as a negative incentive salience in a motivational model, that is, I avoid doing things that might expose me to fearful situations. These concepts can be difficult to disentangle, but they all contribute to a view of behaviour having physiological, and ultimately genetic, underpinnings.

4.3 STRESS AS A BIOLOGICAL PROCESS

In simple terms, the stress mechanism is an evolutionary adaptation for living in a threatening environment. For most of hominid evolution, life has been very threatening. The ancestors of modern humans periodically needed to increase their level of physical performance in order to defend themselves or to get out of harm's way. What we now call the 'Fight or Flight' mechanism is the outcome of that evolutionary process. The mechanism works over two separate time scales. We have an immediate response to an apparent threat to our well-being, which manifests as fear. But there is a slower-acting response, triggered by an ill-defined or uncertain threat, which we call anxiety, but which triggers the same physiological process.

The stress response or general adaptation syndrome (Selye, 1950) can be divided into three phases: the alarm phase, the resistance phase and the exhaustion phase. In the alarm phase, initiated by the detection of a threat or danger, a sequence of responses is triggered in the endocrine and autonomic nervous systems. Adrenaline (epinephrine) and noradrenaline (norepinephrine) are released into the bloodstream. The responses to this hormonal stimulation are summarised here:

- The rate and strength of the heartbeat increase in order to pump oxygenated blood around the body faster.
- The spleen contracts to release stored red blood cells to increase the oxygen-carrying capacity of the blood.
- Sugars stored in the liver are released to provide energy for the muscles.
- Blood is redirected from the skin and viscera to the muscles and brain.
- The bronchi in the lungs dilate and our breathing deepens.
- The pupils dilate, probably in order to improve visual efficiency.

- Lymphocytes and coagulating agents are released into the blood to improve the body's ability to repair wounds.
- Sweat production is increased to improve the cooling effect.

If the stress mechanism continues to be stimulated, then the body moves into the resistance phase. A group of chemicals is released that facilitate the conversion of non-sugars into sugars, and the rate of sugar storage in the liver increases. Growth hormones are suppressed, as are functions associated with sexual reproduction. The body continues to adapt itself to the potential demands of increased immediate workload, but now we start to see an impairment of the immune system. In effect, the resistance phase seems to involve an increased adaptation to the specific threat but a reduced ability to defend against other stressors. Finally, if exposure to stress is prolonged, then the body simply runs out of energy while some of the harmful by-products of the physiological response reach debilitating levels. This is the exhaustion phase, often characterised as 'burnout'.

One consequence of the disconnect between the evolutionary selection pressure and the modern human condition is how environmental conditions will trigger the stress mechanism. Known as environmental stressors, they include heat, vibration and humidity. Most people feel comfortable at 20°C while actual discomfort can be experienced at temperatures above 30°C and below 15°C. Some helicopter cockpits can get up to temperatures of 50°C while cabin attendants working internal routes in some Gulf States report similar temperatures in the back of older aircraft types. High temperatures lead to an increase in heart rate, blood pressure and sweating. Besides, some information processing tasks can be degraded. Low temperatures lead to a loss of feeling and control of limbs. We can moderate the effects of temperature by controlling such factors as insulation (clothing), ventilation and level of activity (metabolic rate). Shivering is one way of increasing our level of activity, but both sweating and shivering consume energy and can result in fatigue.

Vibration includes the direct vibration of the body and the indirect effects of vibrating air; noise. We 'hear' vibrations between 20 and 20,000 Hz. However, the perception of noise is affected by the relationship between the frequency and intensity, measured in decibels. Our ability to detect noise also varies with age. Excessive noise can lead to irritability and, eventually, cardiovascular problems. Low-frequency whole-body vibration can cause a decoupling of sensory inputs, leading to motion sickness. One common source of whole-body vibration in aviation is turbulence. Of course, turbulence can directly lead to injury and death when unrestrained objects and passengers are thrown around the cabin. However, on the flight deck, turbulence can induce a level of stress that impedes information processing while extreme turbulence can make it impossible to read displays accurately. Vibration can interfere with breathing, cause chest pains and lead to speech disorders and headaches.

The extent to which we experience humidity is linked to the ambient temperature. Ordinarily, a comfortable level of humidity, although it varies with temperature, is between 50%–60%. However, the humidity of conditioned cabin air in older generations of aircraft is between 5% and 15%. This can lead to drying of the mucous membranes of the eye, nose and throat and reduced rate of urine production, which, in turn, can lead to additional medical problems. While environmental factors can create a general background level of stress and should not be ignored, it is fear and anxiety that are of more immediate interest as performance inhibitors. I will look at fear in the next section but come back to anxiety when we look at the effects of fatigue on performance.

4.4 STRESS, FEAR AND 'STARTLE'

On 20 May 2002, a Boeing B-727 was approaching East Midlands Airport in the UK (Air Accident Investigation Branch (AAIB), 2003; personal communication). Before their departure from Copenhagen, the crew had received the destination weather forecast, which included a warning of thunderstorm activity. The crew joked that English thunderstorms were nothing compared to the weather systems in their native Canada. The flight from Copenhagen was uneventful, but as they approached the airport, they encountered heavy rain and moderate turbulence. Having been vectored clear of weather, the crew began the approach. In order to intercept the localiser, the captain, who was the handling pilot, had to turn through 120°, descend from 3000 to 2000 ft and decelerate from 225 to180 kts. In a descending 30° banked turn, the captain extended the speed brakes. At some point in the manoeuvre, flap 2 was selected, and the stick shaker activated. Initially, this was interpreted as turbulence but, in fact, was an indication of the aircraft approaching the stall. The captain applied go-around thrust. The aircraft pitched to 7° nose down and, having flown through the localiser, was descending fast.

At 800 ft agl, the captain heard the second officer (SO), who occupied the flight engineer's seat, warning about the speed brakes. By now, the aircraft had accelerated to 235 kts. The speed brakes were stowed, and the aircraft rapidly accelerated further to 250 kts, climbing to 3000 ft before the aircraft was restored to the correct pitch and power. Later, during the investigation, the captain was able to listen to the cockpit voice recording. He was shocked to hear, on the tape, the SO clearly and repeatedly warning of the extended speed brakes from the moment flap 2 was selected. The unexpected severity of the turbulence, the need to execute a complex manoeuvre and the debilitating effects of the rapid onset of stress all played a part in this story.

The personal experience of this captain shows how quickly performance can be impaired. The initial exposure to turbulence was an environmental stressor, and the need to make changes to the planned approach routing represents a plan change, which I will discuss when we look at anxiety. Fatigue was possibly a factor as the crew of the B-727 had flown from Canada to undertake the duty from Copenhagen to the UK. In all probability, they were working off their normal body clock. The rapid, temporary loss of control, together with the captain's momentary impairment is an example of the fear, or 'startle' component of the stress mechanism.

The human 'startle' reflex represents an immediate, often instantaneous, reaction to a situation perceived to be highly dangerous or life-threatening. Startle seems to work through two pathways (European Aviation Safety Agency, 2018; Landman et al., 2017). The first, fastest, pathway is mediated by the amygdala is very old in evolutionary terms and is clearly intended to be a defence mechanism. A slower pathway is routed via the neocortex. The amygdala pathway can often be so fast that we are not really aware of the initiating stimulus; our response is akin to a reflex action. The slower pathway allows us to comprehend the problem. Fear and the associated startle can severely inhibit our ability to reason, interpret and resolve problems. This is usually referred to as the amygdala hijack. The Air France flight AF 447 accident over the Atlantic Ocean in 2009 has become the archetypical example of fear resulting in startle in the aviation industry (BEA, 2012). The example of the B-727 given earlier suggests that fear-induced stress triggering is more widespread than might be thought. Here is the experience of the FO on a B-757 attempting an approach into Puerto Plata, in the Dominican Republic, in 1998 (AAIB, 1999):

> We had broken off a non-precision approach and instead of going around, the captain called for a visual circuit. Although the cloud-base was down to 800ft agl, I had no reason to doubt that we would make a safe landing. I had flown with the captain before and knew him to be a respected pilot. As we turned finals, it became clear that we were going through the centre-line, and we were now manoeuvring a large aircraft close to the ground. I suddenly realised that my field of view had narrowed to a small area immediately in front of my face. My peripheral vision had gone. If I wanted to see the captain, I had to turn around in my seat and look directly at him. To scan the instruments, I had to move my head from one instrument to another. My head filled with voices that I could not get rid of. I snapped out of it when the GPWS alert went off and the captain decided to go-around.
>
> *(Source: personal communication).*

Once again, the most obvious effect of the stress mechanism being triggered was the impairment of the senses with the visual field, ironically, narrowing down to better detect danger. Loss of hearing is also a common symptom in acute stress situations, and we saw earlier that the captain of the B-727 did not initially hear his SO. In evolutionary terms, the adaptive advantage of the stress mechanism (increased chance of survival by avoiding harm) has been offset by changes in society.

Current thinking on stress places less emphasis on its function in avoiding immediately life-threatening challenges. It is now more often taken to be a reflection of an individual's ability to cope with the pressures of everyday life. Specifically, it is the relationship between the perceived demands placed upon us and our perceived ability to cope with those demands. This perspective addresses the problem of more general anxiety rather than fear, and we will come back to anxiety later. First, though, I want to look at fatigue.

4.5 SLEEP AND FATIGUE

Here is a definition of fatigue:

> Fatigue is a suboptimal psychophysiological condition caused by exertion. The degree and dimensional character of the condition depends on the form, dynamics and context of exertion. The context of exertion is described by the value and meaning of performance to the individual; rest and sleep history; circadian effects; psychosocial factors spanning work and home life; individual traits; diet; health, fitness and other individual states; and environmental conditions. The fatigue condition results in changes in strategies or resource use such that original levels of mental processing or physical activity are maintained or reduced (ICAO, 2016).

Hockey (2013) suggests that 'fatigue' has been 'hijacked by the sleep scientists' with important issues around what he refers to as psychological fatigue being overlooked. To further complicate matters, it is not at all clear whether fatigue is an actual condition or simply the symptoms associated with some underlying state. For these reasons, in the rest of this chapter, I want to distinguish between physiological 'fatigue' arising from sleep deprivation or disruption and psychological 'fatigue', which is a longer-term problem resulting from life's challenges. The two aspects do, in fact, interact but warrant separate treatment.

Physiological fatigue, quite simply, is a failure to meet our sleep requirements and results in a state of sleep deprivation. The implications of sleep loss in terms of lost productivity and increased health care costs have resulted in an explosion of interest in, and research into, sleep but, in their review of the proposed changes to the Australian flight time limitation rules, the authors comment that 'fatigue science is very rarely definitive or conclusive' (Civil Aviation Safety Authority (CASA), 2018). The issue of fatigue management can be contentious, even political, in the aviation industry, but the fact remains that the topic is still poorly understood. Most national aviation legal frameworks require operators to implement 'legal' rosters but the regulations also place a responsibility on the individual to report for duty adequately rested and prepared. Reconciling these two positions is difficult.

4.6 WHAT IS SLEEP?

Sleep is, quite simply, a state of unconsciousness. The purpose of sleep seems to be to allow the body to replenish energy stores, repair the wear and tear endured during the day and to organise memory. While we sleep the brain exhibits four distinct types of activity.

- Stage 1, light sleep, involves the brain becoming less responsive to environmental stimuli.
- Stage 2 occurs throughout sleep and represents almost half of the total sleep time.
- Stage 3 is distinguishable by the amount of delta wave activity observable.
- Stage 4 is known as rapid eye movement (REM) sleep and represents about a quarter of the sleep period.

The sleep process is dynamic, and the brain will move through the various stages over cycles lasting approximately 90 minutes. The stages may vary in duration. For example, the first REM episode may last just 10 minutes while the final episode may last an hour. For most people, between 6 and 9 hours of sleep per night is normal although 8 hours of sleep is considered necessary to be fully restorative. It seems that no single 'sleep centre' exists in the brain. Instead, a series of sleep circuits take inputs from the brainstem and the hypothalamus. The sleep cycle does appear to be synchronised with the body's circadian rhythm with periods of wakefulness associated with a rise in the body core temperature while sleepiness coincides with a fall in core temperature. Two physiological processes regulate the need for sleep. First, during the hours of wakefulness, adenosine, a by-product of energy consumption, builds up in the body, and this increase triggers the need for sleep so that energy reserves can be replenished. Second, exposure to light produces melatonin. During the hours of darkness, melatonin production is inhibited and that synchronises the sleep cycle with the period of darkness. Humans are diurnal animals, after all.

The fundamental regulation of sleep – known as the homeostatic model – is disrupted by the very nature of work in aviation. Different types of operation – air taxi, helicopter, regional, low-cost, long-haul, cargo – all have different characteristics but all share some of the following issues:

- The work itself can be fatiguing because of the exposure to environmental stressors. It has been suggested that older pilots, especially those who smoke, could be in a constant state of marginal hypoxia during the cruise because of the relative cabin altitude.
- The work schedule usually has variable start times to the duty period and, especially for long-haul flying, could be classified as a form of quasi-shift work. Both conditions have been linked to health issues.
- Passage through time zones and the subsequent disruption of sleep patterns is a source of fatigue.
- Demands for productivity and disrupted schedules can affect recovery time between tasks.

Sleep deprivation studies have shown that after 24 hours of wakefulness, the brain's metabolic activity is reduced by 6%, with reductions of up to 11% in specific brain centres. Higher-order cognitive tasks are affected earlier than other brain activities. Where tasks require both speed and accuracy, experiments have shown that speed of response decreases before accuracy is impaired. What is more significant is that the performance can be affected by apparently small reductions in sleep. Subjects restricted to 7 hours of sleep a night for 1 week showed degraded performance in tests of response time and in mathematical problem-solving. Tests in driving simulators showed a reduced ability to focus on several tasks at once, and accidents in simulations increased as sleep duration was reduced from 7 to 5 and then 3 hours per night. Sleep deprivation seems to have the greatest impact on motor control whereas psychological fatigue seems to affect the brain's executive control functions (Hockey). Physiological fatigue can be considered to be acute or chronic. Acute fatigue is a sudden onset condition and is mitigated by sleep. Chronic fatigue is an underlying, persistent state not necessarily mitigated through access to sleep.

4.7 ACUTE FATIGUE IN PILOTS

Two scales frequently used in fatigue studies are the Samn–Perelli 7-point fatigue scale(S-PS) (Samn & Perelli, 1982) and Karolinska sleepiness scale (KSS) (Akerstedt & Gillberg, 1990). The S-PS asks pilots to rate their current level of arousal on a 7-point scale while the KSS asks pilots to rate their level of sleepiness on a 9-point scale. These scales, then, tap into the individual's current status and their acute fatigue, and a higher score indicates a worse condition. Both instruments are well established in aviation fatigue studies and have been validated against objective measures.

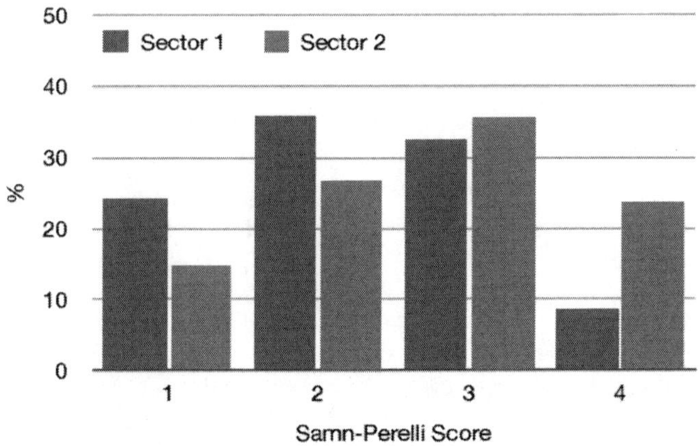

FIGURE 4.1 S-PS scores across two sectors.

I looked at the results of 125 pilots operating a 2-sector day from their home base (Figure 4.1). The working day was fairly routine in terms of start and end times, and no crew flew during the night. I found a moderate negative association between the hours of sleep in the previous 24 (−0.4128 $p < 0.05$) and 48 (−0.4877 $p < 0.05$) hours and S-PS scores. The less sleep pilots had before a duty, the worse was their perception of their state of arousal (higher scores). There was a weak positive association between the hours awake before duty and the S-PS for all pilot groups. The longer pilots were awake before their duty, the more likely they were to feel more fatigue. There was a statistically significant ($p < 0.045$) difference in S-PS scores between the sectors, with higher scores recorded in the second sector. Half of the pilots also completed the fatigue severity scale (FSS), and there was a moderate association between their S-PS and FSS scores (0.42886, p 0.00329). These data suggest that sleep duration, hours of wakefulness and duration of duty affect pilots' acute fatigue state, but the FSS scores indicate that pilots' chronic fatigue state will also be a moderating factor. An S-PS score of 4 is considered the point at which fatigue is starting to become significant. Of the total sample, 16% scored 4 on the SPS while 31.8% scored 1 on the scale, indicating that they felt fully alert. This suggests that under routine daytime conditions a significant number of pilots are operating in a marginal state.

Powell et al. (2008) looked at S-PS scores in a large sample of pilots flying two-crew regional operations. The scores were taken at the top of descent. The highest levels of fatigue were observed in the window of circadian low (WOCL) (0200–0600) and were also associated with the length of the duty day. Fatigue was 0.56 points higher on the S-PS at the end of two sectors compared with a single-sector duty. Aeschbach et al. (2017) found a relationship between the length of the duty, sectors flown and fatigue as measured with the S-PS. Fatigue increased by 0.4 points for every hour awake and 0.3 points for every sector flown. Sallinen et al. (2017) used the KSS to study pilot fatigue and, similarly, found that duty periods that covered the whole local night-time period (i.e. 00:00–06:00 at the domicile) were consistently associated with reduced subjective alertness.

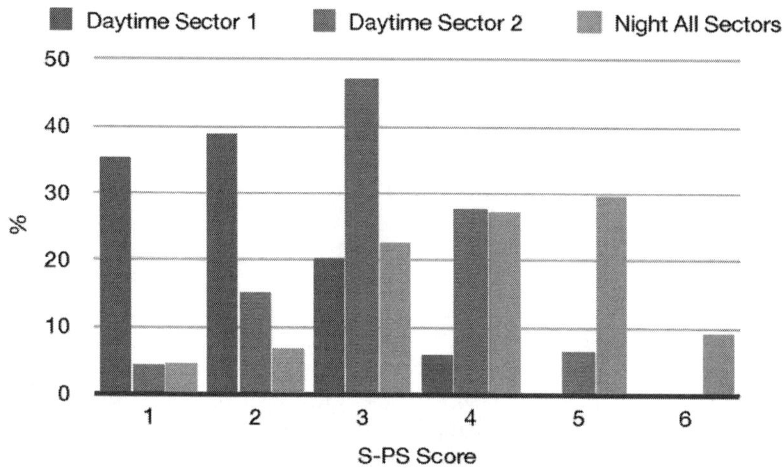

FIGURE 4.2 Types of flying and SPS scores.

Figure 4.2 shows data for a larger sample of regional daytime flights and also includes data for crews flying two-sector night flights. The night flying crews provided SPS data at sign-on (before flight) and after landing for each sector. There was a statistically significant increase in S-PS scores between pre-flight and landing for the first sector but no significant increase across the second sector. These data reinforce the earlier findings that acute fatigue is cumulative and that night-time operations increase the exposure. The apparent plateauing of fatigue in the night flying crews was possibly the effect of a small sample size.

Acute fatigue in pilots, then, is a function of hours of sleep and time awake before a duty, the length of the duty period, the number of sectors flown and the time of day. Furthermore, acute fatigue is cumulative.

4.8 ACCLIMATISATION AND NIGHT FLYING

Long-haul flying and transiting through multiple time zones result in circadian arrhythmia, or 'jet- lag'. In a study of cargo pilots operating in Asia and North America, the crew were asked to complete the S-PS and also to report whether they felt that they were acclimatised to the local time zone. As a very rough rule of the thumb, it takes 1 day per time zone crossed to become acclimatised. In reality, the process is hugely variable, so I relied on each individual to judge their personal state. The first comparison made was of the difference between a sample of acclimatised passenger aircraft pilots flying daytime operations and the cargo crews who typically flew through the night. Table 4.2 shows the distribution of S-PS scores in the two groups. For comparison, data for a regional LCC are included (Yen et al., 2009).

The data suggest that, in a group of acclimatised pilots, 17.6% of pilots operating during daytime and 31.25% of pilots working night schedules experienced excessive acute fatigue (S-PS $\geq$ 4). The comparison with the Low Cost Carrier (LCC) pilots is interesting and might reflect the impact of high rotation, multiple sector days. Different types of flying, then, have different fatiguing characteristics.

TABLE 4.2
Distribution of S-PS Scores (%)

S-PS	Regional	Cargo	Regional LCC
1	13.4	6.25	0.03
2	34.4	18.75	10.08
3	34.4	43.75	64.47
4	16.8	25	24.43
5	0.8	6.25	1

When I looked at the difference between the acclimatised and non-acclimatised cargo pilots, I found that there was a moderate positive association between the hours awake before the duty and the S-PS score in non-acclimatised pilots (0.49, $p < 0.05$). The longer they had been awake, worse was their perceived state of arousal. However, there was no difference between the groups in terms of the average actual hours of wakefulness when compared to their daytime passenger fleet colleagues (passenger 3.34 hours and cargo 3.23 hours). It seems, then, that fatigue builds faster in non-acclimatised pilots. Furthermore, the difference between all-cargo pilots (acclimatised but operating night flights; not acclimatised operating local time night flights while on home base daytime body clock) and the acclimatised pilots flying daytime regional flights in terms of their SPS scores was significant. Pilots who are not acclimatised to the local time zone are starting the day in a worse fatigue state than the acclimatised crew, and night operations seem to be equivalent to flying non-acclimatised.

These results point to the complicated nature of sleep and fatigue. Pilots encounter acute fatigue (as measured by the SPS) routinely in the workplace. Acute fatigue is cumulative across shifts and is related to prior sleep and wakefulness. It is also related to the nature of work (night flying; high rotation) and also the extent to which the pilots are acclimatised to the local time. Having looked at acute fatigue as experienced by pilots during working duties, I now want to move on to chronic fatigue.

4.9 CHRONIC FATIGUE IN AIRLINE PILOTS

This next section is largely based on findings from a survey I conducted of around 350 pilots in a single airline. The sample captured 10% of the pilot population and was representative in terms of age, gender, aircraft fleet and rank. One problem with conventional fatigue risk management models is that they deal with short-term sleep and recovery, which affect acute fatigue levels but fail to take into consideration the endemic, steady-state chronic fatigue experienced by individuals. Most studies of chronic fatigue in the population use questionnaires originally designed to assess recovery from major illness or surgery. As yet, no specific measure of pilot chronic fatigue has been developed although there have been a few studies of pilots that have used the same measurement instrument: The FSS (Krupp et al., 1989). The FSS is a 9-item questionnaire which asks individuals to rate their current state on a 7-point scale. A mean score of 4 or greater indicates excessive fatigue.

Hockey (2013) suggests that about a third of the general population would score above this threshold.

In my survey, 74.4% of the pilots scored 4 or more on the FSS with a mean of 4.6. Pilots in Indonesia (Yuliawati et al., 2015) reported a mean FSS score of 4.66. Two studies of Portuguese pilots (Reis, Mestre & Canhão, 2013; Reis et al., 2016) found that 84% and 90.6% respectively of pilots scored above the threshold as did 68.3% of pilots flying in the Gulf States (Aljurf et al., 2017). Chronic fatigue, as measured by the FSS, would appear to be more than twice as prevalent in pilots as the general population.

To better understand the distribution of chronic fatigue symptoms, a smaller sample of 80 pilots completed the Oldenburg burnout inventory (OLBI) (Demerouti et al., 2003) and the FSS. The OLBI generates two scores: disengagement and exhaustion. The sample specifically included a control group of pilot managers on the grounds that, because of their reduced flying task, they were more likely to be representative of rested pilots. Of the pilots surveyed, 10.71% met the criteria to be considered at risk of burn out on the OLBI (scoring in the upper quartile on both scales). The mean FSS scores for the rested and burnout risk groups were used to create cut-off values which were then applied to the FSS scores for the larger, original survey. This created three groups labelled rested, intermediate and burnout risk. Some 21.89% of the larger sample of pilots scored equal to or better (lower) than the 'rested' cut-off on the FSS while 34.31% scored equal to or worse (higher) than the burnout risk group, leaving 43.79% occupying the middle ground. The FSS cut-off for the rested group was 3.77, which is similar to the standard threshold score of 4. The burnout risk group had a cut-off of 5.14.

Respondents in the original survey also answered questions on sleep quality and daytime sleepiness. The Jenkins sleep quality (JSQ) scale (Jenkins et al., 1988) measures rates of sleep disturbance and difficulty in sleeping, an extreme case of which would be classed as insomnia. None of the rested group scored above the threshold on the JSQ whereas 2.9% of the intermediate and 6.7% of the burnout risk group showed evidence for poor sleep quality.

The Epworth daytime sleepiness scale (ESS) (Johns, 1991) captures symptoms of fatigue manifested as a tendency to doze off during the day. Figure 4.3 shows the distribution of scores for the sample of pilots. Those pilots in the burnout risk group

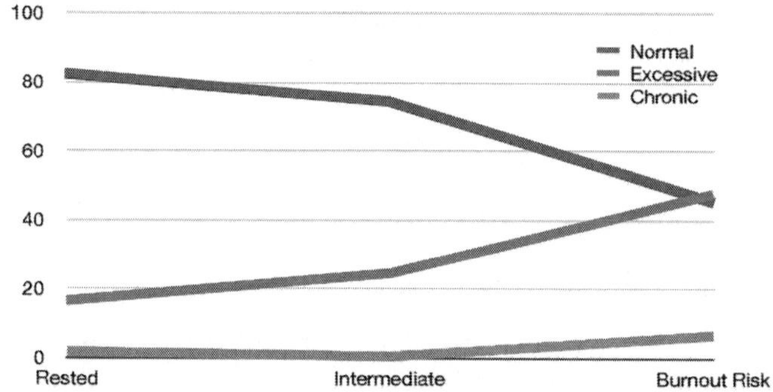

FIGURE 4.3 Distribution of ESS scores (% within the group).

had the highest incidence of abnormal daytime sleepiness and also had a lower rate of normal conditions. However, all three groups contained pilots with issues with daytime sleepiness, and it seems that the transition from a normal state to a chronic state starts slowly but then accelerates approaching burn out.

4.10 RECUPERATION

I said earlier that whereas acute fatigue can be mitigated through sleep, chronic fatigue could not be so easily remedied. This difference may be explained in part by the fact that recuperation is partly a function of the recovery time between periods of work but also involves having the ability to psychologically detach from work. Searle (2012) discussed this need for detachment in a study of the characteristics of crew hotels. Akerstedt et al. (2000) suggested that for workers in demanding roles, such as aircrew, a single day off after a block of workdays offered inadequate time for recovery, and 3–4 days would be required after periods of disturbed circadian rhythmicity.

In addition to the FSS, crews who took part in my survey also completed the need for recovery (NFR) scale (Van Veldhoven & Broersen, 2003). The NFR has a cut-off score of 4, and 80.2% of all respondents scored above that threshold, a number – perhaps unsurprisingly – that almost matches the number of pilots reporting excessive fatigue. Table 4.3 shows the distribution of NFR scores by fatigue group. All groups show a significant recovery need but, interestingly, the intermediate and burnout risk groups suggest that recovery is a significant factor in the fatigue problem.

The sample of pilots used to construct the FSS cut-off values also completed the Whitworth inter-shift recovery scale (Winwood et al., 2005). The output is reported as a percentage (Figure 4.4), and the higher the value, the more recovery has been achieved. The difference between the three groups is statistically significant.

Inadequate recovery from the demands of work would appear to be a significant factor in the prevalence of chronic fatigue in airline pilots. Table 4.3 showed an almost universal NFR among pilots in the upper two fatigue groups, and this is reinforced by the data presented in Table 4.4. A Significant number of pilots in the sample, then, are experiencing chronic fatigue (as measured by the FSS), probably because of an inadequate recovery between the blocks of work. The persistent, debilitating effects of chronic fatigue, a physiological condition, make it a significant contributor to psychological fatigue. Furthermore, there are a group of pilots occupying a space between the 'rested' and 'burnout' who present a different challenge. From an organisational perspective, it would be interesting to know how to stem the drift towards 'burnout risk'. At this point, I want to return to the other elements of the stress response discussed earlier – anxiety – and to try to integrate the process into the fatigue model.

TABLE 4.3
Distribution of Need for Recovery

	Rested	Intermediate	Burnout Risk
Inadequate recovery	66.1	94.8	96.2
Adequate recovery	35.9	5.2	3.8

TABLE 4.4
Inter-shift Recovery

	Rested (%)	Intermediate (%)	Burnout Risk (%)
Need for recovery	67.6	44.64	34.25

4.11 ANXIETY AND PSYCHOLOGICAL FATIGUE

The data presented in the previous section suggest that physiological fatigue is experienced by pilots in both acute and chronic states and that adequate recovery is a significant factor in both conditions. Psychological fatigue is shaped by our physiological state but is manifested through emotional and motivational responses to our environment. We also saw that the stress mechanism was triggered by both fear and anxiety. Whereas fear was an immediate reaction to an apparent existential threat, anxiety was a response to an intangible stimulus. The category of stress triggers associated with anxiety is often labelled 'life stress events'. They are typically problems that we might not know how to deal with or what effect they will have on our lives. The common feature of life stress events is that they represent change and uncertainty, as the following example shows:

> 'My marriage had been falling apart for a couple of years and it got to the stage where I would get up early to be out of the house before my wife woke up. On this particular day I did not manage to get away early enough and my wife and I had a very angry argument, which left me seething with rage as I drove to the airport. I remained in this state throughout flight planning and right up to the incident. I had been warned of a vehicle in the undershoot. I had been given radar vectors to get in between 2 faster aircraft and ATC requested that I land short in order to clear by an early taxiway. Just before touchdown there was a solid bump as I clipped the vehicle. A couple of days later I landed on the same runway and the vehicle was in the same place. It could not have been more obvious but on the day in question I just did not see it.
>
> I realised afterwards that my total loss of concentration was caused by the fact that my mind was full of the row I had had with my wife. We later divorced and I understood for the first time what a strain I had been under. At the time I had been oblivious to the problem and if anyone had suggested that I needed help I would have thought that they were joking'.
>
> *(Source: UK Confidential human factors reporting programme – CHIRP)*

The need to balance the demands of work and home life is reflected in the work/life balance (WLB) concept. The pilots in my survey were asked to assess their own experience of dealing with the demands of home life using the work–home negative interaction scale of the SWING tool (Geurts et al., 2005). This scale assesses the degree to which work has a negative impact on home life and has been hypothesised to reflect the individual's fatigue state. Figure 4.4 shows the percentage distribution of pilots in the top and bottom quartiles on the scale (higher scores reflect better WLB). Chronic fatigue and negative work–home interaction show a strong association. Demerouti et al. (2001) make the claim that burnout negatively relates

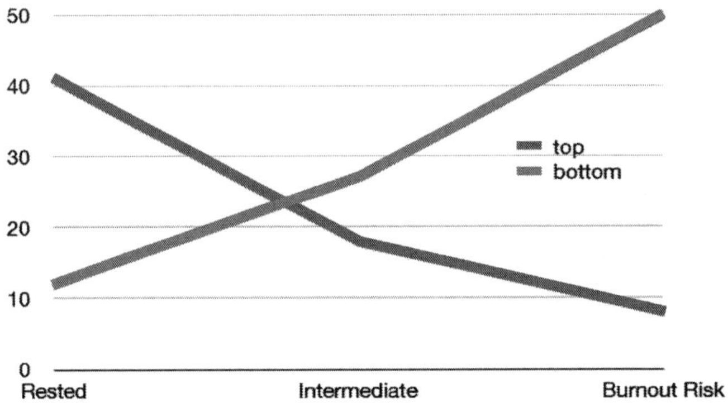

FIGURE 4.4 Negative work–home interaction (percentage of respondents in each quartile).

to happiness and generalises outside of work, impacting on overall life and that of families, too. This relationship is suggested by the survey data.

Hockey identifies three sets of demands we all have to manage: things we must do, such as eat and drink; things we should do, such as comply with instructions; things we want to do, like spend time with friends and family. At any moment we typically have multiple demands to satisfy. He proposes that psychological fatigue shapes the way we cope with multiple demands. In Chapter 3, we saw that Hollnagel's (2009) efficiency/thoroughness trade-off principle proposed that performance can be evaluated in terms of the time invested in completing the task and the degree of thoroughness, or quality, with which the task is accomplished. Demerouti and her colleagues (2019) also looked at the relationship between burnout and performance in the flight simulator during recurrent training. No effect was found, illustrating the complex relationship between fatigue and performance. What the authors did find was that pilots showing high levels of burnout were unwilling to seek additional challenges in training that would improve their skills. Requesting to do more than it is necessary for training is a discretionary task that is happily shed by tired pilots. As well as affecting the way we manage out-of-work demands, psychological fatigue affects the way we cope with the job itself. For the group of pilots in my study, when the data for the disengagement scale were examined, all of the burnout risk group were in the upper quartile on the scale (i.e. less engaged), whereas all of the 'rested' group were in the lowest quartile (most engaged). The data suggest that pilots will shed tasks or elect not to participate in discretionary action when suffering excessive psychological fatigue.

Work stressors include such things as your role in the organisation, relationships at work, concerns about career development and organisational structure and climate. In aviation, the lack of control over one's personal life is often cited as a significant cause of stress. This loss of autonomy sets up a dissonance, contrasting the supposed high level of professionalism associated with the role with apparently inflexible work schedules that conflict with out-of-work demands. In an operational context, constant changes in schedules, alterations to the in-flight service, poor organisational communication, the feeling of too much to do before departure and the behaviour of some passengers are just some of the anxiety-inducing conditions encountered.

Data captured during a LOSA give us some idea of the rate of operational disturbance encountered by the crew. A 'threat', according to LOSA, is something external to the crew that has potential to cause a need to make a change to a plan or be prepared to take appropriate action at some point in the future. A threat can also be a constraint requiring the crew to be aware that their original plan might not be feasible. Threats, then, are perturbations that disturb the normal flow of work, and it is this induced variability or departure from the expected task that adds complexity and increases the workload. Threats can be considered as triggers for anxiety in that they represent problems that have to be dealt with as part of daily life, the implications and consequences of which are not always clear.

In a representative sample of 20 sectors, the average threat rate was 7, with a range of 2–17 threats per sector. The pre-departure phase of flight had the highest threat rate (46.9% of all threats). The need to deal with extreme weather represented 30.61% of the recorded threats of which 'thunderstorms and/or turbulence' was the single largest subcategory. Air traffic control (ATC) comprised 14.28% of the threats, over half of which were ATC clarity. Threats will vary according to season and destination.

Threats are of interest because they influence the work process. For example, the need to repeat ATC transmissions consume time and can lead to rushed actions. Airport issues made up 14.28% of the sample of which three-quarters of the observations were to do with closed runways and taxiways. In terms of crew behaviour, these 'threats' usually require the crew to replan their routing or consider a different arrival or departure. A failure to effectively cope with threats increases the risk of an error being committed by the crew. We will look at error in more detail in a later chapter but, broadly speaking, 20% of errors committed by the crew in a sample of 177 flights were associated with a threat. I should stress that it would be wrong to assume a causal relationship between a threat and an error. Threats absorb attentional capacity, thereby increasing the risk of an error.

In an attempt to quantify the relationship between the interruptions, a commonplace disruption of workflow, and the propensity for error, LOSA observers were asked to note how many times the crew were disturbed during the period from their boarding the aircraft to closing the passenger doors. An interruption was defined as an occasion that required one or both crew members to pay attention to, or respond to, a request from a third party on the flight deck, be it cabin crew, engineers or some other agency. The observers were also asked to note whether the crew subsequently committed an error after the interruption (typically missing an item on a checklist or failing to complete a checklist). In the sample of 177 flights, 80% of departures had an interruption (range 1–22). Crew performance was not affected by interruption rates of up to 5 per departure. From six interruptions onwards, error rates increased and by ten interruptions the probability of an error was 100%. From 11 to 22 interruptions – the maximum observed – the error rate returned to zero: something had now gone wrong, and so the need for care was apparent. These disruptions to the work process, in all probability, represent a low-level anxiety trigger, akin to a moderate surprise. In short, an unexpected, novel event triggers a requirement to reconcile the difference between the planned and the actual situations. Interruptions and distractions occurring at time-critical phases, such as preparing to depart, consume time already allocated to fixed tasks. These low intensity 'surprises' can trigger some of the physiological components of the stress mechanism. In a study of short-haul pilots, Aeschbach et al. (2017) looked at what they called 'hassles', which equate to 'threats' in LOSA terminology. Although 'hassles' had no effect on fatigue, they did have a significant effect on the pilots' subjective assessment of workload as measured by the NASA task load index. Threats, then, increase the workload, and this elevated level of demand increases the risk of error. The following report from a B-777 captain beautifully illustrates how the job of work is, almost by definition, subject to disruptions.

> 'The NOTAMs included a reference to work in progress on a runway that needed a code to be inserted in the take off weight calculation page of the computer. In addition, the Maintenance Log noted an outstanding item, "Semi Gear Lever Sys", which also required a code to be inserted. At this point, I offered our S/O the opportunity to set up the FMC, conduct a preflight flow and brief. I listened and followed along from the middle jump seat while in liaison with the ground personnel regarding cargo loading, refuelling and passenger boarding. The inbound flight arrived significantly late, and we wanted to reduce the outbound delay as much as possible. During the request to pushback we received instructions to expect runway 24L. Takeoff weight was within the parameters to utilise 24R (much shorter taxi), we obtained a new ACARS RTOW and amended the FMC with the new runway. During cruise we noticed that we had accidentally omitted the insertion of the code for the gear issue during the 'amended' final preparation procedure'.

This brief narrative is actually very complex. First, the two codes referred to are changes to the normal workflow, communicated to the crew by different means. The decision to let the SO participate in the preparation for departure was a discretionary act on the part of the captain and is a departure from the prescribed process. The delay, and the subsequent desire to reduce further delay, induced pressure in that the time available for the quantity of work that needed to be done was compressed. Of significance here is that, in all probability, the system was set up correctly in the first instance, but the change in the runway, and by now all the crew were back in their proper seats, would have caused the 'gear' code to be deleted and therefore need reinserting. This is a software design issue. That fact was missed by the captain and the FO as they managed the final checklists during the shorter taxi to the new runway: another time constraint. The 'threats' in the scenario include the changes arising from the two codes, the late inbound flight, the runway change and the interruptions during the final preparation for departure. Most crew reading this example would probably consider it to be a (nearly) normal day at work, and it illustrates the fragmented, disrupted nature of work.

From a system perspective, the boundary is the point at which the aircraft must be correctly configured for take-off. The margin is the space between the crew commencing their flight deck set-up activities to the point at which the aircraft is cleared to enter the active runway. Buffering is the system's capacity to cope with the normal procedures contained in company manuals, the additional work associated with the set-up procedure not necessarily described in procedures, those permitted discretionary acts such as including the SO, contingencies such as operational delays and runway changes and also unanticipated variability, such as interruptions. We saw earlier that interruptions have a measurable impact. In this example, we see fairly innocuous, routine events overwhelming available capacity without the crew being aware, and the system boundary was breached. The implication of the failure was that, on rotation, a change to the geometry of the main undercarriage resulted in reduced ground clearance and an increased risk of striking the runway with the rear of the fuselage. Failure, or tolerance, was graceful. The system still functioned.

It is important to remember that individual differences, such as competence, prior experience and stress tolerance (neuroticism) will shape how we deal with disturbances. The subjective nature of stressors was explored by Vine et al. (2015) using a method that drew on the job demands-resources (JD-R) model illustrated in Figure 4.5. First, pilots were asked to evaluate a series of scenarios according to how demanding they thought the situation would be in real life (demand estimation) and how well they thought they would be able to cope with the situation if they ever had to deal with it (resource estimation). Subjects were categorised based on whether they saw the scenarios as challenges to be met or threats to be dealt with. The pilots then flew an exercise in a simulator that included an engine failure just after take-off, while wearing eye-tracking technology. Pilots who evaluated the initial scenarios as challenges tended to perform better in the simulator (as measured by accuracy of flying and assessment by instructor pilots) while those who fell into the 'threat' group demonstrated reduced ability to inhibit distractions, a more random scanning of displays and a higher search rate (less structured scan). Coping with the disruptive nature of work, then, depends, in part, on our disposition towards variability in life. Those with poor coping skills are more prone to experience anxiety.

The JD-R model (Bakker & Demerouti, 2007) brings together out-of-work and in-work factors as inputs and generates two outputs, exhaustion and cynicism, that then shape performance. Psychological fatigue captures how individuals respond to anxiety-inducing stressors and then act in response to task demands. It is one output

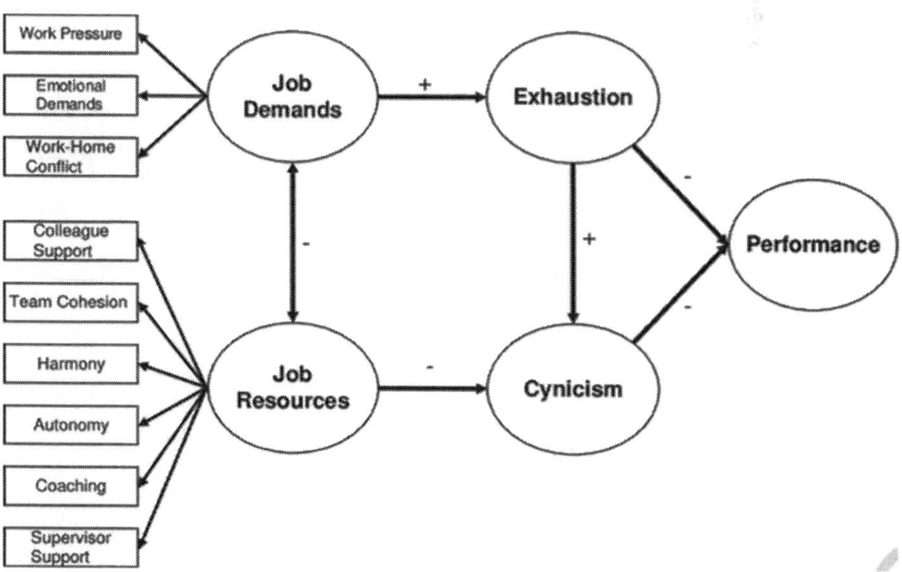

FIGURE 4.5 Job demands-resources (JD-R) model.

from the JD-R model, and at high levels of psychological fatigue, it is manifested in the effort we apply to a task such that efficiency and/or thoroughness are minimised or the task itself is abandoned: we engage in load-shedding.

The factors that trigger anxiety seem to be instrumental in shaping our level of psychological fatigue. Furthermore, physiological fatigue similarly reduces our resilience. However, I said earlier that it is not exactly clear if 'fatigue' is a condition or a symptom. The implication of this is that you do not have to 'feel' fatigued (that is, experience, or be manifesting the symptoms of, physiological fatigue) to still be demonstrating psychological fatigue-related behaviour. This next example illustrates the point.

A B-777 aircraft was on approach to San Francisco Airport after a 12-hour flight. The captain was the flying pilot. The weather was fine with good visibility. Having disconnected the autopilot, and as they were descending through 300 ft, the crew then heard ATC give clearance for another aircraft to depart on the crossing runway. This is a commonplace event at San Francisco. The crew of four on the flight deck could see the other aircraft as it lined up on the runway. At 200 ft, ATC instructed the B777 to go-around, the FO acknowledged but the captain countermanded the instruction. In effect, the captain has just disobeyed a lawful command from ATC. At 98 ft the aircraft was given clearance to land. San Francisco ATC filed a complaint. When questioned, the captain said that he had rested well before the flight, had taken adequate rest during the flight and did not think that fatigue played a part in his decision. When asked why he did not comply with the instruction, he said that he was concerned that there might be a risk of a collision between his aircraft and the departing B757 over the airfield. When asked what might have happened if the B757 had rejected the take-off, the captain was shocked: it had never occurred to him that this was a possibility. A cursory search of the ASRS database found several instances where, in identical situations to this one, the departing aircraft had, in fact, rejected the take-off causing the landing aircraft to go-around.

There are several aspects of this event that are of interest. In Chapter 2, we talked about risk. In aviation, in trying to avoid one risk, we can often set up a countervailing risk. So, instead of meeting over the airfield, a risk the captain wanted to avoid, the two aircraft could have met on the crossing runways. We saw earlier that sleep-deprived subjects have a reduced ability to assess risk but, in this case, the captain reported good sleep. Load-shedding seems to be the most likely explanation for this event. Executing a go-around in busy airspace with a very powerful, relatively light-weight aircraft at the end of a 12-hour flight is demanding. Landing the aircraft was the simplest thing to do despite the fact that it constituted a violation, resulted in legal action and could have ended in a collision on the runways.

Job demands and coping abilities are not absolute; they are estimations made on the part of each individual. As a result, we can see differences both between and within individuals in the extent to which they believe they can cope with specific events. Many people experience stress when, in fact, the situation is well within their level of competence. Figure 4.6 illustrates the concept. At low levels of fatigue, personal and task goals will be aligned, but as fatigue increases, those goals become disconnected until a point is reached at which behaviour will be influenced by those actions needed to sustain a belief in one's ability to cope.

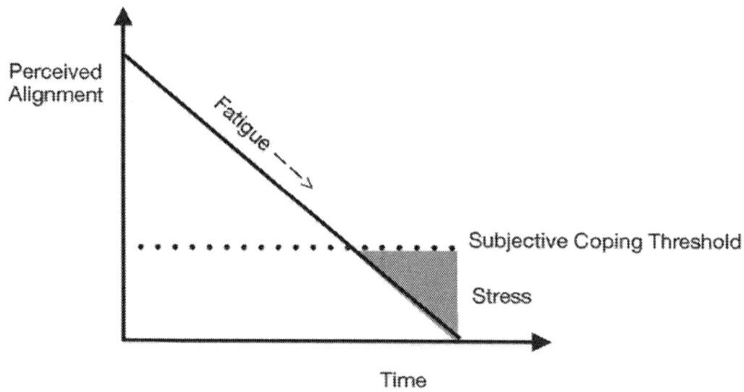

FIGURE 4.6 The relationship between fatigue and stress. (After Phillips, 2014.)

4.12 FATIGUE MITIGATION

We saw earlier that acute fatigue can be mitigated by sleep. The crew do attempt short-term mitigations such as the use of caffeine (coffee or energy drinks), taking a leg stretch out of the seat and taking a break that does not include a nap. Napping on the flight deck is known as controlled rest (CR). In an online survey of 218 pilots, Stait (2019) found that 69% of respondents had used CR. Many airlines make provision for crews to take naps in flight, and the mitigation is bounded by quite strict rules. The first problem we find is that the actual rate of CR in aviation is not really understood. In one LOSA, we looked at the actual incidence of CR. During the study period, nine instances of CR were taken, representing 6% of possible flights. We found that if the captain took rest then the FO might, but FOs never took rest if the captain had not already done so. So, it seems that there is a social dynamic that might inhibit FOs from taking rest even if they needed it. Resting on the flight deck is just one-way crew can tactically manage fatigue (recognising that the provision is not always used appropriately). Stait also reported that 65% of the crew made use of the autopilot as mitigation, and 33% had elected not to act as the pilot flying because of tiredness. These are examples of using technology or procedures to cope with fatigue in the workplace.

If we now look at who takes CR, we see that there are, in fact, three categories. The intended users of CR are those who find, at that moment, they are in need of rest. The assumption is that the crew member reported fit for duty but has experienced such a rapid onset of fatigue that they need to take action. However, there are two other groups that take advantage of the rules: those who fail to prepare appropriately for duty and use CR to manage their lifestyle, and those suffering from chronic fatigue and are unable to recover between shifts and so need to rest in flight.

Chronic and psychological fatigue typically need more than simply additional sleep to effect a recovery. It has long been recognised that humans have a tendency to resort to symptom-directed coping to mitigate stress, which includes using alcohol as a distraction. Sleep medication can also be used as a stop-gap fatigue management tactic. In a study of 435 Portuguese pilots (Reis et al., 2016), the authors found that the percentage of individuals taking sleep medication or using alcohol as a sleep aid was

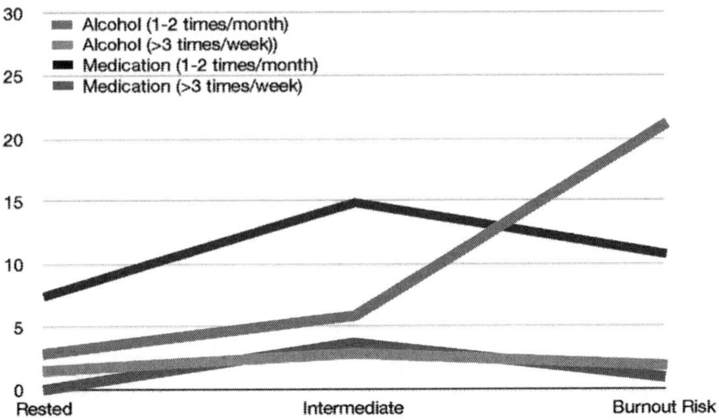

FIGURE 4.7 Use of alcohol and sleep medication.

very low (1.4% for each). There was a positive significant relationship between the use of alcohol and medication and sleep complaints (measured using the JSQ questionnaire), but not with fatigue or daytime sleepiness. The sample comprised pilots mainly involved in short and medium-haul flying (72%), and mitigations seemed to be responses to disrupted sleep patterns. Stait (2019) found that 6% of respondents in her survey reported frequent (>3 times a week) use of sleep medication. In my survey, 305 pilots provided data on their use of alcohol and medication. The results are shown in Figure 4.7. Of interest here is that it is the intermediate group that seems to be resorting to aids to sleep, the exception being the (relatively) moderate use of alcohol by the burnout risk group.

It should come as no surprise that the most fatigued pilots feel that they do not recover sufficiently between work periods, have poor sleep quality and experience serious levels of daytime sleepiness. However, the data suggest that all pilots suffer from inadequate recovery, and even the least fatigued group has individuals who suffer from sleep problems. Of particular interest is the 40% of pilots in the sample comprising the moderate fatigue group. The data suggest that, rather than occupying the middle ground, they tend more towards the highest fatigued group in terms of characteristics. If we look at their attempts at mitigation, it seems that the moderately fatigued pilots are most likely to resort to alcohol, initially. Alcohol consumption within 3 hours prior to sleep actually degrades the sleep experience: it is counter-productive as fatigue mitigation. Once physiological fatigue reaches a certain level, it seems that a self-sustaining feedback cycle is set up that, ultimately, will lead to long-term health effects.

4.13 FATIGUE AND PILOT HEALTH

Pilots in my survey were asked to complete a standard health status questionnaire, the SF-12 (Ware, Kosinski & Keller, 1996). The survey generates a physical and a mental health score. The pilots all had access to private medical insurance and were subject to periodic medical checks in order to retain their license to fly. Therefore, it should be expected that the group would be in good physical health. A score of

<50% is indicative of poor health and indeed, only 7.5% of the pilots score less than 50% on the physical scale. On the mental health scale, 29.2% of pilots scored less than 50%, but Figure. 4.8 reveals a markedly different distribution of scores on the two scales. As we would expect, the group was heavily skewed towards the upper end of the scale for physical health but the mental health scores follow a more normal distribution. The respondents also completed the FSS, and when we factor in chronic fatigue, the effect is quite stark. Figure 4.9 shows the percentage of pilots in each fatigue category scoring more than 50% on each component scale and, thus, reflecting a better health state. Pilots in the poorest fatigue group still make up the majority of those scoring lowest for *both* mental and physical health.

Fatigue, then, has a significant impact on both physical and mental health. In Figure 4.5, we saw that emotional demands and work–home conflict were factors in

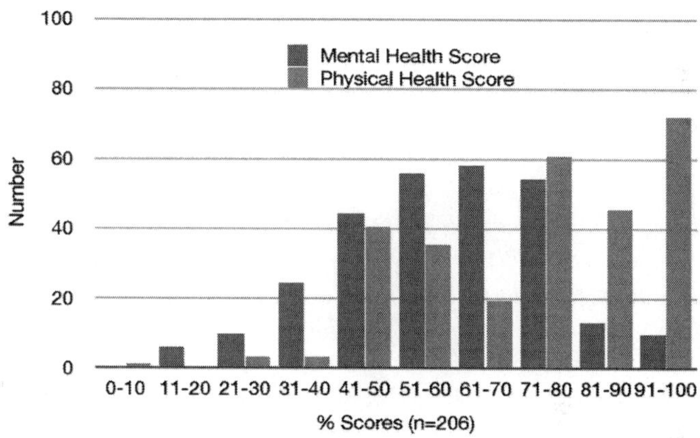

FIGURE 4.8 Pilot physical and mental health.

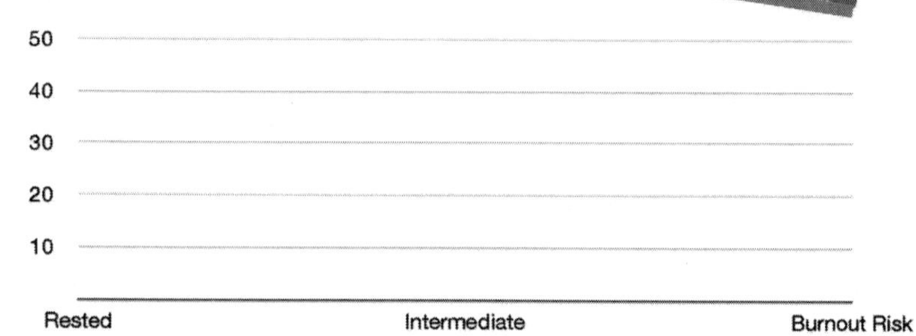

FIGURE 4.9 Pilots' health and fatigue.

the JD-R model. The model operationalises the concept of stress as being a dissonance between expectations and capability but also has exhaustion as an output. Job demands are defined as those physical, psychological, social or organisational aspects of the job that require sustained physical and/ or psychological effort. Job resources are those physical, psychological, social or organisational aspects of the job that are functional in achieving work goals, reducing job demands and the associated physiological and psychological costs. They can also stimulate personal growth, learning, and development. The findings for the relationship between fatigue and mental health suggest that a pilot's capacity to cope with emotional demands will be diminished. Effective support from friends and family is considered an important factor in stress coping, but, again, high levels of fatigue appear to be associated with degraded WLB.

Sustained inadequate recovery can lead to 'burnout', a concept that describes an extreme condition often associated with a degraded health state. There is no agreement about the specific condition we call 'burnout', its development and the way it presents in terms of symptoms. Nonetheless, it remains a useful indicator of a problem. In a study of 1147 UK airline pilots using the OLBI, 20% of the sample reported clinical levels of burnout (Demerouti et al., 2019). A benchmarking study of 13,000 employees from several different occupations suggests that 4% had the symptoms of burnout (Bakker et al., 2000). We saw earlier that in my study of 80 line pilots 10.71% scored sufficiently high on the OLBI to be classed as burnout candidates. The Germanwings crash on 24 March 2015 (BEA, 2016), in which the mental state of the FO appears to have been a cause, has focussed more attention on mental health issues in aviation. In one survey of 328 pilots (Aljurf et al., 2017), 34.5% scored above the cut-off on a measure of depression. A larger survey of 1848 pilots from the United States of America (Wu et al., 2016) found that 12.6% of respondents showed evidence of mental health issues. Of interest, symptoms of depression were higher in pilots using sleep medication more than three times a week (33.3%). In Figure 4.7, a small percentage of pilots in the intermediate and burnout risk group fell into that category. O'Hagan et al. (2017) found that pilots who spent long hours on duty per week were twice as likely to report feeling depressed or anxious.

The arc traced in this chapter ultimately arrives at medical impairment, either in terms of degraded general health or a psychiatric condition. I think that it is important to recognise that among 'rested' individuals and those living with permanent tiredness, there is a hinterland of pilots struggling to cope with the demands of life but are not yet at the point of suffering a collapse in their mental health. They appear to be more reliant on alcohol and sleep medication to arrest their worsening condition than the other groups. Demerouti makes the point that 'the job of a pilot is not only detrimental for health-related outcomes, i.e. exhaustion (because of high or badly designed job demands) but also for motivation due to lack of job resources' (Demerouti et al., 2019).

4.14 FATIGUE AND SAFETY – CAUSE OR RISK FACTOR?

This chapter has looked at factors that both shape and impair performance. While individual differences can undoubtedly affect the efficacy of our interventions, the question we want to answer, of course, is what effect does stress and fatigue have on safety. Fatigue as a causal factor has been demonstrated in some contexts, such as road traffic accidents, but the same relationship in aviation has been more difficult to establish (Williamson et al., 2011). The authors of the CASA review of aviation fatigue rules (CASA, 2018) make the point that there are simply too few accidents at the extremes of duty periods to draw sound conclusions and, in any case, fatigue is likely to be just one contributory factor. Thomas and Ferguson (2010) suggest that fatigue effects are masked because the crew protect the safety of the operation by modifying their behaviour.

We saw earlier that relationships can be found between hours of sleep and wakefulness before duty and subjective measures such as the S-PS and KSS. Working with LOSA data, Thomas and Ferguson (2010) looked at hours awake before a duty, hours of sleep, error rates and error management. They identified a threshold value of 6 hours of sleep before a duty below which error rates increased, and error management was degraded. No effect was found for hours of wakefulness. However, a second LOSA study (Clockwork, 2014) found a significant relationship for both sleep and wakefulness. If both the crews had been awake for more than 8 hours before the start of the duty then error rates were higher, and if one or more crew members had less than 6 hours of sleep before a duty, error management was impaired. Unfortunately, increased error rates do not necessarily mean reduced levels of safety. It is not possible to predict exactly what errors will be made or how significant they might be to the operation. In another LOSA study, the crews with the highest S-PS scores committed least errors, the difference being that these were daytime operations. The relationship between fatigue and error is complicated.

The effect of hours of wakefulness was evident in a 2017 sample of cargo flights. The aim of the study was to investigate fatigue effects in two-sector operations through the night. The crew would be operating in the WOCL, which is the phase of the circadian cycle when the body is really supposed to be asleep. This circadian effect has been observed in aircraft maintenance where Hobbs et al. (2010) found that skill-based errors (using Rasmussen's SRK taxonomy) were significantly more frequent during the early hours of the morning. In a study of Brazilian pilots (de Mello et al., 2008), error rates were found to increase by 50% for the early morning period.

The bulk of the observations of the cargo crews were done on triangular patterns. The first sector was operated by a crew who took the aircraft from home base to a major cargo airport. They then handed the aircraft over to a second crew who operated two sectors through the night, back to home base. Data were collected on all three sectors. In addition to the normal LOSA data, the crew were asked to provide information about sleep, hours awake before the duty and to complete the S-PS and KSS. Using the LOSA data from a previous, mainly daytime, study as a benchmark, I compared the error rates for the two-sector WOCL flights. I need to add a caveat here: there is no such thing as a 'normal' error rate, and we will explore error in more detail in Chapter 6. The benchmark error rate across all aircraft types was 6.25/sector. The error rate for the cargo crews operating through the night was 10.6/ sector. However, the error rate for the crew on the initial sector was also above 10 and yet this sector, technically, was not through the WOCL. The explanation seems to lie in hours of wakefulness. The WOCL sector crews had, on an average, been awake for 2.5 hours before starting the duty. The crews operating the first sector had been awake, on an average, for 7.5 hours before the duty: a value very close to that found in the 2013 study to be associated with increased error rates. Hours awake before a duty, then, is a predictor of performance, and its effect is comparable to that of operating through the WOCL.

It seems, then, that there is a relationship between the two sleep parameters (prior sleep and wakefulness) and performance (error rate and error response), but that is still not an incontrovertible proof of a threat to safety. In Chapter 2, I discussed the concept of risk, and so it might be useful to view fatigue through a safety management system (SMS) prism. An SMS risk matrix compares the probability of an event against its severity to arrive at a 'risk level'. De Matos Alceu (2015) looked at the distribution of fatigue risk using S-PS data. The S-PS intervals were mapped onto the 'severity' scale of a risk matrix while the frequency of S-PS scores in a sample of flights was mapped onto the 'likelihood' scale. A problem with this approach is that there is no universal agreement on the grading of risk and severity. However, a risk assessment of night-time operations into Queenstown Airport in New Zealand's South Island recommended that crews with an S-PS of 4 or more should not be permitted to fly the approach (Navigatus Consulting, 2017). The risk was too high. Figure 4.10 illustrates the risk distribution of SPS scores for cargo flights. The crew were sampled before flight (BF1 and BF2) and after landing (AF1 and AF2) across their two-sector pattern. The rate of scores of '4' was normalised, and then the percentage of scores above and below '4' were calculated. The data for the daytime B-777 regional flights (S1 and S2) are included for reference. We can see that the fatigue scores for the daytime regional flights suggest a low-risk profile. However, for the two-sector operation through the night, the risk profile shifts progressively through medium towards high risk. In an SMS, the output from the matrix is an instruction to management to take appropriate action in response to the specific risk.

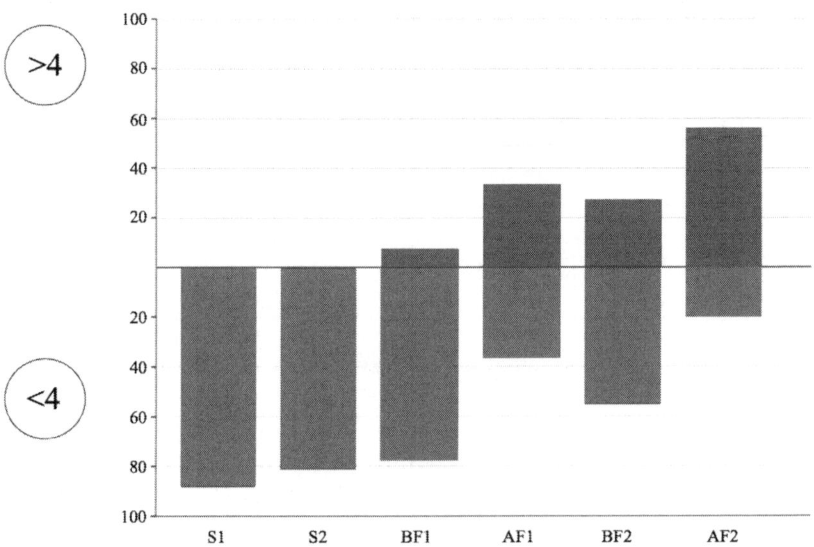

FIGURE 4.10 Distribution of S-PS scores.

Whereas hours of sleep and wakefulness are direct measures, the other fatigue scales we have looked at in this chapter are indirect measures. A pilot's self-perception of fatigue and its subsequent manifestation in terms of performance is complicated. It has been suggested that fatigued pilots fly more conservatively, but commercial aircraft are usually operated by a crew. Therefore, the fatigue status of the individual crew members might not be an adequate measure of the risk given that the performance reflects the inputs of two or more crew members. This final example illustrates the problem.

In November 2018, the crew of a B-747 made two attempts to get airborne for a night-time departure without first selecting take-off flap (personal communication). After their second attempt, they realised their error and finally got safely away. The story starts at pushback. The captain was waiting for the ground engineer to indicate that pushback was complete, that he had disconnected his headset and that the crew were clear to taxi. There had been some discussion, generally, among pilots that this process seemed to be taking longer and longer. Once the engineer had disconnected, the next step in the process was for the take-off flap to be selected. As usual, on this evening, the engineer was taking his time, but once he had he finished, the captain turned back to the task in hand. Under normal circumstances, the pilot monitoring, in this case the FO, would have had his hand on the flap selector waiting for the instruction to select the flap. On this occasion, for whatever reason, the FO had moved his hand and a visual cue that would normally act as a place marker in the procedure, triggering the captain's next call, was missing. The checklist was continued but without that crucial step. There was one more opportunity to check whether the flaps had been set, which was when the crew verified that the take-off checklist had been completed. This step has some discretion as to when it is undertaken. A new requirement had been introduced to verify the correct runway for take-off, and so the task was usually delayed until the crew had arrived at the nominated runway entry point. On this occasion, because traffic was light, ATC was gently pushing the crew along and had already given them clearance to take-off The crew turned onto the runway and brought up the power, having omitted to finish the take-off checklist. When they pushed the TOGA buttons, they got a warning. They did a slow-speed rejection of the take-off, turned off at the next intersection and come back to start again. As they were turning off the runway, while investigating the reason for the warning, the crew did a quick check that autothrottles had been selected. As the relevant switch was in the correct position, they assumed that it must be a TOGA switch malfunction. The crew knew that manual selection of thrust for take-off was permissible so the apparent malfunction was no impediment, and they did not need to call an engineer. Cleared for take-off a second time, they brought up the power and immediately got a 'CONFIG FLAP' warning. The problem was now clear. We will look at error in a later chapter but, technically, the crew made two lapses – or forgetting to do something – and one mistake. They forgot to select flap in the first place, and then they forgot to complete the checklist. They then misdiagnosed the reason why the TOGA buttons did not work: that was the mistake. Both crews had positioned to operate the day before so neither was fully acclimatised. The time zones crossed by each pilot prior to their duty differed, one travelling five zones west (supposedly the more beneficial direction) while the other travelled eight zones in an easterly direction.

One pilot had managed to sleep until about 3 hours before the reporting time while the other pilot had woken at his normal time as though he were at home. He had been awake for more than 14 hours when the event happened.

We cannot say that fatigue 'caused' this crew to perform well below their usual, professional standards. What we can argue is that fatigue is a risk factor: the presence of a fatigued state – and we have seen more than once in this chapter that hours awake before a duty correlates with error – increases the risk that something contrary to expectations will happen. We cannot predict what error will be committed, just that the performance of both individuals and crews as a unit is likely to be more erroneous when fatigued.

4.15 CONCLUSION

Differences between individuals, both physiological and psychological, will shape behaviour and, under certain circumstances, impair performance. Stress and fatigue are controlled by completely different mechanisms but they are, nonetheless, linked. Prolonged exposure to stress will lead to the exhaustion phase, at which point our energy reserves have been used up: stress, then, can lead to fatigue. Extreme sleep disruption adds to stress by undermining our sense of being able to cope. Psychological fatigue seems to have some similarities with anxiety in that it reflects our belief in our ability to cope with the demands of life. Stress and fatigue may be different physiological processes, but they clearly contribute to a common lived experience on the part of the individual. This chapter has attempted to throw light on the extent of, and inter-relationships between, some of these factors. My argument is that the physical and mental state of a pilot will have an effect on his performance, even though direct cause and effect can be difficult to demonstrate.

Life stressors and fatigue state appear to be related but establishing causality is a significant problem. For example, do issues at home result in, say, disturbed sleep resulting in increased fatigue? Having children at home has been shown to correlate with fatigue in several studies (Stait, 2019). Or does workplace fatigue result in degraded home life because of poor engagement, which then affects sleep patterns? I have said nothing in this chapter about the role of emotion. The complex interplay of personality, motivation and self-esteem, moderated by our state of stress and fatigue, gives rise to outward qualitative displays that, in turn, generate responses from others. Maintaining equilibrium while sustaining task-directed activity is effortful.

Stress and fatigue will shape workplace performance. In the rest of this book, we will look at normal workplace behaviour. Every aspect of normal operations will be influenced by the factors discussed in this chapter. The management of startle, stress and fatigue also represents the first area of competence we need to consider. We will pick this up again in Chapter 11, but self-management is key to resilient operations.

REFERENCES

AAIB. (1999). Report No: 3/1999. Report by the Dominican Republic Authorities into the accident to Boeing 757-200, G-WJAN, at Puerto Plata Airfield, Dominican Republic on 1 January 1998.

Aeschbach, D., Vejvoda, M., Mendolia, F., & Tritschler, K. (2017). Additive effects of the number of completed flights and time awake on fatigue in short-haul airline pilots. *Sleep*, 40(Supplement 1), A58.

Akerstedt, T. & Gillberg, M. (1990). Subjective and objective sleepiness in the active individual. *International Journal of Neuroscience*, 52, 29–37.

Akerstedt, T., Kecklund, G., Gillberg, M., Lowden, A., & Axelsson, J, (2000). Sleepiness and days of recovery. *Transportation Research Part F: Traffic Psychology and Behaviour*, 3(4), 251–261.

Aljurf, T.M., Olaish, A.H., & BaHammam, A.S. (2017). Assessment of sleepiness, fatigue, and depression among Gulf Cooperation Council commercial airline pilots. *Sleep and Breathing*, 22, 411–419.

Bakker, A.B. & Demerouti, E. (2007). The job demands-resources model: State of the art. *Journal of Managerial Psychology*, 22(3), 309–328.

Bakker, A.B., Schaufeli, W.B., Sixma, H., Bosveld, W., & Van Dierendonck, D. (2000). Patient demands, lack of reciprocity, and burnout: A five-year longitudinal study among general practitioners. *Journal of Organizational Behavior*, 21, 425–441.

Behrend, J., Dehais, F., & Koechlin, E. (2017, September). Impulsivity modulates pilot decision making under uncertainty. *The Human Factors and Ergonomics Society Europe (HFES 2017)* (pp. 1–12), Rome, Italy, Hal ID: hal-01739838.

BEA. (2012) Final Report On the accident on 1st June 2009 to the Airbus A330-203 registered F-GZCP operated by Air France flight AF 447 Rio de Janeiro – Paris.

BEA. (2016). Accident on 24 March 2015 at Prads-Haute-Bléone (Alpes-de-Haute-Provence, France) to the Airbus A320-211 registered D-AIPX operated by Germanwings (English version).

CASA. (2018). Independent review of aviation fatigue rules for operators and pilots. Canberra.

Clockwork Research. (2014). Unpublished LOSA Report.

Demerouti, E., Bakker, A.B., de Jonge, J., Janssen, P.P., & Schaufeli, W.B. (2001). Burnout and engagement at work as a function of demands and control. *Scandinavian Journal of Work and Environmental Health*, 27(4), 279–286.

Demerouti, E., A. Bakker, A.B., Vardakou, I., & Kantas, A. (2003). The convergent validity of two burnout instruments. *European Journal of Psychological Assessment*, 19(1), 12.

Demerouti, E., Veldhuis, W., Coombes, C., & Hunter, R. (2019). Burnout among pilots: Psychosocial factors related to happiness and performance at simulator training. *Ergonomics*, 62(2), 233–245.

De Matos Alceu, P.M. (2015). Managing Fatigue in a Regional Aircraft Operator: Fatigue and Workload on Multi-segment Operations. Unpublished thesis. Lisbon University.

de Mello, M.T., Esteves, A.M., Pires, M.L.N., Santos, D.C., Bittencourt, L.R.A., Silva, R.S., & Tufik, S. (2008). Relationship between Brazilian airline pilot errors and time of day. *Brazilian Journal of Medical and Biological Research*, 41(12), 1129–1131.

Dunbar, R. (2014). *Human Evolution*. Harmondsworth: Cambridge University Press.

EASA. (2018). Final Report No: EASA_REP_RESEA_2015_3. Research Project: Startle Effect Management.

Graziano, M.A. (2013). *Consciousness and the Social Brain*. Oxford: Oxford University Press.

Geurts, S.A.E., Taris, T.W., Kompier, M.A.J., Dikkers, J.S.E., Van Hooff, M.L.M., & Kinnunen, U.M. (2005). Work-home interaction from a work psychological perspective: Development and validation of a new questionnaire, the SWING. *Work & Stress*, 19(4), 319–339.

Hobbs, A., Williamson, A., & Van Dongen, H.P. (2010). A circadian rhythm in skill-based errors in aviation maintenance. *Chronobiology International*, 27(6), 1304–1316.

Hockey, R. (2013). *The Psychology of Fatigue*. Cambridge

Hollnagel, E. (2009). *The ETTO Principle: Why Things That Go Right Sometimes Go Wrong.* Farnham, UK: Ashgate.

ICAO. (2016). Doc 9966, Manual for the Oversight of Fatigue Management Approaches. Montreal.

Jenkins, C.D., Stanton, B.A., Niemcryk, S.J., & Rose, R.M. (1988). A scale for the estimation of sleep problems in clinical research. *Journal of Clinical Epidemiology*, 41(4), 313–321.

Johns, M.W. (1991). A new method for measuring daytime sleepiness: The Epworth sleepiness scale. *Sleep*, 14(6), 540–545.

Klinect, J.R. (2005). Line Operations Safety Audit: A Cockpit Observation Methodology for Monitoring Commercial Airline Safety Performance. Unpublished PhD thesis. University of Texas.

Krupp, L.B., LaRocca, N.G., Muir-Nash, J., & Steinberg, A.D. (1989). The fatigue severity scale: Application to patients with multiple sclerosis and systemic lupus erythematous. *Archives of Neurology*, 46, 121–1123.

Landman, A., Groen, E.L., van Paassen, M.M. (René), Bronkhorst, A.W., & Mulder, M. (2017). Dealing with unexpected events on the flight deck: A conceptual model of startle and surprise. *Human Factors*, 59(8), 1161–1172.

Mesarosova, K., Plouff, R.A, Saklofske, D., Smith, M.M., Tremblay, P.F., & Siegling, A.B. (2017). Personality measurement and profile in a European sample of civil airline pilots. *European Journal of Psychological Assessment*. 35(6), 791–800.

Navigatus Consulting. (2017). Queenstown Airport Night Operations Foundation Safety Case.

O'Hagan, A.D., Issartel, J., Nevill, A., & Warrington, G. (2017). Flying into depression: pilot's sleep and fatigue experiences can explain differences in perceived depression and anxiety associated with duty hours. *Workplace Health & Safety*, 65(3), 109–117.

Phillips, R.O. (2014). What is fatigue and how does it affect the safety performance of human transport operators. Fatigue in Transport Report I, TOI Report No: 1351/2014, Oslo.

Powell, D., Spencer, M.B., Holland, D., & Petrie, K.J. (2008). Fatigue in two-pilot operations: Implications for flight and duty time limitations. *Aviation, Space, and Environmental Medicine*, 79(11), 1047–50.

Reis, C., Mestre, C., & Canhão, H. (2013). Prevalence of fatigue in a group of airline pilots. *Aviation Space Environmental Medicine*, 84(8), 828–833.

Reis, C., Mestre, C., Canhão, H., Gradwell, D., & Paiva, T. (2016). Sleep complaints and fatigue of airline pilots. *Sleep Science*, 9(2), 73–77.

Sallinen, M., Sihvola, M., Puttonen, S., Ketola, K., Tuori, A., Harma, M.I., Kecklaund, G., & Akerstedt, T. (2017). Sleep, alertness and alertness management among commercial airline pilots on short-haul and long-haul flights. *Accident; Analysis and Prevention*, 98, 320–329.

Samn, S.W., & Perelli, L.P. (1982). Estimating aircrew fatigue: A technique with application to airlift operations. Brooks Air Force Base. San Antonio, TX. Report No: SAM-TR-82-221.

Searle, B.J. (2012). Detachment from work in airport hotels issues for pilot recovery. *Aviation Psychology and Applied Human Factors*, 2, 20–24.

Selye, H. (1950). Stress and the general adaptive syndrome. *British Medical Journal*, 1(4667), 1383–1392.

Stait, V. (2019). In it for the Long Haul: Perceived Occupational Fatigue and Pilot Roles over a Long-haul Career. Unpublished MSc thesis. Coventry University.

Thomas, M. & Ferguson, S. (2010). Prior sleep, prior wake, and crew performance during normal flight operations. *Aviation, Space, and Environmental Medicine*, 81(7).

van Veldhoven, M. & Broersen, S. (2003). Measurement quality and validity of the "need for recovery scale". *Occupational and Environmental Medicine*, 60, 3–9.

Vine, S.J., Uiga, L., Lavric, A., Moore, L.M., Tsaneva-Atanasova, K., & Wilson, M.R. (2015). Individual reactions to stress predict performance during a critical aviation incident. *Anxiety, Stress & Coping*, 28(4), 467–477.

Ware, J.E., Kosinski, M.M., & Keller, S.D. (1996). A 12-Item Short-Form Health Survey: Construction of scales and preliminary tests of reliability and validity. *Medical Care*, 34(3), 220–233.

Williamson, A., Lombardi, D.A., Folkard, S., Stutts, J., Courtney, T.K., & Connor, J.L. (2011). The link between fatigue and safety. *Accident Analysis and Prevention*, 43(2), 498–515.

Winwood, P.C., Winefield, A.H., Dawson, D., & Lushington, K. (2005). Development and validation of a scale to measure work-related fatigue and recovery: The Occupational Fatigue Exhaustion/Recovery Scale (OFER). *Journal of Occupational and Environmental Medicine*, 47(6), 594–606.

Wu, A.C., Donnelly-McLay, D., Weisskopf, M.G., McNeely, E., Betancourt, T.S., & Allen, J.G. (2016). Airplane pilot mental health and suicidal thoughts: A cross-sectional descriptive study via anonymous web-based survey. *Environmental Health*, 15, 121.

Yen, J-R., Hsu, C-C., Yang, H., & Ho, H. (2009). An investigation of fatigue issues on different flight operations. *Journal of Air Transport Management*, 15(5), 236–240.

Yuliawati, I., Siagian, M., Abudi, T., & Basuki, B. (2015). The number of sectors and other risk factors related to fatigue among short-haul commercial pilots in Indonesia. *Health Science Journal of Indonesia*, 6, 69–75.

Zuckerman, M. (2012). *Psychobiology of Personality*. Cambridge: Cambridge University Press.

5 Doing Normal Work – Processes at Level 1

5.1 INTRODUCTION

In Chapter 2, I proposed an organisation that will serve as a framework for structuring the rest of this book. I have offered the view that what we call crew resource management (CRM) is actually the application of expertise to the task of controlling the action. Aviation is a collaborative, distributed process that aims to achieve the desired goal within the bounds of tolerable risk. Finally, humans are frail and our capacity to do work is moderated by a suite of individual differences. In this chapter, we start to look at the behaviours associated with the systems model. When I outlined the model (or, possibly, my organising framework), I listed some of its characteristics, one of which was emergence. Each level in the hierarchy will exhibit properties that cannot be explained simply based on the functioning of the individual components at that level.

The chapter starts with an elaboration of activity as thought. All observed behaviour is the output from mental processing but rarely do we have the opportunity to trace a pilot's line of thought as events unfold. Next, I elaborate on the concept of goals, boundaries and margins because this hypothetical 'space' is the arena in which individuals act. Sense-making is fundamental to action, and this is examined in some detail before we look at decision-making as option selection. A key theme of this chapter is that individual action is both shaped by and, in turn, shapes the environment. The problem we face, though, is captured by Johnson (2004), who suggests that things fail in ways that we simply cannot imagine (remember non-ergodicity and radical uncertainty from previous chapters). As an aside, because of this, 'safety management', Sisyphus-like, is forever doomed to fail. Chater (2018) argues that, in any case, because the future is unknowable all we can ever do is respond to events as they happen. In Chater's view, because humans are inveterate storytellers, we create narratives that rationalise the relationship between the actions and outcomes and identify 'causes' that are no more than fictions. From this perspective, the concept of 'proaction' becomes absurd. These two positions imply that, even under normal conditions, any intervention we make to achieve a goal will have a degree of uncertainty attached to the outcome and, thus, we will be constantly having to adjust to circumstances as events unfold.

Controlling complex systems is an act of collective computation. Fundamentally, individuals engage with the world and then respond to outcomes. They share cognitive resources to make choices about future actions. Expertise makes us more sensitive to the cues that exist in situations as they develop. Research now suggests that brain activation precedes conscious thought and, from an evolutionary perspective, it makes sense for the brain to be leading consciousness rather than lagging. But the fact remains, we are still largely simply responding rather than predicting. This idea that we are always reacting to the present drives the rest of this chapter.

Having said that aviation is a collaborative and distributed process, work, by definition, is shared across individuals. As a result, it can be hard to separate those specific properties of a situation that are exclusive to a single person. For this chapter, though, I will try to limit the discussion to those key internal processes that are the necessary precursors to shared activity. Ironically, the idea of single-pilot CRM has seemed to be an oxymoron ever since the CRM was first specified as a training requirement. However, if you look at single-pilot accident reports you will see many features that have parallels with large commercial aircraft accidents. In all probability, then, it is in single-pilot commercial operations that we might see some clues to functions at Level 1.

Doing a job of work involves initiating activity, monitoring the conduct of that activity and assessing progress towards the desired goal. A key part of work is making decisions about which activities to modify, and how to modify them, to ensure continued progress towards the goal. Once the goal is achieved then the job can be considered complete and we can terminate the activity. From a systems model perspective, at this level, control is manifested in action. Individuals require a schema of some sorts to control action or a process for resolving uncertainty. Feedback is provided by the responses to inputs observed in the world. Effective evaluation of feedback requires the individual to have adequate mental models of the world and the processes being managed. Figure 5.1 illustrates Level 1 in my hierarchy. Part of the task is to correctly configure the aircraft to cope with disturbances while ensuring progress towards the operational goal but, of course, work involves more than just manipulating the aircraft. We will look at errors in more detail in the next chapter but, typically, things are deemed to have gone wrong when pilots either fail to execute an action plan within acceptable boundaries or adopt a plan that was inappropriate for the situation.

The focus of this chapter, then, will be on the individual operator engaged in a task. Although some of the discussion will involve multi-crew operations, I will not be addressing the social nature of work but will focus, instead, on what individuals do. That said, we will touch on the idea of distributed processing: two or more actors combining their cognitive attributes to get the work done. I propose that safety is an emergent property at this level. It is only as a result of the way we act that outcomes are either of increased risk or most likely to succeed. Safety is a product of how individuals act in the world.

5.2 WORK AS THOUGHT

On 17 January 2008, a British Airways (BA) B-777 was on final approach to London Heathrow Airport's Runway 27L after a flight from Beijing (AAIB 2010; personal communication). The aircraft was fully configured for landing and passing 1000 ft on the approach, at which point the captain handed control to the flying officer (FO) in accordance with their briefing and the company procedures. It was around 12:40. The wind was blowing from the south, over Terminal 4 and the engineering hangars. The captain of a Turkish Airlines aircraft landing before the BA flight reported that it

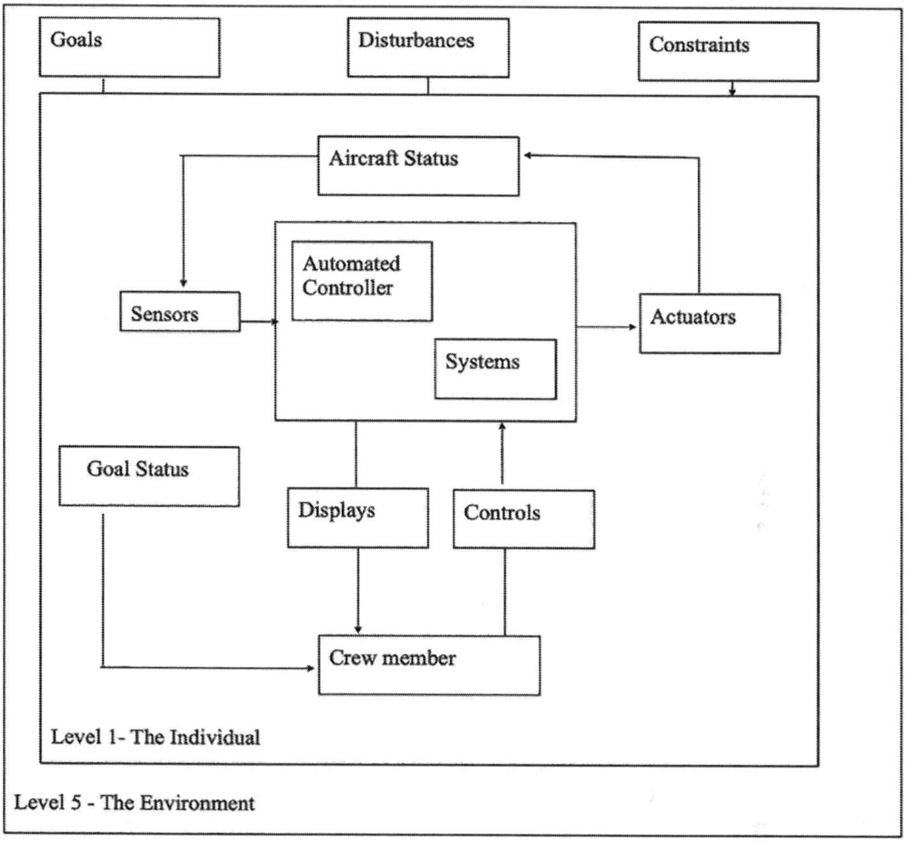

FIGURE 5.1 Systems model levels 1 and 5. (Adapted from Leveson.)

was quite turbulent on the approach (personal communication). Having handed over control, the captain began scanning to see whether their assigned gate was clear. The FO had intended to disconnect the autopilot and auto-throttles at 800 ft for a manual landing. Just after he took control, the FO noticed that the auto-thrust system, first, commanded an increase in thrust, and then one lever reduced power, followed soon after by the other thrust lever. Finally, the levers split, adopting different positions. The FO asked, 'was it doing this when you were flying?'. The captain, still searching for the gate, assumed that the FO was referring to the fact that the thrust levers had been gently moving to compensate for the turbulent, gusty conditions. Nothing in the FO's comment suggested that he might be referring to something different. At 500 ft, the automatic radio altimeter call was made and the landing clearance was given. The captain called 'stable', and the FO responded 'just'. This comment was sufficiently odd to cause the captain to, now, direct his attention back into the flight deck. Observing the strange condition of the thrust levers and engine displays, the captain wondered whether the aircraft warning systems had failed. The relief FO, sitting on the centre seat, commented, 'it looks like a double engine failure'. The captain thought that it could not be a failure because the engines were still operating, just not producing sufficient output. At that point, he switched his attention to the aircraft panels to confirm that everything was in the correct position, that they had not knocked something off by accident.

Having been distracted by the behaviour of the thrust levers, the FO had not disconnected the autopilot, which was now trying to maintain the instrument landing system (ILS) glideslope. With insufficient power from the engines, the airspeed was reducing as the autopilot raised the nose of the aircraft to try to maintain the glideslope. As the speed reduced to 115 knots, the 'Airspeed Low' warning was triggered. The captain realised that they had to reduce drag. The aircraft was fully configured for landing, and so he briefly considered raising the landing gear. This idea was rejected because it would, initially, increase drag as the doors opened before the gear came up. He also felt that they might need the undercarriage as a protection, as it was not clear that the aircraft would make the runway. He then tried to recall the speed margins associated with different flap positions and felt that they could probably reduce the flap setting by one stage. As the flaps retracted, the captain felt that they would now clear the perimeter fence, at least, and avoid the buildings beneath the flight path. Ten seconds before touchdown the stick shaker operated and the FO pushed the nose forward, disconnecting the autopilot. The captain considered making the emergency call 'brace for impact' but reckoned that, as the crew were already seated for landing this might just cause panic. He also considered taking over control for the landing but recognised that handing over control so late in the flight was not sensible. In any case, the FO had almost as much time on the B-777 as he had. The aircraft touched down on the grass 330 m short of the runway and 110 m inside the boundary fence.

What is interesting in this example is that, in the space of 34 seconds, the captain, having had his attention drawn to the unusual situation, needed to address several issues: 'why was the aircraft not giving a warning?'; 'was the relief FO's diagnosis correct?'; 'had the crew accidentally deselected a system?'; 'how can we reduce drag?'; and 'shall I take control of the aircraft?'. There was very little interaction between the crew members during the event. Instead, the captain was engaging with the situation, and his responses were the output from the internal dialogue. At Level 1, in my hierarchical model, we are dealing with how individuals engage with events through internal processing. It is these internal processes that result in action.

5.3 GOALS, BOUNDARIES AND MARGINS – THE STRUCTURE OF TASKS

In Chapter 3, I outlined some properties of the system, such as boundaries and margins. I now want to elaborate on the idea of a task and to incorporate the concept into a systems model. Any commercial aviation task can be described as a trajectory from assembly to dispersal. Assembly is the process of bringing together the resources needed to accomplish the task. This will include the airframe, crew, fuel, payload, etc. Dispersal is the reallocation of resources at the end of the specific task. A trajectory will comprise a set of goal states. A goal state is a specific configuration of the aircraft and resources relevant to some aspect of the task and can be described in terms of the set of constraints to be satisfied. Once a goal state has been satisfied

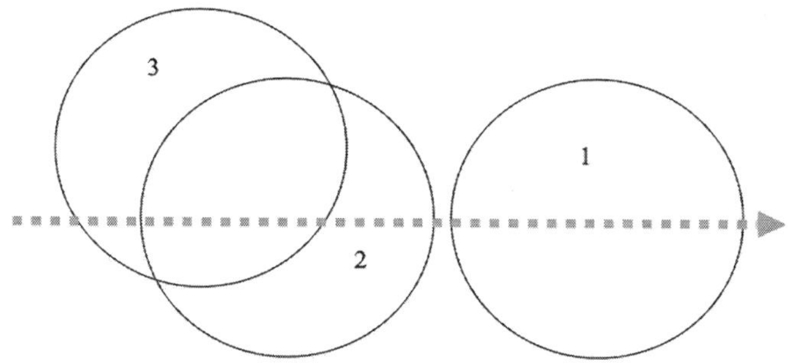

FIGURE 5.2 A goal state model.

the task continues along its trajectory to the next goal state. The status of the system reflects the degree to which the actual state and planned state are congruent.

Task management, then, involves establishing a future goal and then initiating actions that will allow that goal to be achieved. Figure 5.2 shows the main components of a dynamic goal state model. As the trajectory unfolds (represented by the hatched line in Figure 5.2), we move from the current goal state (2 in Figure 5.2) towards the desired future goal state (1). The status (3), as we have said, is the current configuration. If the task management is successful, status and state will be congruent until the next state change is required. As the trajectory transitions through one goal state to the next, the status will diverge until it can be brought back into equilibrium.

Some goal states can be specified in fairly specific terms. Landing the aircraft requires a fairly limited, specific set of constraints to be met if the landing is to be deemed a success. However, other goal states can be better represented as a volume of space and time. For example, we can draw a notional 'top of descent' as a mark on a chart, but, in reality, it is an event in time rather than a point in space. The actual descent point will vary according to several factors, such as weather, other traffic and air traffic control (ATC).

The primary role of the crew is to manage the aircraft through these transitions from one goal state to the next, as illustrated in Figure 5.3. For example, during the taxy phase, the runway in use is usually identified by ATC together with the route to follow to arrive at the point at which the aircraft can enter the active runway. The movement from the gate to the runway is more than simply a physical relocation of the aircraft. First, the crew must successfully depart from the terminal area and any associated ground support. Systems are tested and configured for departure. The actual take-off configuration and route are validated through the monitoring of ATC transmissions from the departing aircraft; contingencies are discussed, and crew actions allocated. There are probably four distinct goals buried in this stream of activity: push back, test and configure, manoeuvre and, finally, confirm the next step.

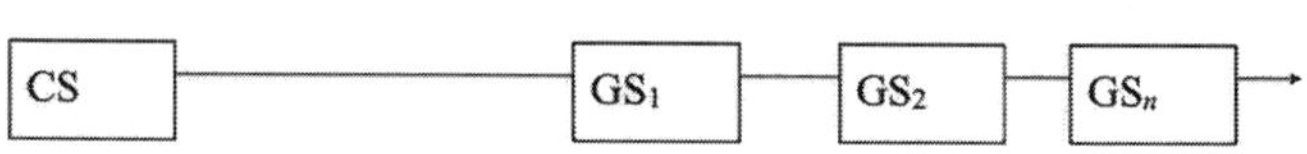

FIGURE 5.3 An idealised trajectory.

The nature of the transitions between the goal states in terms of complexity and time available varies throughout the trajectory. A very short taxy route from the gate to the runway requires an increased work tempo whereas a long taxy at a busy airport, with other aircraft waiting to depart ahead of you, can create a different set of problems. Achieving the immediate goal state will trigger the next set of actions necessary to accomplish the subsequent goal.

Desired goal states are usually associated with a critical point (CP), which is a notional point on the trajectory by which a particular goal state must be achieved before the aircraft can safely (or efficiently) proceed to the next goal state (see Figure 5.4). Convergence between the current status and the desired goal state will vary in terms of rate and the extent to which convergence will occur before the CP. Failure to achieve a goal state by the CP will have implications downstream. Looking at the example of taxying the aircraft, the crew cannot start the journey to the runway until they have successfully finished all the activities associated with the push back. Aircraft have struck ground equipment that had not yet been moved injuring personnel and dragged still-attached equipment and collided with other manoeuvring aircraft during this phase. Taxying to the runway cannot commence until a variety of other activities have been done. Equally, the take-off cannot commence until after the aircraft has been correctly configured. Each element of the move from the terminal to the runway will have a CP by which tasks must have been accomplished.

It is possible to look forward to successive goal states and if a downstream state becomes illegal, all preceding states becoming irrelevant. As an extreme example, when the magnitude 9 earthquake and subsequent tsunami struck Japan on 11 March 2011, aircraft in the area were told that all airports were closed. For those aircraft in the descent or on approach, their goal state instantly became one of 'where can I land before I run out of fuel'. Everything else was redundant.

I need to distinguish between a system boundary and CP. The boundary is the point at which the system, as a whole, becomes unstable, shifting from a safe to an unsafe state. A CP, on the other hand, is a notional point in a stream of work by which the goal state constraints should be met. A really simple example would be the need to select gear down sufficiently in advance of landing in order to allow time for the undercarriage to deploy. To bring control and predictability to the flight process, airlines develop procedures, many with associated notional 'gates', such as a stabilised approach point. These are often, in fact, pseudo-goals in that, although easy to define, they do not necessarily map onto the operator's cognitive model of the current process, but they, nonetheless, represent CPs set sufficiently in advance of the boundary. The landing system boundary is a broader concept, but in some circumstances, a goal state CP and the boundary may be identical.

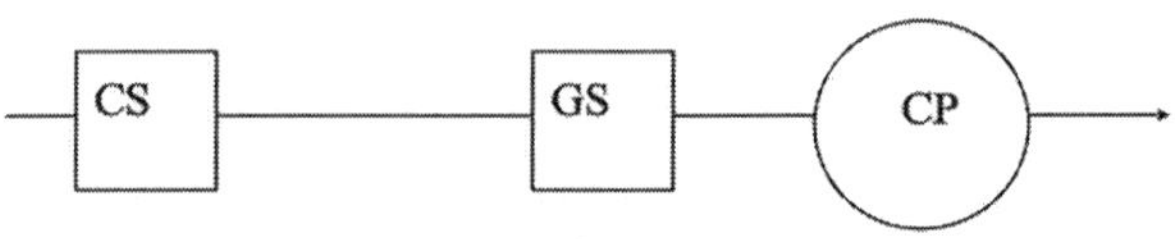

FIGURE 5.4 Desired condition.

The margin is the space available to satisfy the goal state constraints and can be considered synonymous with the use of the term in relation to the broader system. In both cases, it is a volume of space in which action is required. Margins are not constants and are created as the trajectory unfolds. In the example of taxying to the runway, the taxi route length will affect the time available to complete the necessary tasks without incurring a penalty, such as delaying take-off. In the case of the approach, ground speed and the point of interception of the inbound course (height in relation to desired glide slope, distance from touchdown) will shape the margin available. Satisfying the requirement to align the status and goal state by the boundary requires the crew to anticipate and act in such a way that states, transitions and constraints can be satisfied.

5.4 BUFFERING AND EFFICACY – THE MANAGEMENT OF GOAL STATES

Buffering describes the range of behaviours the system can cope with. If we consider goal states in relation to CPs, we can identify situations where events become compressed or the attainment of the goal state is deferred. Figure 5.5 illustrates some of the possible relationships between the goal states and CP. Failing to achieve the appropriate status by the CP will complicate subsequent activity but will not necessarily lead to breaching the boundary. However, in situations such as these, the workload will increase and both load shedding and forgetting will increase the risk.

Because of the dynamic nature of both atmospheric processes and the operating environment, pilots must continually verify the validity of the goal state. Changes in the circumstance will alter the values of certain attributes, and this will, in turn, affect the extent to which the constraint set can be satisfied. It may even render the desire goal state illegal. For example, on an approach to a runway in winter, a report of runway braking action made by an aircraft landing ahead of you may contain new information that could trigger a change in your plan. Changes in a goal state, therefore, may require additional activities in order to satisfy any updated constraints or even the selection of a new goal state. Buffering also reflects the adaptive capacity of the system.

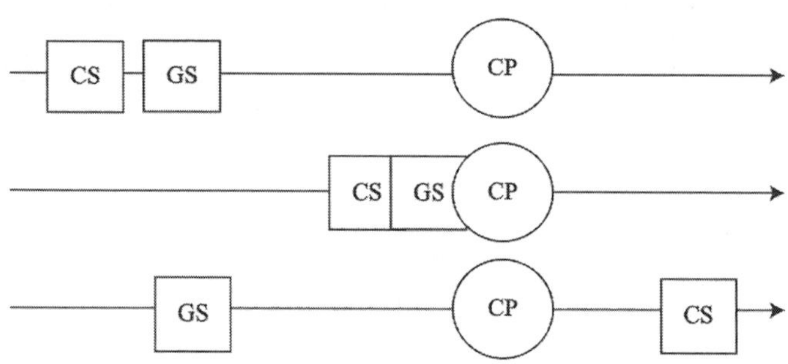

FIGURE 5.5 Possible conditions.

Efficacy describes the degree to which interventions will satisfy goal constraints within the margin available and lies at the heart of CRM. Efficacy is sustained by effective collaboration, but it is a reflection of the contribution of individuals. It involves us in continually comparing the current status, in part captured by the device sensors and displayed to the pilot, with the desired goal state to verify convergence or congruence. Unfortunately, here we have several potential problems. First, reflecting on the experience of the Saab 340 crew in Chapter 3, the interface between the aircraft and the crew can be remarkably opaque. Second, the system status is a construct based on the relationship between the condition of the actual world and the operator's understanding of the world. The problem is that, as we will see in the next section, the individual's view of the current state is often based on a plausible interpretation of available information rather than an absolute version of reality (Weick, 1995; Hutchins, 1995). Operators work with representations of the world around them, not with absolute truths. The degree to which operators can construct valid representations is, in part, a function of the feedback received from the system and the environment.

Pilots can modify the trajectory by bringing forward, deferring, deleting or inserting new task elements. Buffering will reflect the capacity of the system to cope with these task modifications, especially improvisations, whereas efficacy reflects the fitness of the chosen action. Task management skills include the array of techniques to sustain or recover the desired trajectory, the processes by which all team members maintain a shared view of the task and the techniques used to resolve the implications of task modifications, all underpinned by fundamental knowledge of the performance and system's behaviour. In the next section, we will look at how individuals make sense of the work and organise work. The signals we use to confirm that a constraint has been satisfied will vary in significance and salience. Expertise is a measure of the richness of our constraint sets, and our ability to detect the cues that suggest that constraints will not be satisfied.

5.5 GOAL STATES AND RESILIENCE

In order to negotiate a successful path through the various goal states comprising the aircraft's trajectory, individuals must construct an understanding of the prevailing state of the world, but, as was stated above, we often work with approximations of reality. Returning to the B-777 landing at Heathrow described earlier, the notional plan for the last few seconds is illustrated in Figure 5.6. Unfortunately, events unfolded rather differently, and the revised approximate sequence is illustrated in Figure 5.7.

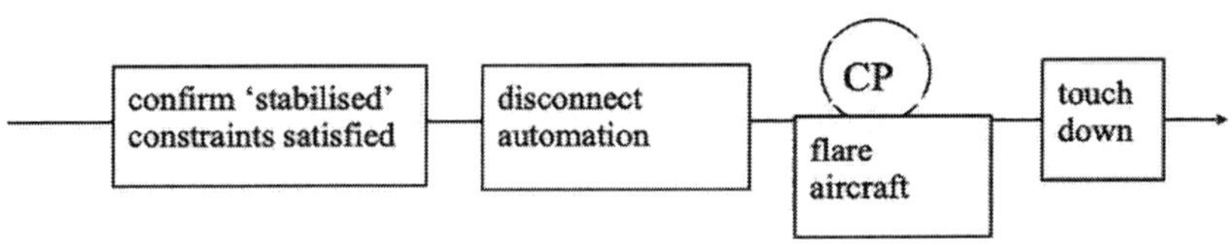

FIGURE 5.6 Heathrow landing goal states.

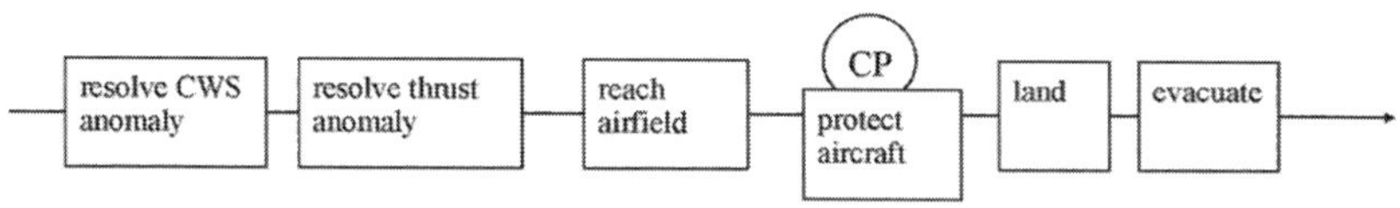

FIGURE 5.7 Heathrow revised trajectory.

These two representations illustrate how the space occupied by the activity can be both concrete and abstract. Under normal circumstances, the transition from the descent to touchdown is concrete. It can be witnessed, and the implications of a failure to manage the transition are known. The set of abnormal goals illustrated in Figure 5.7 are largely abstract, and they are attempts to resolve ambiguity or to increase the probability of the desired outcome, 'reach the airfield', for example. This latter goal points to the nature of the action. By intervening, the captain changed the status of both the aircraft and the world, as it was presented to him. His action was to reduce the flap setting in order to reduce drag. Because of the condition of the aircraft, which was established by the inputs of the FO to the controls, both the drag states, the path of the aircraft and the point of touchdown, were modified. The captain now had a different view of the world and, as a result, could anticipate a different set of outcomes. The point I want to make here is that the world is not a static entity to which individuals respond. Action shapes the world thereby offering changed sets of possibilities. This concept will be important when we come to explore sense-making.

5.6 HUMAN INFORMATION PROCESSING

In Chapter 3, I introduced the concept of cognitive load and its three components: information processing, time on task and task switching. These components lie at the heart of performance at the individual level in a system. The efficacy of actions is a measure of the success of the individual in meeting the demands posed by balancing cognitive load. In this next section, I want to address the first component: information processing. Psychologists have a long tradition of studying the way humans receive and handle information from the outside world. The classical model of information processing, illustrated in Figure 5.8, has six basic components: inputs, perception, attention, processing, storage and outputs. The model argues that the first stage in

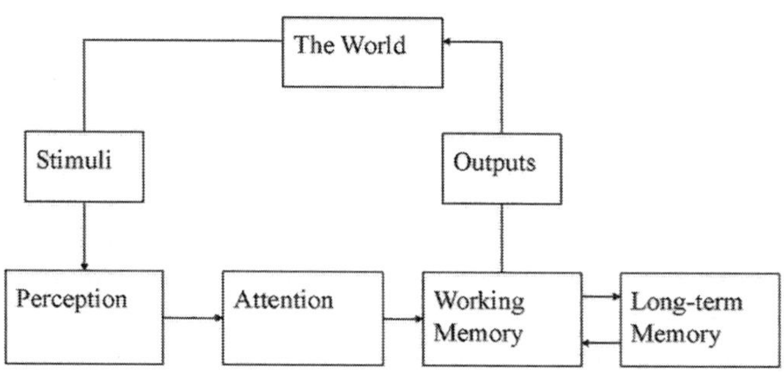

FIGURE 5.8 Human information processing.

making sense of the world is perception. Stimuli from the external environment generate sensory inputs that are detected by our sense organs. These stimuli are held in sensory registers and processed to varying degrees. The processing of inputs takes two forms. One mechanism, called bottom-up processing, entails the direct analysis of the stimulus and is constrained by the availability of sensory inputs. What this means is that if the signal is of insufficient strength or is not detected in the first place, then it will be overlooked, irrespective of its significance. The second type of processing, top-down, involves a search for expected features in a stimulus based on past knowledge and the current context. This first stage of information processing, then, is constrained by the physical properties of the sense organs and shaped by our prior experience. This can lead to erroneous perception of events, but at the same time, it allows us to fill in the gaps if a stimulus is of very brief duration or degraded in some way. Signal salience describes the extent to which a signal stands out from the general noise. The significance of a signal suggests its importance in terms of what we are trying to achieve, but the significance is a construction, not necessarily a reality. Stress and fatigue will affect perception, as we saw with the B-727 at East Midlands Airport.

The inputs available to the captain of the B-777 on approach to Heathrow would have included:

Visual: Views of the outside world (planned gate, buildings under the approach path, airfield dimensions), instrumentation and annunciators, the position of the thrust levers, the position of flap and gear levels.

Aural: Engines, transmitted signals and messages, warnings and crew utterances.

Kinaesthetic: Acceleration detected by semi-circular canals (buffet and turbulence), stick shaker and reaching for controls.

These sensory inputs were perceived and attended to. As the aircraft descended, the captain was processing stimuli concerning a variety of tasks. Initially, processing was framed by expectations derived from a model of normal procedures – searching the visual field to see whether the assigned gate was free. Once the captain became aware of the unusual situation, his attention was directed towards, first, resolving ambiguities and then, second, to the relationship between the current path of the aircraft and the likely point of contact with the ground. The specific stimuli attended to were relevant to the task goal in play at that time. Initially, the goal was related to planning the arrival on the parking stand. Stimuli from inside the aircraft were not being attended too. The FO's initial comment – 'was it doing this when you were flying' – was processed in a top-down manner'. The ambiguity in the message was resolved by comparing it with the captain's own prior experience of the turbulence on the approach and its effect on the thrust levers. The outcome had sufficient congruence such that the comment did not trigger any additional information-seeking behaviour. The second comment by the FO, however, was dissonant and triggered additional behaviour. The captain then went through two cycles of searching for information about aspects of the aircraft's condition. At this point, the captain started to process in a bottom-up manner. He scanned the available information and tried to make sense of it. None of the information matched his prior experience or his understanding of the aircraft's systems. Finally, the captain attended the stimulus associate with the performance of the aircraft in relation to the ground.

Only the stimuli reaching a particular signal strength are attended to. Throughout the episode, some available data were not processed. For example, the fact that the autopilot was still engaged was missed and was only disengaged when the stick shaker was triggered. The aircraft's speed was slowing because of the reduced power from the engines and the fact that the autopilot was trying to maintain the ILS glideslope. Although there was some attempt to understand the cause of the power loss, the task was traded off against the more salient task of reaching a safe place to land. The implication of speed was not appreciated until the stick shaker was activated. Attention, then, is a process of recognition, of isolating out certain aspects of the environment. It is possible to attend to multiple stimuli by allocating attention according to the demands of signal processing. We can manipulate attention to a degree. Focussed attention allows us to concentrate on a specific stimulus but at the expense of processing other stimuli. Divided attention allows us to track multiple stimuli. The effectiveness of divided attention is a function of the difficulty of the tasks being monitored and the degree of practice we have had in those specific tasks. In the example we have just explored, the crew were at the end of a long flight from Beijing, and we need to accept the insidious nature of psychological fatigue in relation to information processing. It is an understatement to say that stress was also probably elevated given that the situation they were presented with was outside of any procedural or training framework they had ever encountered. These factors all had an impact on the attentional capacity and attention management tactics of the crew.

Returning to our model of information processing, inputs attended to are brought into the working memory (WM) and manipulated. The WM is where we extract order from the mass of data presented to us. It is where we analyse inputs, make decisions about what to move into storage, prepare responses to inputs and generally manage tasks. Once processed, some information is then retained in the long-term memory (LTM). If we consider the captain's analysis of his options for reducing drag, thereby extending the track of the aircraft, we can see the interplay between the WM and LTM. Having processed the available stimuli, he identified a possible course of action, which was to retract the undercarriage. However, knowledge of the operation of the undercarriage, its associated doors and the drag implications, held in LTM, was recalled, but the applicable constraint – no action should increase drag – could not be met. He then considered the implications of raising the flaps, to reduce drag, in relation to the aircraft's performance. Because the aircraft speed was below the standard target speed, the margin between the minimum speed for the new flap setting, held in LTM, and the actual speed would be narrower than usual. In fact, we have another trade-off. In this case, we see the need to extend the range being traded off against the performance margins associated with the aircraft's condition. The output from this processing activity was an act that reconfigured the aircraft: it was a decision.

To summarise then, a pilot interprets signals from the environment, some of which are mediated through aircraft sensors, systems and indicators, and makes inputs to the aircraft to reconfigure the device as appropriate for the local needs. The role of the pilot is two-fold: first, it is to execute a plan determined in advance and, second, it is to respond to perturbations in the environment in real time. Accurate information processing is underpinned by robust knowledge structures. Fundamental to safe operations is the importance of sense-making. We need to remember that the model of information processing outlined above was heavily influenced by early work on computers. Unfortunately, humans are not computers.

5.7 SITUATIONAL AWARENESS, DISTRIBUTED COGNITION AND SENSE-MAKING

The concept of situational awareness (SA) was developed to describe the way in which fighter pilots establish 3D models of the airspace around them during combat (Endsley, 1989, 1993). The project was triggered by a finding that, in studies of air combat, over 50% of pilots never saw the aircraft that shot them down. The concept was later applied to civil aviation and incorporated into CRM (Endsley, 1995). It has been defined as:

> 'The perception of the elements in the environment within a volume of time and space, the comprehension of their meaning and the projection of their status in the near future'
> **(Endsley, 1989)**

Unfortunately, this description of SA seems to be no more than a reframing of the classical model of information processing. Baxter and Bass (1998) observe that, in reliable systems, operators do not generally need a high level of SA, and that there does not appear to be a direct causal link between the level of SA held by an operator and the general level of performance of the system as a whole: high SA can still lead to an accident. Flach (1995) points out that the failure to separate the cause and manifestation will lead to circular arguments in which accidents result from incomplete SA which, in turn, was the result of an earlier loss of SA. Despite its prevalence and the fact that it is widely, and confidently, used as an assessment marker, SA is a contentious subject. Parasuraman et al. (2008; see also Banbury and Tremblay (2004); Endsley et al. (2003)) make a robust defence of the concept stating that it is underpinned by strong empirical evidence whereas Dekker et al. (2010) take a different view, claiming that SA is no more than 'folk psychology'. At the heart of the debate lies the nature of the experienced world. The traditional approach to information processing, and remembering that the brain as a digital computer metaphor prevailed at that time, assumed that information processing resulted in a version of the outside world generated inside the head which had a high degree of verisimilitude. Parasuraman et al. (2008) considered SA to 'represent a continuous diagnosis of the state of a dynamic world'. As such, there was 'a 'grounded truth' against which its accuracy can be assessed. The authors claim that SA is an estimate of the accuracy of the relationship between the 'objective' world and the world as understood by the actors. That said, others have argued that SA is both a condition – our understanding of the current status of the world – and a process – the methods by which we clarify that understanding.

More recent conceptualisations see the creation of meaning as a two-way process between the human and the outside world. Clausen and Hutchins (1996), in describing the performance of a crew in a simulator, observe that an aircraft flight deck represents a distributed cognitive system. Information processing in such a context can be characterised as propagation of representational states across representational media. Representational media include the values presented by displays and indicators, and also any other aspect of the workplace through which we can gain an understanding of the status of a process. For example, some pilots use the position of the landing lamp switch to remind them about the granting of permission to land. The switch is not moved until after landing has been authorised by ATC and, thus, if the switch is 'on' then permission must have been given. The switch conveys information about the legal status of the approach. As the task progresses, the flight deck is progressively reconfigured to reflect the current status of the system. From this perspective, aspects of the aircraft's configuration as represented by the flight deck is a surrogate for SA. However, holding an accurate representation of the world is of little use unless it supports sense-making. Weick (1995) identifies seven properties of sense-making:

- Focussed on and by extracted cues
- Driven by plausibility rather than accuracy
- Grounded in identity construction
- Social
- Enactive of sensible environments
- Ongoing
- Retrospective

This view suggests that 'reality' is socially constructed rather than some sort of objective internal representation of the world 'outside of my head'. Weick's 'extracted cues' are the flags that draw our attention to those aspects of the situation that support future actions. To save cognitive effort, and we need to remember that the brain consumes 30% of the energy produced by the body, we tend to stop processing as soon as we reach a plausible explanation of events. The relief FO on the Heathrow B-777 stated that the problem 'looked like a double engine flame-out'. This 'plausible-but-wrong' analysis was framed by experiences of failure drills in training. Sense-making is also influenced by the social niche we occupy. For example, the captain's sense-making was shaped, in part, by the broader responsibilities he held as the commander of the aircraft. So, his analysis of the situation took into consideration the need to either take back control of the aircraft or leave the FO as the flying pilot, by the need to prepare the crew and passengers for a possible adverse outcome. Finally, and probably the most important, sense-making is continually refreshed as the world unfolds, and as a result, we can only know what has already passed and what we face now; we cannot know the future.

The 'environment' we work in, then, is created by us, and, in turn, we respond to that environment as it changes through time. Furthermore, we can see that the interpretation of external information is a subjective process and is driven not by what information is objectively available but, rather, by what is accessible to individuals and what they chose to sample. Sense-making is not static but, rather, it is a dynamic process that changes as events unfold. The extent to which our mapping of facts onto reality is successful is a measure of the way we employ information-seeking behaviour.

5.8 PERFORMANCE AS APPROXIMATION

This next case study illustrates how cognition is distributed across systems and how operations can progress despite the crew having significantly flawed awareness of the situation. Again, the case study is a two-crew operation, but I want to focus, initially, on aspects of the captain's performance. On 8 August 2003, an ATR-42-300 en route from London Luton Airport to Galway, in the Irish Republic, experienced an in-flight shut down of the right-hand (RH) engine as a result of fuel exhaustion (AAIU, 2005). Fog, earlier in the day, had delayed the departure of the flight from Galway. As a result, the rostered crew could not complete the four planned sectors in the remaining duty time, and an intermediate landing was made in Dublin to change crews. Inbound to Dublin the aircraft left-hand (LH) fuel gauge malfunctioned and was reading empty, an intermittent snag that the aircraft had been carrying for several days. Given the delay and the fact that the operating crew were confident that the maintenance response would be no more than to sign off the defect without rectification, no technicians were called and the aircraft was handed over to the new crew. Having, himself, encountered the fuel gauge problem with this aircraft, the on-coming captain was comfortable with the situation but, nonetheless, he decided to refuel in Dublin. The 962 kg of fuel requested were not recorded in the technical logbook. The flights to Luton and then to Waterford were uneventful, and the pilots found that by recycling the fuel panel they could occasionally get the LH gauge to give a fairly accurate reading. Fuel was taken on at Waterford, the process being supervised by the captain. The penultimate sector was back to Luton where the captain, again, supervised the fuelling. A combination of factors resulted in the captain inadvertently diverting all of the fuel taken on at Luton into the LH fuel tank. The crew detected a lateral imbalance after take-off, and an attempt to rectify the problem was made by feeding both the engines from the LH tank for 15 minutes. At the same time, the RH fuel gauge showed an unexpectedly low value, eventually dropping to empty. This was dismissed as simply the RH gauge failing as the LH tank gauge had done. At 23:06, 40 miles west of Dublin, the RH engine stopped, and a precautionary landing was made at Shannon. The crew later discovered that the RH engine had, in fact, flamed out because of fuel starvation. The RH tank was empty.

This brief description of events serves as an introduction to a very complex episode. Although fuel was taken on during the crew handover at Dublin, the quantity of fuel loaded was not recorded. The on-coming captain, therefore, started his duty without an accurate fuel state for the aircraft. Technically, this was not an issue until the aircraft completed its second scheduled sector and landed at Waterford. At this point, the aircraft needed to take on additional fuel to complete the remaining two sectors. The process of loading fuel is controlled through the electrical refuelling control panel (ERCP) located on the LH wheel fairing. The panel allows the operator to insert the desired final fuel total into a numerical totaliser selector. The ERCP will distribute fuel based on the quantity in each tank at the start of fuelling, as measured by the fuel gauge, and the planned total quantity required at the end of fuelling. Blue lamps on the ERCP indicate that the valve to the respective tank is open, and fuel is distributed in such a way that both tanks will contain equal quantities. A refuelling point is located on the underside of the port (left) wing.

Because the fuel uplift at Dublin had not been recorded and given the problems with the LH fuel gauge, the captain had probably estimated a sensible quantity to load rather than planned an accurate figure based on prior consumption and the fuel required to complete the task. He entered 500 kg into the ERCP intending that to be the volume of fuel to be put into each tank (i.e. total uplift would be 1000 kg). Instead, the ERCP, as we have just seen, was configured to put up to 250 kg into each tank (i.e. the starting quantity plus the additional fuel) resulting in a final on-board quantity of 500 kg. The system was, of course, dumb and did not 'know' the captain's intentions.

During the fuelling process, the captain noticed that the RH valve light quickly went out, indicating that the valve had closed. This was because the total quantity in that tank had reached 250 kg (i.e. half the 500 kg planned quantity). The LH valve light was still illuminated. He tried to open the valve using the electrical override switch on the ERCP, but this did not work. The captain then showed the fuel truck operator how to open the valve using the manual trigger located by the fuelling point on the wing. The captain was able to observe that a blue valve light illuminated. This episode is significant. The fuel tank valves are controlled by the fuel tank gauges. The failed LH gauge meant that, as far as the automatic system was concerned, the tank was permanently empty. As a result, the tank valve would remain open with the valve lamp lit irrespective of the actual tank contents. The RH tank was functioning correctly and, so, when the captain tried to override the system with the valve switch, it did not operate. The system logic dictated that it should be closed, given that the RH tank contents were already at the planned value. A total of 600 kg of fuel was uplifted, less than the planned total of 1000 kg. Once refuelling was finished, the captain estimated the quantity in each tank by inspecting the drip sticks. The drip stick is an electro-mechanical device that gives a numerical value representing the quantity of fuel in each tank. This value is used to enter a table in the aircraft manual to derive the actual fuel quantity. The value has to be factored for lateral inclination (the extent to which one wing might be higher than the other on uneven ground) and to do this, the pilot has to take a further reading from an inclinometer installed on the aircraft. Poor maintenance of the device made it difficult to read, but the captain estimated that the LH tank contained 600 kg of fuel, and the RH tank contained 720 kg. A post-flight reconstruction of the fuel consumption suggests that the actual contents were 618 and 689 kg.

If we deconstruct this episode, we see that the captain started from an uncertain position, not knowing the exact quantity of fuel on the aircraft after landing at Waterford. Given that the aircraft should have had a balanced fuel load on departure from Dublin and that the fuel consumed by the two engines en route was similar, the tank contents at this point should have been more or less equal. His attempts to manage the refuelling process represent a set of goal-directed behaviours. He directly intervened in his working environment, first, by inserting a value into the ERCP and, second, by directing the action of the refueller. These interventions shaped the expectations of future actions and both coalesce around the behaviour of the valve lamp on the ERCP. I will have to make an assumption, at this point, but I believe that the captain was expecting both the valve lights to extinguish almost simultaneously, which suggests that he was unaware of the effect of a failed gauge. The use of the drip sticks was a formal, procedural requirement but, nonetheless, was goal-directed, and its purpose was to confirm the current status of the fuel state of the aircraft. The output from this activity was an approximation of the actual state but one that was sufficiently plausible not to trigger any concern on the part of the captain.

The next sector was from Waterford to Luton where the aircraft was again refuelled. The captain assumed that the quantity of fuel in each tank was the same (325 kg), and he planned to uplift 960 kg of fuel, sufficient for the final sector, with 480 kg to go into in each tank. Again, he supervised the process but did not change the 500 kg previously set in the totaliser selector panel. At the start of refuelling, the LH lamp was illuminated, for the reason discussed earlier, but the RH lamp was extinguished because the actual tank already contained more than 250 kg (actually estimated to be 277 kg). By now, it was getting late, and the visibility was fading. Because of his earlier experience with the electronic valve switch at Waterford, the captain chose to not bother with the electronic override switch, instead of moving straight to the manual triggers. This episode illustrates how prior experience shapes the way we choose to engage with the world: action is created afresh with each encounter with the world. He initially asked the fuel truck driver to operate the trigger, but this was declined because it was not something the driver had been trained or approved to do. The captain then climbed onto the refuelling vehicle platform and opened the valve himself. The triggers are not labelled but are simply located on either side of the refuelling point in relation to the tank they operate, orientated in relation to the nose of the aircraft. The captain operated the valve on his RH side. Facing aft and using his right hand, he inadvertently pulled the trigger for the LH valve, which was already open. As a result, the entire fuel uplift was directed to the LH tank. Again, the captain observed a blue valve light, but, at such a distance and in the fading light, he could not see which tank is related to. It was later estimated that the distribution of fuel on departure from Luton was 1166 kg in the LH tank and 277 kg in the RH tank. A set of steps was needed to reach the drip sticks but none were available at Luton. Because finding a set would only add to the already considerable delay, the captain did not bother checking the contents of the tanks before departure. The crew noted that the RH fuel gauge was now indicating less than they were expecting, but this was explained away as being a second gauge failure.

The FO was the handling pilot on the final sector from Luton to Waterford. Soon after rotation, she detected that the aircraft was out of lateral balance, feeling left-wing heavy. She commented on this to the captain who took the controls to check for himself. In fact, at take-off, the difference in contents between the two fuel tanks was 889 kg, 339 kg above the maximum permissible lateral imbalance. There were three aspects of the operational environment relevant at this stage. First, the imbalance was immediately noticeable through the controls. Second, the imbalance was within the command authority of the autopilot, possibly indicating that the problem was not excessive. Finally, a light on the overhead panel was indicating that the starboard spoiler was occasionally being deployed to counter the imbalance. In response to this fragmented repertoire of clues, the crew decided to run both the engines from the LH tank for 15 minutes to reduce the imbalance. About 30 minutes after take-off, the RH tank was indicating empty, and both the fuel low warning lights were illuminated. These warning lights are triggered by the relevant fuel gauge indication and are not linked to an independent measure of tank contents. Thus, presumably, the LH light had been illuminated for most of the flight. One hour and 21 minutes after leaving Luton, the RH engine stopped, and the crew diverted to Shannon.

The captain attributed the failure to an engine technical problem and chose to remain slightly high on final approach in case the LH engine also failed. Meanwhile, the FO was attempting to reconstruct the aircraft's fuel state and concluded that the RH gauge was accurate and that, in fact, the tank was empty. Not even after it landed on one engine at Shannon did the captain believe that lack of fuel was the problem.

5.9 ACTING IN AN UNDER-SPECIFIED WORLD

In the previous discussion of SA and sense-making, I suggested that views of sense-making predicated on the need for a 'truthful' internal representation of the outside world might be wrong. Hoffman (2019) goes as far as to suggest that striving for accuracy is inefficient in evolutionary terms. When you open your eyes, what you actually see are a collection of icons, emojis and GIFs that represent attributes of the world. If I look at a bowl of fruit what I initially see is an icon for 'something good to eat'. When I look at a group of people, I see a collection of emojis that allows me to establish an understanding of the mood of the group without investing too much effort. Having surveyed the world in shorthand, I then focus attention on those elements of interest and render them in finer detail. The ATR 42 case study shows how this might work in an aviation context. On shutdown in Shannon, the aircraft had 480kg remaining in the LH tank. Fuel planning and in-flight monitoring are not just routines but fundamental pilot skills. Given the degree of randomness in the refuelling at both Waterford and Luton, on another day the aircraft could have landed safely at its destination by chance alone. It seems that rather than seeing an accurate representation of actual fuel states, the captain saw a Hoffmanesque icon for 'fuel state sufficient for the remaining journey' based primarily on the contents of the RH tank. But it was not just the actions of the captain that created the situation he engaged with on that day. To fully understand this event, we need to place the captain's actions into a system's context.

Airworthiness requirements mandate criteria for the design and performance criteria of fuel systems. These were translated into a solution by the aircraft manufacturer who, in turn, communicated with the airline through the embodied technology, incorporating controls and indicators, and the associated documentation, which contained system descriptions and operating instructions. As part of the solution, the manufacturer provided redundancy. In this case, the electronic fuel switches were backed up by manual overrides. The design of the system afforded opportunities to create novel solutions to problems.

Company Standard Operating Procedures (SOPs) are designed to facilitate work; they are formal work instructions that guide activities but also represent embodied knowledge about how to do a specific task. The aircraft's fuel gauge problem had been known for at least 5 weeks before the flight incident. The company operations manual stated that *defects* are to be written up *after flight* (author's italics). Because the problem was intermittent, it might be construed that the crew did not see the need for reporting a *defect* as the gauge had not failed completely. Furthermore, the landing at Dublin was an unscheduled, intermediate stop and might not have been understood by the first crew as the *end of the flight*, which was destined for Luton. A flight crew instruction circulated to remind crews of the appropriate action in the event of a defect simply talked about accepting an aircraft with open entries in the tech log but does not refer to aircraft with known defects that are not entered in the log. Where maintenance assets are limited (either in terms of manpower or spares), it was once common for faults to be written on notepaper and only those that could be easily remedied during turn round were, in consultation with maintenance, written up in the tech log.

If an aircraft was at an outstation, the procedure in this company was for the relevant page in the log to be faxed to base maintenance who will make a decision and fax the annotated page back to the aircraft for insertion in the log. The report noted that it was often difficult to find a fax machine at some destinations (Asynchronous drift resulting from developments in technology?), and so crews operated with open maintenance items rather than incur the delays typically associated with compliance with the instruction in the operations manual.

Although the company's minimum equipment list contained advice on how to estimate fuel tank contents in the event of a gauge failure, there was no formal requirement to do so, and although the company procedures did require fuel checks to be completed, there were no specific additional requirements in the event of unreliable or faulty fuel gauges. Therefore, the captain's failure to check the tank contents before the final take-off from Luton was not necessarily a breach of procedure. Formal rules, then, can also provide sufficient latitude for local solutions to real-time problems to be implemented. Work very rarely proceeds as planned, and variations and anomalies are a normal part of the task. The fact that individuals are required to exercise discretion in order to get the job done renders the work process fallible. We can see this in the way in which the ATR 42 crew managed the faulty fuel gauge.

In this episode, we saw two examples of aircraft refuelling. Each iteration comprised different component parts as the captain tried to negotiate solutions to problems as they arose. His performance was framed by company procedures and shaped by prior experience but also involved collaboration with, in this case, the refuellers. Although one agent was willing to operate the manual trigger at the captain's request, the second declined. Presumably, both agents were constrained by similar rules, but each acted in a manner that satisfied their interpretation of their freedom of action.

This section has attempted to illustrate how individuals construct meanings in support of their actions and how that action, in turn, modifies the world. Some would say that the captain 'lacked SA', but I have attempted to show that sense-making is a dynamic process. Organisational factors and aircraft design rules all shaped the way the captain behaved. However, effective metacognition requires performance to be matched against some anticipated set of outcomes. The incident report referred to 'common sense, prudence and good airmanship' in suggesting that taking a manual reading of tank contents at Luton would have been advisable under the circumstances, a classic example of hindsight bias. In fact, the report's authors have captured the extent to which processes are generally under-specified, relying on operator expertise to elaborate on the process should circumstances dictate. Dekker (2002, 2005) discusses the difference between the work 'as-imagined' and work 'as-done'. Shorrock (2016) goes further, identifying 'work-as-disclosed' and 'work-as-prescribed' as additional artefacts of the process of describing an activity. This incident reveals how the process of refuelling comprised a task that was represented in documents (imagined, prescribed and disclosed) but also a task was understood by various actors (done). Importantly, it also existed as a task as enacted (outcome). The outcome was clearly not the captain's intention, but the result was almost inevitable given the way the activity was undertaken in a system's context based on a buggy construction of the world. Furthermore, the action was the resultant of a solution generated in response to a version of the world created anew on each encounter. Importantly, the captain's actions flowed from the decisions he made.

5.10 DECISION-MAKING AS TASK MANAGEMENT

In the next two sections, I want to look at decision-making in terms of actions associated with the active goal state and then decision-making as adopting a new goal. Having just looked at information processing, the other two components in the cognitive load model, time on task and task switching, relate to the activity associated with the current, active task. Doing work involves individuals engaging with the world and making interventions appropriate to the needs of the task. Because of the inherent variability of normal operations – non-ergodicity – individuals have to adapt their performance within the framework of policies and procedures. Selecting the most appropriate action, therefore, is an act of decision-making. Task management involves making decisions about initiating, sequencing and controlling actions. Task switching is the act of shifting between specific target activities depending upon an assessment of priorities. Crucial in all of this, of course, is the timing of interventions: delay reduces downstream opportunity, which is what we mean by margins being eroded.

Trained and competent individuals have a set of expectations about how a task is to be undertaken based on prior training, the applicable procedures and any relevant experience. As we saw in Chapter 1, these factors are encapsulated in a 'plan' or 'script' for how the flight will unfold, including the relevant constraint sets that must be satisfied, all stored in LTM. There will be several different scripts for different flight segments and probable common contingencies. Scripts comprise aggregations of schemata relevant to the task. Thus, at a mental level, tasks can be considered as sets of component activities with triggers that initiate an activity and feedback loops that allow for monitoring, all compiled into routines. Tasks will vary in their complexity, the demands they make upon specific skill sets or the competence of individuals, the level of accuracy demanded and the time available for completion. We can identify three criteria for assessing task accomplishment: the quality of task activity, its completeness and its validity (efficacy, or fitness) in relation to goal achievement. Under normal conditions, work involves drawing on the following knowledge structures:

Declarative Knowledge
- What to do – what are the steps to the next goal?
- Why do it – what is the rationale for this specific course of action?

Process Knowledge
- How to do it – what is the procedural activity required to achieve the goal?

Work, of course, is messy and involves enacting bundles of tasks: rarely will a skilled operator be engaged in a single activity. Task elements will be conducted concurrently, in series and interleaved. For example, imagine a light aircraft flying the downwind leg in the traffic pattern. Keeping the aircraft straight and level represents two concurrent, albeit integrated, tasks. Configuring the aircraft for landing (taking flap and lowering the undercarriage) represents tasks conducted in series. Making radio calls is representative of an interleaved activity. This suite of behaviours requires management if the whole performance is to be orchestrated successfully.

Task management activity comprises initiating, monitoring, interrupting, resuming and terminating task elements, all of which are outcomes of decisions made about the current state of progress towards a goal. The scale of task management problems was shown in a study by Chou et al. (1996) who found that 76 out of 324 accident reports (23%) and 231 out of 470 incident reports (49%) contained cockpit task management errors. Task switching is effortful, and most of the problems associated with normal task management stem from poor prioritisation and forgetting. We will look at the error in more detail in the next chapter, but 'forgetting' is a specific type of error called a lapse. Initiating new tasks and tracking those tasks that are in play places a burden on cognitive capacity. Deviations from the script, such as changes, loops (repetitions), deferrals or omissions, will increase the cognitive demands, requiring additional attention. That task management is effortful is captured in the concept of workload. Task management is the process of organising work while cognitive load is the controlling of the mental framework and workload is the subjective assessment of the effort involved. The workload is a function of the complete assemblage of activities being undertaken by operators over a unit time and includes both formal activities associated with an immediate task and other informal activities. Social conversation between individuals requires attention and, thus, contributes to workload. The workload can be physical as well as cognitive. One commonly-used tool for assessing workload is the NASA task load index (NASA TLX) (Hart and Staveland, 1988). It comprises six scales that look at different aspects of workload. In an attempt to better understand the cognitive demands of flying departures and approaches, I asked the crew to complete the NASA TLX at the top of climbing and after landing. Table 5.1 shows the mean scores of the cognitive effort for the pilot flying (PF) and the pilot monitoring (PM) The difference between the roles was not statistically significantly, but the difference for the PM between the two stages of the flight was significant. In short, the monitoring role requires more cognitive effort during the arrival phase. We also saw in the last chapter that the threat rate, which

TABLE 5.1
NASA TLX Scores

	Top of Climb	After Landing
PF ($n = 21$)	20.3	22.8
PM ($n = 20$)	21.5	27.0

is indicative of the rate of perturbations encountered by the crew, adds to workload as measured by the NASA TLX (Aeschbach et al., 2017).

Prioritisation is the process of making decisions about sequencing activity. Colvin (1999) identified three factors associated with prioritisation: status, procedure and value. The status of the task describes the extent to which the operator believes the conduct of the current task is satisfactory. Colvin's use of 'status' is similar to my use in relation to goal states. The factor labelled 'procedure' captures the degree to which the operator believes the activity is correct within the overall plan. Value refers to the importance of the task within the current activity and the perceived consequences of not completing the task. Colvin's 'value' maps well onto the concept of efficacy. Prioritisation, of course, may require switching between the concurrent tasks, and there is evidence that the interruption associated with switching impedes task resumption. Furthermore, competing tasks induce forgetting, which is not simply a function of workload. Forgetting is representative of poorly executed skill-based routines and is symptomatic of an overloaded WM.

Where a task is well defined, the problems of management are typically associated with the rates at which inputs can be received, processed and responded to. The number of tasks processed per unit time does become an issue as capacity limits are reached. However, ill-defined tasks are associated with a different range of problems. Unfamiliarity can result in a poor structuring of the situation with a resultant fixation on certain features of the problem. Tunnel vision (focussed attention) limits the range of options considered and cognitive lock-up results in aspects of the problem being treated in sequence as opposed to in order of importance. Poor prioritisation, generally, and incorrect sequencing are also features of coping with an ill-defined task.

We saw in the ATR 42 fuel scenario that individuals exercise agency: they act with a degree of autonomy. In a systems model, SOPs represent a form of control but individuals' acceptance of constraints on the work process is a variable, not a given. This property of 'agency' has a bearing on other aspects of task management. Once the primary goal of work and the associated cognitive requirements have been satisfied then operators can consider other performance modifiers, such as:

- perceived risk of failure
- current workload
- time remaining
- perceived social acceptance of task accomplishment

The control of the current task, then, requires the activity to be tracked and evaluated in terms of its validity in relation to the current active goal. Task management requires action sequences to be modified and this increases the workload. The process can fail when the capacity is overloaded or prioritisation is suboptimal. Attempts to control tasks through procedures can fail if individuals elect to work outside of prescribed bounds. Ultimately, though, action at the level of the individual is a stream of continuous decision-making directed at achieving the planned goal.

5.11 DECISION-MAKING AND GOAL STATE MODIFICATION

Because of the variability in the operating environment, task modification may also require the selection of a new goal state. Orasanu and Martin (1998) propose a framework of decision types characterised by the nature of the information available and the responses available. These are summarised in Table 5.2. The simplest course of action is a 'codified' strategy where an event triggers a rule-based response. As a decision type, a rule-driven situation has a single response, and the go/no go decision in aviation is a classic example of this. Successful performance is related to the experience of the decision-maker, and thus the richness of the response sets available, and to the quality of the information available. Typical failures involve not being able to retrieve the correct response from memory.

The second strategy, associative, is a situation in which multiple options are available and involve trade-offs. Various suggestions have been offered for how humans choose between the alternatives. The pairwise comparison suggests that we set up alternatives and delete one that fails to meet the required or certain criteria, working through the list until one option remains. The 'first past the post' strategy suggests that we implement a plan and stick with it until it no longer works. The 'dominant search' model proposes that unsuitable options are discarded early, and the most obvious solution applied. The solution will remain in use until either the task is achieved or additional information renders the solution no longer suitable. Again, experience and the availability of information are significant variables in choice decisions as well as the effectiveness of simulation based on stored mental models. The final strategy is the analytic or creative approach and is characteristic of ill-defined situations. This is the domain of problem-solving and will be discussed later in this chapter.

In dynamic contexts, a decision generates a future action. That action, in turn, may require a further decision. Rule-based problems tend to function in a feed-forward manner: the decision is made and action is taken. The execution of intervention then brings about a state change. However, in situations requiring more creative solutions, action will feedback to the decision-maker and may well change the understanding of the situation as well as the freedom of action for any responses. Reason (1990) identifies two intervention conditions: reactive and multiple dynamics. In the former, the action of the decision-maker influences the state of the task in a lock-step fashion. In the latter, the problem space is influenced by the actions of the decision-maker but also, independently, by external influences.

TABLE 5.2
The Course of Action Strategies

Strategy	Strengths	Weaknesses	Characteristics
Codified (rule)	Fastest, most resilient	Least flexible, cannot generalise to new situations	Single response
Associative (choice)	Faster, efficient, resilient to effects of stress.	Requires good mental models.	Multiple choices with trade-offs. Well defined.
Analytical (creative)	Most flexible	Demanding attention and STM Susceptible to stress	No suitable option readily available. Ill-defined.

We have already seen that sense-making is fundamental to successful intervention in the world, and studies of decision-making have added to our understanding of the process. Quite often, when presented with a problem the tendency is to consider 'have I seen this before and what did I do last time'. If our analysis of the situation is wrong then the solution we apply will not work. One study (Fischer, Orasanu & Montalvo, 1993) found that more effective pilot decision-makers spent longer clarifying the situation, worked through checklists to completion and took time to prevent misdiagnosis. Fischer, Orasanu and Wich (1994) found that the situation structure was not constant across team roles. They found that captains tended to view a situation from a severity and flight safety perspective whereas FOs considered situations in terms of the response required – what will they be required to do. Both groups did consider the time available as a factor. The authors also found that better decision-makers balanced ongoing tasks, avoided cognitive lock-up (fixation) and also retained options for as long as possible.

Goals can also differ in terms of their structure, which can affect situation comprehension. Goals can be either open or closed. A closed goal would be easily defined: land on the correct runway, for example. An open goal might be something like 'how to efficiently and safely dissipate excess energy in the descent in order to achieve a stable approach while negotiating storm cells'. The structure of the goal might be visible or hidden. So, the relationship between the inputs and outputs is clear in a situation where the structure is visible but is less clear where systems are opaque. Time pressure can be high or low, and the information available may be dense – and therefore difficult to process – or sparse – leading to ambiguity. These factors, especially time pressure and structure, will affect the ability to prioritise, which is simply the act of selecting the next best action sequence. Breaking goals down into meaningful sub-goals and establishing a purpose for each goal has been found to be an effective behaviour (Dörner and Schaub), but the oversimplification of situations is, actually, a recognised bias. Simplification is a counter to complexity but can result in a flawed analysis. Pilots often use checklists to frame situations but this runs the risk of falling victim to the 'availability' heuristic. Just because a checklist appears to match the situation does not mean that it is the correct checklist.

Two further issues that relate to situation assessment are bounded rationality and the structure of the actor's mental model. Bounded rationality relates to the fact that, unlike supposed rational decision-makers, ordinary people tend to work within a subset of available information. They do not interact with an absolute model of the world but, instead, work with a skewed, biased, approximate version. Furthermore, because of flaws in training or through aberrant schemata modification as a result of misinterpreted experience, individuals' mental models may be incomplete or inaccurate. Therefore, their ability to run simulations of events will be jeopardised.

Work unfolds by individuals enacting sequences of behaviour directed at achieving a goal state, each sequence triggered by prior action or by some change in the environment. In aviation, work is often conducted in conditions of ambiguity and with high risks involved. Often, crews have to reconcile conflicting goals, and their actions can have unintended consequences. Selecting between the courses of action often involves incurring known costs in order to mitigate unknown risks. Unlike normative models, where there is usually a single 'best' solution, this is not the case

TABLE 5.3
Metacognition and Task Management

	Cognitive	Metacognitive
Knowledge	Systems goals and safety conditions Regulatory constraints Running mental models to generate new rules	Proper ranking of production and safety constraints Making trade-offs between available courses of action Constraint set satisfaction
Rule	Selecting actions relevant to the current goal Choices between alternative routines	Balance speed, workload and risk Cue-action set fitness
Skill	Sensory-motor patterns	Detection of signals that modify rules Error detection

Source: After Rasmussen and Svedung.

in aviation: a choice is 'good enough' in that it works under the circumstances. Our behaviour can be classed as satisficing in that it is acceptable although probably not always optimal. A course of action, with hindsight, can often be criticised, but at the time. decisions are usually judged acceptable if 'nothing went wrong'. These 'decision-action' sequences occur in real time as the flight progresses.

We can now see that effective performance comprises both action and control of that action. Metacognition is the term we use to describe this controlling function. Table 5.3 elaborates the skill/rule/knowledge framework, discussed in Chapter 1, in terms of metacognitive functions.

To be successful, the work needs to be monitored. Monitoring requires us to be able to visualise the task being controlled and, thus, is a function of the completeness of the mental model held by the operator. Monitoring requires the current status of the aircraft to be established with reference to information distributed around the aircraft and shared between the aircraft and outside agencies. We will look at monitoring in more detail in Chapter 7.

5.12 THE SPECIAL CASE OF PROBLEM-SOLVING

It is customary to differentiate between decision-making and problem-solving as somehow separate processes. A decision can be seen as a choice between two or more known alternatives. With a problem, it is not necessarily clear at the outset what the current situation entails and, therefore, what alternative solutions are available and which one will have the greatest efficacy. Decision-making is often fast and intuitive whereas problem-solving is often slow and laborious (Kahneman, 2011). Problems are what Orasanu and Martin called 'creative' or 'analytic' decision types, often require the creation of novel solutions. On closer examination, however, what we call problem-solving is actually a process of making decisions about decisions.

Problem-solving is usually triggered by either an impediment to progress or a realisation that there is no obvious course of action available. In the event of a non-normal or emergency situation like the B-777 on approach to Heathrow, the fact that the crew were presented with a 'problem' was apparent. In Chapter 3, I suggested that the margin was often the zone in which discrepant signals emerged, and in the discussion of SA, we saw that the final stage in Endsley's model was 'projecting ahead'. Given Chater's caution that the future is unknowable, a key element of task management is tracking the current status of the task and validating the future goal state constraint set. This prompts two key questions: 'will the conditions that prevail now remain applicable until the next goal state?' and 'are there any signals in the current situation that suggest either a significant change to the prevailing conditions before the next goal state or a change to the goal state constraints?'. If the answer to either question is positive then there is a need to consider a change to the current, active plan or to select a new course of action. From a systems perspective, we once again come across the issue of scale effect. Quite often, as conditions deteriorate (at least, from the perspective of the outside observer), pilots continue to apply skill-based and rule-based behaviour to sustain performance within bounds: they are engaged in what I call 'trajectory trimming' in order to remain safe and legal. Their actions are directed at the immediate situation. Unfortunately, the system will be operating at a larger scale and maybe changing in more fundamental ways. By the time they recognise the severity of their situation, their range of options has narrowed significantly. At some point along a notional trajectory, the boundary has shifted, and the change in conditions has reduced both the margin and the buffering capacity. Because of the inadequate tracking of the current status of the operation, actors often fail to consider the time and space required for a manoeuvre or goal state transition. Furthermore, where actions are deferred, actors rarely anticipate that any further change in circumstance might impact the ability to perform the delayed action. In part, this is reflected in task forgetting.

The traditional view of problem-solving is that it is a normative process in which the decision-maker has identifiable, stable expectations. The choice is then between a set of known alternatives, each of which has a specific value, efficacy or perceived benefit. Prescriptive models, derived from such an approach, are common in aviation, and Table 5.4 presents some popular examples. Although referred to as 'decision-making models', the first five rows in the table actually describe what we might call the steps in a problem-solving process. The last three rows describe the decision point and subsequent control activity. The 'problem-solving' phase comprises a set of decisions about recognising a need for additional activity, about how and where to search for information and about what future actions have a tolerable probability of achieving the desired goal. Since the early 1990s, attempts have been made to offer descriptive models of naturalistic decision-making (NDM) that better capture real-world performance. It is best to think of NDM as a cluster of views about decision-making rather than a fully developed theory.

Table 5.5 presents a model of problem-solving drawn from work using business simulations (Dörner & Schaub, 1994). The table lists the steps and the potential error types. The model offers an elaboration of the first five rows in Table 5.4 and, importantly, some of the key reasons why problem-solving, especially under stress, can fail. The authors make the point that we have limited capacity for thinking and, as we have seen, are prone to forgetting. They also stress that humans tend to guard feelings of competence and efficacy: we want to feel that what we are doing is serving a purpose. Kahneman has demonstrated that we tend to be overconfident: we think we know more

TABLE 5.4
'Decision-Making' Models

DECIDE	FORDEC	SADIE	DODAR	GRADE	CLEAR
Detect change	Facts	Share information	Diagnose	Gather information	Clarify the problem
Estimate significance		Analyse information		Review information	Look for information
Choose a safe outcome					
Identify alternatives	Options	Develop the best solution	Options		
	Relative advantages			Analyse alternatives	Evaluate alternatives
	Decide		Decide	Decide	
Do	Execute	Implement decision	Assign tasks		Act
Evaluate	Check	Evaluate outcome	Review	Evaluate	Review

TABLE 5.5
Stages in Problem-Solving

Step	Error Types
Goal definition	Inadequate time invested; failure to balance contradictory or competing goals; oversimplification
Information collection and hypothesis testing	Channelling or selective use of information; action deferred pending more information, dogmatic refusal to change; abbreviated searches under time pressure
Prognosing	Linear thinking in dynamic situations; inability to cope with rapid changes in direction (situation reversals)
Planning	Disregarding side effects or long-term effects; stress induces an apparent need for action (sit on your hands); overconfidence in the effectiveness of actions
Monitoring	Feedback delay – failure to monitor long-term changes
Self-reflection	Premature abandonment of metacognitive tasks

than we do. Furthermore, our natural loss aversion makes us more likely to take risks in order to avoid a potential loss, and our inability to recognise a 'sunk cost' means that we are likely to continue to throw good money after bad in order to avoid feelings of regret.

One solution proposed for creative situations is critical thinking or narrative creation. In situations like this, actors attempt to create dynamic descriptions of events – a narrative or story – that they can then use to generate future actions. As long as the outcomes observed can be explained by the narrative, then the model is deemed adequate to support future action. If subsequent events cannot be explained by the narrative then the story must be reviewed and a new narrative is created.

5.13 SAFETY DRIFT AT THE INDIVIDUAL LEVEL – BEHAVIOURAL TEMPLATES

We saw in Chapter 3 that systems are capable of drifting towards an unsafe state. At the level of the individual, similarly, we see behaviour that represents drift. Because we all act with a degree of autonomy and have different experiences of the world, we develop habits that result in work being managed in a riskier fashion. As with the system-level effects, drift at the individual level constrains the system's buffering capacity. Drift represents a subversion efficacy in that the activity is considered in terms of personal benefit or identity creation. I propose four categories of drift at Level 1 in my model:

- Consolidation: The reshaping or simplification of action sequences, often seen in situations where delays are built-in to work processes, and individuals chose not to interrupt the workflow. For example, not waiting for the required cooling period after an event (brakes and engine shut down) or completing checklists in advance of the intended point of execution.
- Homogenisation: The standardisation of actions, results from adapting generic approaches to an operation rather than follow the specific procedural steps. For example, contract pilots flying for multiple clients sometimes follow a generic process, or pilots trained on multiple types sometimes standardise work processes rather than maintain two repertoires.
- Person/Task fit: Choosing to do what is easiest, reflecting the fact that we all differ in our capabilities, and some people work within their limits.
- Role drift: Creating interpretations of the task that reflect self-perception and perceived latitude to act.

While recurrent training and periodic standardisation checks attempt to limit drift, it seems to be a fundamental process. Procedural migration is often a product of social interaction in the workplace as workers share experiences or find new ways to achieve a task.

5.14 CONCLUSION – WORKING AT LEVEL 1

At Level 1, in my hierarchy, individuals engage with the world and act in order to accomplish a task-related goal. Because of the uncertainty and variability in the real world, the crew are continually having to respond to disturbances and create solutions. Action is predicated on a robust understanding of the job and an ability to interpret and apply rules that, by design, are under-specified. Action is directed at accomplishing a goal that serves a purpose in the broader task assigned to the crew. Control of the situation is a function of the individual's interventions, and feedback is provided by the extent to which actions are successful. Task failure, or error, is discussed in detail in the next chapter. Actions concerning events are the product of decisions made by individuals. All activities are conducted in a world that is created by the individual and is shaped by the information needed from that world.

I want to end this chapter with a case study that, to my mind, nicely illustrates the speed with which plans can unravel. On 29 December 2010, an American Airlines 757 ran off the end of the runway at Jackson Hole (JAC), Wyoming (NTSB, 2012). The FO was PF and there was a light snow falling at the time. JAC is at an elevation of 6491 ft, and the runway in use was 6300 ft long. Being a ski resort, aircraft are often operating at maximum weights during the winter as passengers bring their skiing equipment as well as their normal baggage with them. The captain, as PM, had considerable experience of landing at JAC, and the crew were diligent in checking the weather and running landing performance calculations. An aircraft that landed an hour before the 757 had reported that the braking action was 'good' for the first two-thirds of the runway but 'poor' for the final third. Runway surface friction readings were obtained 18 minutes before landing. The FO stated that he planned to touchdown within the first 1000 ft of the runway, and the crew set the spoilers for automatic deployment and the wheel brakes were set to 'MAX AUTO'. The landing was 'firm', about 600 ft from the approach threshold. So far, the performance of the crew was exemplary.

After the aircraft touched down, two things were supposed to happen. First, the spoilers should deploy automatically, an action monitored by the PM, and, second, the PF was supposed to deploy the thrust reversers. Both crew were also required to ensure that the aircraft deceleration was in accordance with their expectations. In order to manage this sequence of events, the pilots, each independently, needed to observe, or perceive, signals in the array of information available inside and outside of the flight deck, attend to specific signals and fit events into a sequence coded as a 'script' or the action plan as defined in procedures. While doing their allocated tasks, pilots had to be aware of what the other crew member was doing and whether that activity supported the achievement of the planned goals. What all this means in terms of the events of that day is that, having armed the wheel brakes and speed brakes, one action required of the captain was to monitor that the automated systems activated as required and make the appropriate call-outs.

Similar to the pilot activity, aircraft systems have their own logic that shapes activity. In his determination to land slightly short, the FO's touchdown was firm. Immediately after touchdown, the FO tried to deploy the thrust reversers. Unfortunately, the aircraft bounced on touchdown, and the air/ground sensing momentarily transitioned to 'air' before returning to 'ground'. This sequence coincided with the FO's selection of thrust reversers. Unfortunately, the sync-lock mechanism, designed to prevent the thrust reversers deploying in the air, was triggered as the aircraft bounced. Once the air/ground switch returned to the ground mode, there was a 5-second lag before the reversers could be deployed a second time. Unfortunately, on this occasion, the system froze.

Simultaneously, the spoilers started to deploy. The captain's role was to monitor both the spoilers and the thrust reverser deployment. The speed brake (spoiler) lever started to move, and this prompted him to call 'deployed' before he then shifted his attention to the thrust reversers. Unfortunately, because of the slight bounce, the speed brake lever, unnoticed by the captain, returned to the 'armed' position.

Because a component in the thrust reverser clutch mechanism had been improperly installed, the system failed in the automatic mode, and it took both pilots a further 18 seconds to deploy the reversers. Ironically, a standard manual deployment of the speed brakes would have worked, but the crew were too busy trying to get the automatic system to work to think of this as a fallback. Despite the best endeavours of the crew to do the best job they could, things worked against them on that day. This chapter starts and ends with two examples of how the world can go from normal to catastrophe in less than 30 seconds. Luckily, in both cases, there were no injuries.

REFERENCES

AAIB. (2010). Report on the accident to Boeing 777-236ER, G-YMMM, at London Heathrow Airport on 17 January 2008. AAR 1/2010.

AAIU. (2005). AAIU Formal Report No: 2005-014: Serious Incident to ATR 42 EI-CBK near Dublin, 8 August 2003. File No: 2003/0043.

Aeschbach, D., Vejvoda, M., Mendolia, F., & Tritschler, K. (2017). Additive effects of the number of completed flights and time awake on fatigue in short-haul airline pilots. *Sleep*, 40(Supplement 1), A58.

Banbury, S. & Tremblay, S. (Eds.) (2004). *A Cognitive Approach to Situation Awareness: Theory and Application*. Aldershot: Ashgate.

Baxter, G.D. & Bass, E.J. (1998). Human error revisited: Some lessons for situation awareness. In *Proceedings of the 4th Symposium on Human Interaction with Complex Systems* (pp. 81–87). Los Alamitos, CA: IEEE Computer Society Press.

Chater, C. (2018). *The Mind Is Flat: The Illusion of Mental Depth and the Improved Mind*. Harmondsworth: Allen Lane.

Chou, C., Madhavan, D., & Funk, K. (1996). Studies of Cockpit task management errors. *The International Journal of Aviation Psychology*, 6(4), 307–320.

Clausen, T. & Hutchins, E. (1996). Distributed cognition in an airline cockpit. In Engström, Y. & Middleton, D. (Eds.), *Cognition and Communication at Work* (pp. 15–34). Cambridge: Cambridge University Press.

Colvin, K.W. (1999). Factors that affect Task Prioritisation on the Flight Deck. Unpublished PhD thesis.

Dekker, S.W.A. (2002). Reconstructing human contributions to accidents: the new view on error and performance. *Journal of Safety Research*, 33(3), 371–385.

Dekker, S.W.A. (2005). Why we need new accident models. *Human Factors and Aerospace Safety*, 2, 181–198.

Dekker, S.W.A., Hummerdal, D., & Smith, C. (2010). Situation awareness: Some remaining questions. *Theoretical Issues in Ergonomics Science*, 11(1–2), 131–135.

Dörner, D. & Schaub, H. (1994). Errors in planning and decision-making and the nature of human information processing. *Applied Psychology: An International Review*, 43(4), 433–453.

Endsley, M.R. (1989). *Tactical Simulation 3 Test Report: Addendum 1 Situational Awareness Evaluations* (81203033R). Hawthorne, CA: Northrop Corporation.

Endsley, M.R. (1993). A survey of situation awareness requirements in air-to-air combat fighters. *International Journal of Aviation Psychology*, 3(2), 157–168.

Endsley, M.R. (1995). Toward a theory of situation awareness in dynamic systems. *Human Factors*, 37(1), 32–64.

Endsley, M.R., Bolte, B., & Jones, D.G. (2003). *Designing for Situation Awareness: An Approach to Human-Centred Design*. London: Taylor & Francis.

Fischer, U., Orasanu, J., & Montalvo, M. (1993, April). Efficient decision strategies on the flight deck. In Jensen, R. (Ed.), *Proceedings of the 7th Symposium on Aviation Psychology* (pp. 238–243). Columbus, Ohio State.

Fischer, U., Orasanu, J., & Wich, M. (1994). Expert pilots' perceptions of problem situations. In *Proceedings of the 8th International Symposium on Aviation Psychology* (pp 777–782). Hillsdale, NJ: Lawrence Erlbaum Associates.

Flach, J.M. (1995). Situation awareness: Proceed with caution. *Human Factors*, 37(1), 149–157.

Hart, S.G. & Staveland, L.E. (1988). Development of NASA-TLX (Task Load Index): Results of empirical and theoretical research. In Hancock, P.A. & Meshkati, N. (Eds.), *Human Mental Workload. Advances in Psychology* (vol. 52, pp. 139–183). North Holland: Elsevier.

Hoffman, D. (2019). *The Case against Reality: How Evolution Hid The Truth from Our Eyes.* Harmondsworth: Penguin.

Hutchins, E. (1995). *Cognition in the Wild.* Cambridge, MA: The MIT Press.

Johnson, C.W. (2004). Human error and the failure of imagination. In Johnson, C.W. & Pelanque, P. (Eds.), *Human Error, Safety and Systems Development* (pp xi–xvi). Boston, MA: Kluwer Academic Press.

Kahneman, D. (2011). *Thinking, Fast and Slow.* Harmondsworth: Penguin.

NTSB. (2012). AAR12/01. Runway Overrun American Airlines Flight 2253 Boeing 757-200, N668AA, Jackson Hole, Wyoming. 29 December 2010.

Orasanu, J. & Martin, L. (1998). Errors in aviation decision making: A factor in accidents and incidents. *Proceedings of the HESSD '98* (pp 100–107). Seattle.

Parasuraman, R., Sheridan, T.B., & Wickens, C.D. and his colleagues (2008). Situation awareness, mental workload, and trust in automation: Viable, empirically supported cognitive engineering constructs. *Journal of Cognitive Engineering and Decision Making*, 2(2), 140–160.

Reason, J. (1990). *Human Error.* Cambridge: Cambridge University Press.

Shorrock, S. (2016). *Varieties of Human Work.* Humanistic Systems Blogpost.

Weick, K. (1995). *Sensemaking in Organisations.* London: Sage.

6 Error as Performance Feedback

6.1 INTRODUCTION

In Chapter 2, we saw that the actions of individuals constitute a class of hazards. In the previous chapter, I addressed the question of how individuals engage with the world in order to get jobs done. I was interested in what constitutes normal work. Ironically, my examples all illustrated a failure of some kind. In two cases, the crew were presented with very unusual situations, under extreme time constraints. The third example showed how a flawed performance, unfolding over time, can still look 'normal'. The cases were chosen because they illustrated how work comprises responses, typically within a procedural framework, to signals from the outside world. I also suggested that those responses, in turn, shape the world. Actions create signals that feedforward to bring about change which, in turn, feeds back to shape further action. In this chapter, I want to specifically address the fact that normal work is buggy: work is enacted in an imperfect manner. These imperfections are labelled 'errors'.

The focus of this chapter, then, is on things going wrong. Such is the potency of the term 'pilot error' that I need to recalibrate the discussion. Things go wrong because the efforts of the individuals engaged in the activity did not meet the requirements of the task, which is to say that the constraint set of the intended goal was not satisfied. By default, we subsequently frame the debate in terms of failure. The Safety-II paradigm, correctly, claims that most work 'goes right', and that, simply focusing on what goes wrong distracts attention from exploring 'why' things go right so often. By looking at error in detail, I hope to uncover some properties of the world of work. First, I am interested in the conditions, or risk factors, that increase the probability of a mismatch between the actions of individuals and those needed to satisfy the goal state constraints. Second, what are the symptoms of, or precursors to, erroneous behaviour? What signals exist, especially in marginal states, that suggest that buffering is being overwhelmed? Finally, how does the system respond under conditions of mismatched behaviour and goal state constraints?

This last point is important. Once an error occurs, crew either adapt their future actions to the changed circumstances they now face or they fail. We said in the last chapter, though, that interventions change the environment. Error 'management' is not a case of 'doing it again but right this time'. Error management is almost an oxymoron. Instead, we sustain control in the new context that flows from the prior insufficiency. If you accept this argument, then it suggests a possible flaw in the 'Safety-II' position: failure remains an object of interest. In fact, doing things right is often all about nudging 'failures' back on track. In effect, the control of the current task has drifted such that operators recognise that something is now clearly not right, and action is called for to restore equilibrium.

This chapter will draw on evidence from field observations using the line operations safety audit (LOSA) methodology. Given that LOSA is an audit, the benchmark of performance is contained in company documentation, and the findings are shaped by the taxonomy used to classify observations. Because error is an emotive term and, as we will see, the vast majority of negative events are inconsequential, I prefer to think of errors in the context of LOSA as examples of quality non-conformity. The evidence gathered by observers is primarily of departures from the prescribed task. One airline's 'error' would not exist in another airline having different procedures. By design, then, the evidence base is constrained. That said, LOSA data are a rich source of (fairly) standardised information about workplace performance.

Error, in reality, is the trace that the process of work leaves behind. Errors are like archaeological remains that can be excavated and used to reconstruct what the individual was trying to do at the time. As with fossils and bits of pot, a degree of interpretation is needed. With these caveats in mind, this chapter will look at categories of error, linking error to performance and, finally, regaining control. We start with an example that left many observers bewildered at the time.

6.2 THE HELIOS B-737 CRASH NEAR ATHENS, 2005

On 14 August 2005, a Boeing 737-315 took off from Cyprus, heading for Prague with an en route stop at Athens (AAIASB, 2006). Having left Larnaca at 06:07, the aircraft passed over Athens and entered a holding pattern at 07:38. No radio calls had been received from the crew for most of the flight. At 08:24, as the aircraft made its sixth orbit, it was intercepted by F-16 fighters from the Greek Air Force. The military pilots were able to see that the captain's seat was vacant and the occupant of the first officer's (FO's) seat was slumped over the controls. The oxygen masks had been deployed in the cabin. At 08:49, a person was seen entering the flight deck. A minute later, the left-hand engine flamed out due to fuel starvation. During the descent, two attempts were made by someone on board – thought to be a member of the cabin crew – to broadcast a Mayday call. The right-hand engine also flamed out passing 7000 ft, and the aircraft crashed at 09:03, 33 km north-west of the Athens International Airport. There were no survivors. The subsequent investigation attributed the accident to the incorrect configuration of the pressurisation system. Those passengers not killed in the final crash had probably already died of hypoxia.

6.3 WHAT IS AN 'ERROR'?

The media reaction to this tragic event was one of disbelief, and yet, aircraft pressurisation problems, while not common, do regularly occur. The NASA ASRS database contains 171 reports relating to the Boeing 737 alone, over the period 1994–2004. Of these reports, 58 were deemed critical. Boeing reported that the rate of pressurisation events was 8.5 per million departures for the B-727, 2.7 for the B-737, 4.9 for the B-757/767 and 6.3 for the DC-9/MD-80 (AAIASB. 2006). For the aircraft pressurisation system to work, three elements need to be correctly configured. The engine bleed valves must be open, the cabin conditioning pack valves must be open and the pressurisation control panel must be correctly set. Issues with meeting these system constraints provide us with the opportunity to clarify what we mean by an error. The Threat and Error Management framework underpinning LOSA defines errors as 'actions or inactions' that lead to 'deviations from *organisational* or flight crew intentions' (authors italics). The definition is normative in that it suggests that the error is a 'deviation' and that a third party – the organisation – had an expectation of the way the work will be conducted. In short, it is possible to distinguish between the correct version of events (organisational intentions) and an incorrect enactment of plans (crew actions). Reason (1980) defines an error as an 'action not as planned'. Here, we see that error is manifested in behaviour which departs from intention. Both definitions share the idea that the 'error' is an observable state, condition or action that is not what was expected. Importantly, 'error' is seen as an outcome. Reason's work triggered several attempts to develop error classification systems or taxonomies. However, the behavioural precursors to the erroneous outcome are actually of more interest than the final observed output. Hollnagel (1983) suggests that, if work has an inherent variability as a result of the constant need to adapt to circumstances, then, error is an integral part of that work. We, therefore, have no need to categorise elements of work as being somehow 'other'. Furthermore, if apparently erroneous performance is integral to normal work then we also have no need for pathologies and taxonomies of error. The argument developed in the previous chapter leads to a similar conclusion. Performance flows from underlying knowledge structures, their application to solving tasks and their subsequent control. Errors are simply symptoms of a performance: neither a cause of a problem, nor an outcome.

We saw in Chapter 1 that skills, rules and knowledge structures can be categorised as either declarative or process knowledge. While I might see a skill being performed, though, rules and knowledge are only observed indirectly through the behaviour of individuals (although they may be accessed through questioning). These memory structures represent the inputs to activity. The outputs can be considered, according to Reason (1980), as slips, lapses and mistakes. A slip is the inadequate control of motor performance. When you knock over a glass, drop an object thrown to you or trip over your feet, these all count as slips. An incorrect read back of an ATC instruction would also be a slip. Lapses are, simply, forgetting to do something. We have already seen that an interruption can lead to the subsequent forgetting to complete a task and prospective memory failure is the forgetting to do some planned future action. A mistake is a decision error: what you chose to do was not sufficient for the task in hand. Slips and lapses map onto encoded skills in that they flow from flawed enactment of behaviour. Mistakes flow from a flawed application of rules and knowledge. Violations represent a subset of mistakes (or 'intentional non-compliance' in LOSA terms). Whereas mistakes reflect action choices that fail to fully map onto task demands, a violation occurs when an individual elects to act outside of prescribed bounds. The language around violating is pejorative but, as we will see, violations are often well intentioned. I now want to return to the Helios accident to explore these categories in more detail. I will use other aircraft pressurisation events from around the time of the Helios crash to elaborate on performance types. My goal is to show that the Helios accident was not really an exceptional catastrophe but a tragic example of the probable outcome from a normal performance.

6.4 SLIPS – EXECUTING TRAINED BEHAVIOUR PATTERNS

When the Helios crew boarded the aircraft, they completed the checklists associated with preparing the flight deck. This included scanning the various panels on the flight deck to check the position of switches and controls. Neither pilot detected that the pressurisation controller was in the wrong position. Scanning arrays and controls and conducting routine checks are a fundamental part of the job and represent classic examples of skill-based behaviour. It is the high level of automaticity associated with well-rehearsed skills that gives rise to slips.

Consider, for a moment, the way pilots move their hands around the cockpit space. With increasing familiarity, a pilot can reach for an object in the cockpit without necessarily moving their eyes to track the position of their hands and to locate the target object. Kinaesthetic feedback from the position of their limbs allows them to roughly position themselves within the workspace and the sense of touch facilitates the final, accurate location of the control. In October 2004, a pilot knocked off the conditioning pack switches of an MD-83 when he attempted to put the tech log back on the glare-shield after writing up some items. In this case, the degree of muscle motor control, the execution of a skill, was inadequate and the pilot exhibited 'clumsiness'. The control of our limbs in space is controlled by structures in the cerebellum. This region of the brain is larger in Highland Gorillas that any other primate (including humans), reflecting the fact that these large animals negotiate complex arboreal habitats that require considerable planning of movement and thus require additional computing power.

In information processing terms, the cue represented by the pressurisation controller in the wrong position on the Helios aircraft lacked salience: it was not distinctive enough to draw the crew's attention to the condition. Having missed the configuration of the controller on first boarding the aircraft, the pilots had four further opportunities to check the system according to the aircraft checklist: before engine start, cleared for start, after start and after take-off. There were two cues that should have triggered the correct behaviour in relation to the cabin pressurisation system: the green caption indicating that the pressurisation controller was in manual mode and the position of the selector switch. When the Helios pilots did their visual scan, for whatever reason, these were missed. The design of the working environment can reduce the effectiveness of cues. The intensity of the annunciators on the aircraft can be set to two levels, bright and dim. The light switch on the Helios aircraft was found to be in the bright position. In March 2004, the crew of a 737-300 were doing their pre-flight checks for an early morning start. With the rising sun in their eyes and the lights set to 'dim', the pilots reported that the manual annunciator was invisible. Switches and annunciators are not equally visible from both seats and are often embedded in a complex array of other controls and indicators. The ergonomics of the cockpit can impede detection of a miss-set item, especially if inadequate attention is paid to the task.

Signal detection can be shaped by our expectations. In January 2003, the crew of a B-737-300 also got airborne with the controller in manual mode. The pilot had done a bleeds-off take-off and, having selected the bleeds ON (which would have no effect with the controller set to manual), he verified that the cabin was pressurising. In fact, the cabin was climbing at the same rate as the aircraft, but the captain reported his expected visual cue rather than the actual cue.

Prior learning can interfere with new skills. For example, in February 2006, an Embraer 170 landed with the parking brake applied, causing its tyres to burst (SIBF, 2006). The captain had only recently completed his type conversion course. Because the aircraft was too fast on the approach, the captain had selected what he thought was the speed brake. However, he was using a behavioural routine ingrained from flying his previous type of aircraft. On the new aircraft, the parking brake is in the same location as the speed brake was on his former aircraft. The captain joked about his error and then set the brake to OFF. Unfortunately, a brake circuit failure resulted in the brake remaining partially applied. The Embraer does have a warning light for inadvertent brake selection in flight, but its inconspicuous design means that it is easily overlooked. The crew did not notice that the park brake did not appear to have been released. (The same error was made by the crew of an EMB 145 on landing at Bristol Airport, UK on 22 December 2017. The report (AAIB 1/2019) refers to two previous events involving the ERJ 135/140/145 family of aircraft. The first was in August 2000 and the second was in August 2008. The 170 is a derivative of this type). Stress, as well as recent experience, can result in the reversion to prior learned habits.

Finally, in September 2002, the pilot of a Boeing 737-204 demonstrated a bleeds-off take-off to an inexperienced FO. After take-off, the FO turned the bleeds on and then inadvertently turned the air conditioning packs off. The demonstration was an interruption of the normal sequence but the outcome was a classic example of how slips occur. The action of moving one set of switches was associated in the mind of the FO with a similar action involving a second set of switches. However, in this case, the air conditioning packs had already been turned on and the two linked behavioural responses resulted in the switches –already in the correct position – being moved back to the off position.

Slips, then, are the outcome of control actions that go awry. Novice pilots are less likely to make skill-based errors than expert pilots simply because they are concentrating more on the control task than an expert would. It is typically the reduced allocation of attention to a well-rehearsed task that increases the risk of inadequate control. Sequences of skill-based behaviour are linked into event chains, each sequence triggering the subsequent actions in the chain. Elements can fail to be triggered, or steps can be triggered out of sequence or in reverse. The dynamic interplay between the environment and the individual is worked out in terms of the quality of the final performance.

6.5 LAPSES – FORGETTING AS A PART OF WORK

Chou and Funk (1990) found that 39% of task management errors related to failing (forgetting) to initiate task elements at the appropriate point in the sequence. In a study of prospective memory (that is, the memory function that keeps track of future intentions and the failure of which results in a lapse), Marsh and Hicks (1998) found that the difficulty of one task increased the probability of forgetting another task, the value increasing from 28% under conditions of low workload to 57% under high workload. Lapses are the result of forgetting to do something.

The Helios aircraft had undergone maintenance the night before the flight, and one task undertaken by the technicians was a check of the aircraft pressurisation. When they had finished their work, it was the view of the investigation board that the engineers had left the pressurisation mode selector in the 'manual' position (although this was contested by the technician involved). If it was, indeed, an oversight, then it is an example of a lapse: they forgot to do the appropriate action for some reason. In February, and again in March of 2005, the same problem was reported to the ASRS. The first involved an MD-88 and the second a Boeing 737-300. The completion of the maintenance task should have triggered the behaviours necessary to return the flight deck to its normal state. Such actions may sometimes be prescribed in procedures but, more often, 'work' comprises procedural elements embedded in a generic process, usually referred to as 'common sense'. This episode is indicative of the true nature of the complex work undertaken by skilled practitioners. Because we simply cannot specify every single action to be taken at every stage throughout the day, we rely on 'expertise' to fill the gaps. Increased expertise allows for greater autonomy in the way work gets done. However, autonomy also means variability and increased risk of a breakdown in control. Control and autonomy are traded off in hazardous systems in order to achieve flexibility, efficiency and safety.

The use of checklists to manage activity is variable across the aviation industry. Although aircraft and equipment manufacturers develop checklists as part of the airworthiness certification process, the complexity of checklists will vary according to the nature of the aircraft. For smaller aircraft, like the Cessna Caravan we looked at in Chapter 1, it is not uncommon for pilots to work completely from memory once they have mastered the aircraft's – admittedly succinct – checklist. For more complex aircrafts, checklists are used in a variety of ways. In some cases, they are used in a 'challenge and response' sense in that one pilot calls an item off the list, and another verifies that the action is complete. In other cases, the pilots complete the checklist actions from memory and then use the document to verify that actions have been completed. In some multi-crew aircraft, each pilot is responsible for completing their assigned portions of the checklist, which they do in a 'silent' manner. More modern aircraft have electronic checklists that incorporate locks or warnings that protect against incorrect or incomplete checklist action. The checklist process can also differ between normal and abnormal or emergency situations. These different approaches to the same task are an embodiment of decisions about job design and task management. A common response to a previous failure is to add additional steps to the checklist in an attempt to prevent a repetition.

We saw in the previous chapter that work comprises bundles of interwoven tasks often involving interaction with other agencies. Concurrent tasks compete for attention, and this can lead to lapses. In February 2005, an MD-88 was climbing after take-off, having been initially cleared to 6000 ft. The captain, who was the pilot monitoring (PM), noticed that the aircraft was accumulating rime ice and so he turned on the wing anti-icing system. At this point, the crew received a further clearance to climb to 13,000 ft. Soon after recommencing the climb, the pilots received a report from the cabin of a squealing noise coming from the rear galley, quickly followed by the cabin altitude aural warning. By now, the aircraft was at 10,000 ft, and the crew realised that the cabin altitude control selector was in the manual position, something that should have been picked up in the after-take-off checklist. At that point, the captain remembered that, because of having to deal with the icing conditions, the checklist had not been completed. Coincidentally, the selector was in the manual position because the maintenance personnel had conducted a pressurisation check during the previous night, exactly the same task that had been conducted on the Helios aircraft.

Because, as we have just seen, complex control requires several tasks to be running concurrently, the nature of work often requires us to interrupt tasks while we attend to another, more important, task. The fact that we work in a social context also means that other team members who need our attention, momentarily, as part of their work process sometimes interrupt us. This is the essence of prioritisation. When we interrupt a routine, we run the risk of failing to complete the action sequence or of resuming the process at the wrong point in the sequence. In December 2000, a B-737 crew were doing their cockpit checks prior to departing from Cork in the Irish Republic (AAIU, 2001). The captain noticed that his passengers were attempting to board an empty aircraft parked next to theirs. He tried to alert ground operations to the problem, but a line technician luckily intervened. During the engine start sequence, the crew encountered a problem, which needed the support of an engineer. Next, while completing the after-start checklist, the crew were interrupted by the cabin supervisor who informed them that the front left cabin door was making a strange gurgling noise. The FO went to investigate and, on his return, the crew agreed that it was probably water in the seal. At that point, the forward toilet fire alarm was triggered by smoke from the galley ovens where the cabin crew were heating the passengers' breakfasts. Once they finally got airborne, it is not clear if the crew completed the after-start checklist, but what we do know is that they forgot to set the pressurisation. We will return to this crew later but, for now, theirs is a classic illustration of the consequences of being interrupted.

The crew of the Cork aircraft were dealing with external factors, but we also have to recognise that tasks can be deliberately interrupted or modified. In January 2003, the FO on a Boeing 737-300 was giving a briefing to airport fire service personnel, which included a discussion of how to configure the aircraft for an evacuation. During the demonstration, the FO showed the visitors how to turn the pressurisation to manual. After his guests had departed, the FO forgot to reset the panel. An event reminiscent of the Helios tragedy, it shows how quickly task requirements – resetting the panel – can be forgotten. Interruptions and alterations to work routines represent sources of variability.

Skill-based behaviour, the precursor to slips and lapses, involves the execution of action routines associated with the configuration and control. Errors arise from the manner in which control is exercised. Of course, in complex manipulative tasks, like hand-flying an aircraft, individual competence and experience will have an effect on the sophistication of the control being exercised. Specific consequences associated with skill-based behaviour are:

- Failure to start a sequence
- Failure to complete a sequence
- Omitting steps in a sequence

- Incorrect sequencing of steps
- Inserting fragments from other sequences
- Doing the wrong sequence
- Inadequate manual dexterity

Slips and lapses flow from the execution of skill-based behaviour. Because skill-based behaviours have become highly automated through extensive rehearsal, attention is allocated in proportion to the perceived load. Ordinarily, as demand increases so should the degree of attention allocated to the task but the irony of being highly skilled is that attention management can lag behind demand.

The behaviours we have been looking at capture the flow of events that we see on the flight deck under normal conditions. Well-rehearsed work routines are executed according to the specific demands of the task. When discussing task management in Chapter 5, I proposed that decision-making was a part of the normal management of tasks. Occasionally, though, there was a need to consider the suitability of planned goals. Changed circumstances dictated a need to adopt a new goal or even to create an innovative solution to a problem. And that exposes the operation to the risk of mistakes being made.

6.6 MISTAKES – THE FAILURE OF RULES

Whereas slips and lapses are a function of inadequate attention applied to skill-based behaviour, mistakes flow from the selection of future courses of action. Mistakes are the result of both rule-based and knowledge-based behaviour. In this section, we will look at the application of rules to the management of work. Quite simply, rules are repertoires of 'IF<x>THEN<y>' statements acquired in training or through direct experience. Unfortunately, rule selection can be problematic. We are now moving into the realm of conscious, premeditated activity (although rule-based behaviour can also be incorporated into fairly automatic routines). In October 2004, a B-737-300 failed to pressurise after take-off and the crew elected to land and change aircraft. Later, the captain was surprised to discover that, based on the output from the flight data recorder, the aircraft pressurisation had not been turned on in the first place. On reflection, the captain recalled that he had checked the pressurisation switches and had seen that they were both in the off position. However, as the checklist was underway and it called for the switches to be moved one at a time, he assumed that this action was pending and did not think to query it. In effect, he had sampled the work process at what he thought was too early a stage in its progression. After speaking to his FO, he was surprised to find that the FO, too, had seen the switches in the 'off' position but had assumed that the captain had moved them for some reason. Both pilots had created a 'rule' to explain the configuration of the aircraft. The behaviour of the FO also points to the role of social hierarchies in shaping action.

The crew of the Helios aircraft were engaged in more formal rule application. As the aircraft climbed through 12,000 ft, the cabin altitude warning horn sounded. The cabin altitude warning horn alerts the pilots if the cabin altitude exceeds 10,000 ft, the point at which the partial pressure of atmospheric oxygen starts to fall below that required to support normal human functioning. From the flight data recorder, we can see that the crew, first, disengaged the autopilot and then disengaged the autothrottles. The thrust levers were briefly retarded and then the automatics were reset and the aircraft continued its climb. What the crew did not do was recognise that the cabin was not pressurising. Instead, they called dispatch for advice.

The warning horn actually serves two purposes. At lower levels, it warns the crew if the aircraft is incorrectly configured for take-off or landing. The usual context for hearing this horn is on approach with the power levers at idle and the undercarriage still up. The use of the same warning for two different conditions is an example of an engineering solution (Level 3) to satisfy an airworthiness requirement (Level 4), the assumption being that the differences between these two circumstances will avoid confusion. The crew of the Irish aircraft departing Cork also experienced the horn sounding at 10,000 ft but the captain chose to disregard it as spurious. He had misidentified the cause of the horn, presumably because of its stronger association with the undercarriage on approach. The actions of the Helios crew in moving the power levers suggest that they also associated the horn with a configuration problem rather than the cabin altitude.

If we return to the information processing model discussed in a previous chapter, cues from the environment are detected and attended to. They are then brought into working memory and processed. The crews of both the Helios and the Irish aircraft mapped the cue onto sets of examples of when the horn would sound held in long-term memory. They then applied the strongest association – configuration warning – and disregarded the input as spurious. The use of a wrong rule (IF <horn> THEN <configuration problem>) caused them to stop searching for other possible associations. We can see the role of bias in decision-making at work in these situations. In October 2004, the crew of a 737-300 experienced problems with the auto throttle and the transponder in the climb. Suspecting a failure of the air/ground sensing system, the crew was fault-finding when they got an intermittent sounding of the warning horn. Assuming that this was the configuration warning, they were able to partly rationalise the symptoms they were encountering. Although, in this case, the cabin pressurisation had been set correctly, the system had actually failed. While the crew were distracted, the cabin altitude had been slowly rising until it set off the horn. This is an example of 'confirmation bias': information is being used to confirm the current assessment of the situation when, in fact, the new information was completely unrelated. The captain of the aircraft later commented, 'after a few minutes the FO remembered that the take-off warning horn doubles as the cabin altitude warning. He checked the cabin altitude and saw that it was above 10,000 ft and that we were slowly losing pressure'. In this case, the two pilots were individually processing cues from their environment. They were fitting information into their model of aircraft systems in order to develop a plausible explanation for the symptoms they were encountering. However, the FO then recalled an additional piece of information that offered a completely different narrative. Whereas the initial response was to look for commonality between the existing problem and any new information as it arrived, the correct solution was arrived at when the FO treated the horn in isolation and sought other associations that could be made. It is important to recognise that breaking out of one mode of sense-making and developing an alternative strategy is effortful. We also need to consider what we referred to as 'startle' in Chapter 2. In November 2003, a captain reported, 'I was amazed at how long – 10–15 seconds – it took us to fully realise what was happening. I always thought I would instantly recognise the take-off warning as the cabin altitude warning once I was airborne. Not so!'

These examples have all involved crew applying the wrong rule to a situation in order to make sense of a warning. I suggested in the previous chapter that decision-making is a stream of responses to the environment typically driven by formal and informal rule sets. This section has looked at how rules shape interpretation – the rule was used in retrospect – but wrong rules can be used prospectively to select the next course of action in a sequence. The fitness of the rule, given the circumstances, will determine the effectiveness of the action.

6.7 BRINGING KNOWLEDGE INTO PLAY

So far, we have looked at crew applying rules that are mostly contained in procedures or are the result of training directed at specific contingencies. Rasmussen's knowledge-based behaviour describes activity undertaken once rule sets are exhausted: what do you do when there is no rule to apply? In the previous chapter, I suggested that 'problem-solving' was a process of making decisions about seeking information and evaluating alternative courses of action. Knowledge-based behaviour allows us to run mental 'what if' scenarios from which we can identify gaps in data and derive new rules that could work in the current situation. This second category of 'mistakes' is the result of the application of underpinning knowledge to a current state but with the result that does not satisfy the applicable constraints.

Returning to the Irish crew departing from Cork, we saw that, after the disrupted passenger boarding process and engine start sequence, there was a possibility, according to the investigation report, that the after-start checklist might have been missed. The next anomalous event to occur was the illumination of the Master Caution light while the crew were completing the after-take-off checklist. The warning was cancelled, and the captain informed the FO that the Auto Fail warning light on the pressurisation panel was on. The FO later reported that he did not understand the significance of this warning. The Auto Fail light will come on under one of the four conditions:

- Loss of DC power
- Controller failure
- A rate of cabin pressure change in excess of 1800 ft per minute
- Cabin altitude in excess of 13,875 ft

The captain assumed that, as the pressurisation must already be on, the warning was related to a controller failure, and so, the 'standby' mode was selected. This action will have no effect if the pressurisation is not, as in this case, turned on in the first place. The FO's comment illustrates how full participation in problem-solving is dependent upon robust technical knowledge. The episode also illustrates my description of 'problem-solving' as being a process of making decisions about decisions. The captain held a belief about the status of the aircraft. The triggering of the Master Caution was the result of one of the four possible conditions applying at the time.

I think we can assume that the crew could safely reject the first and last conditions. However, depending on the aircraft's rate of climb, the third condition could still be true. In this case, the decision to disregard one possible cause constrained the subsequent information search. Instead, the action was driven by what the captain considered to be the most plausible cause.

The Helios crew, too, had to deal with confusing symptoms. As their aircraft passed 17,000 ft, and while they were discussing the warning horn with the company dispatch office, the master caution light illuminated. In fact, two events occurred in close proximity, both of which would have triggered the warning, but we cannot say, based on the evidence, which occurred first. Because of the reduced density of air, the equipment bay cooling system was not functioning effectively, and so, the warning light came on. At about the same moment, the passenger oxygen light illuminated to indicate that the cabin masks had dropped. Communication between the cockpit and the ground at this stage revolved around the 'configuration warning' – the horn – and the status of the equipment bay cooling warning lights. From the evidence of the ground engineer who had spoken to the captain, it seems that the interpretation was that the equipment bay cooling fans had failed. The warning light indicated a loss of airflow mass, which can result from either the failure of a fan or from the reduced air density at altitude. The captain's diagnosis of the initial problem – the warning horn – as a configuration warning was absorbing his attention when the second warning occurred, the Master Caution. Of the two conditions that could have triggered the caution, the cooling fan indicator warning took priority. The passenger oxygen light, strangely overlooked, and the warning horn were direct indications of the actual problem, whereas the cooling fan warning was an indirect symptom. In both of these cases, the crew response was shaped by the way signals were integrated into an understanding of the world. We also need to accept that crew cognitive functioning may already have been affected by hypoxia by this stage.

Providing access to knowledge is a decision made, primarily, by manufacturers but moderated to some degree by the company. Checklists and manuals are artefacts of knowledge embodiment and serve to communicate background information and preferred actions to the crew. Checklists cover both normal routine activity and also a subset of non-normal situations. Because they are, themselves, the result of prior decisions, checklists carry an inherent risk. Consider the case of a helicopter taking workers to an oil platform off the coast of Scotland (AAIB, 1998). At 0729 hours, while cruising at 120 kts, at a height of 3000 ft amsl, there was a sudden onset of severe airframe vibration. The captain immediately instructed the FO to reduce main rotor pitch. The FO lowered the collective lever to reduce the main rotor pitch from 15.6° to about 12.2° and, in addition, applied left yaw to maintain the heading. The captain reported a 'major vibration' to Aberdeen radar and requested an immediate return.

Both crews thought that the vibration was associated with the main rotor, and the captain decided to follow the FREQUENCY ADAPTOR FAILURE checklist and read aloud the following extract from the drill:

> 'Reduce collective pitch slowly to the setting which reduces the vibration to a level acceptable for continued flight – approximately12 to 14 degrees pitch if vibration level does not decrease to an acceptable level – land/ditch immediately'

He made several small pitch adjustments in an attempt to find a setting that reduced the level of the vibration, but this appeared to make no significant difference, although the captain did think that the vibration had moderated. The crew then reviewed the situation. They noted that the vibration appeared to be constant, that the engine indications were normal and that there was no fluctuation in the torque readings. As the flight progressed, the crew became less convinced that it was a Frequency Adapter failure, but in the absence of any other evidence, they continued to associate the vibrations with the main rotor. The fact that there was no 'buzz' through the yaw pedals appeared to support that conclusion. The crew brought the aircraft back to the mainland, the passengers were all disembarked and the aircraft shut down. It was only then that they discovered that the problem was with the tail rotor.

The initial association of the vibration with the Frequency Adapter may have been triggered by the fact that it was the only drill in the Flight Reference cards relating to vibration. At this stage of the aircraft's in-service history, there had been one failure of a main rotor gearbox, in July 1987, and a checklist procedure was subsequently developed. At the same time, the Frequency Adapter design was modified. At that point, the global fleet had not suffered a tail rotor failure, and so, no emergency drills had been developed for this contingency. Although reducing the collective pitch to 12°, the first action in the drill, did appear to reduce the vibration slightly, there was little or no lateral component present. It may be that the drill misled the crew rather than assisted them in their attempted diagnosis. Because the cyclic, collective and yaw pedal controls act through hydraulic servo units, any feedback felt through the controls would have been dampened. In fact, the vibration felt at the controls seemed to be in sympathy with that felt generally through the airframe. With the controls appearing to function normally, the only source of evidence was the frequency of the vibration and its relation to main rotor (265 rpm) or tail rotor (1279 rpm) rotational speeds. The main rotor blades rotate 4.4 times per second, and therefore, a 1R vibration would have a frequency of 4.4 Hz and a 4R vibration, a multiple, would have a frequency of 17.6 Hz. A vibration associated with the tail rotor (1T) would have a frequency of 21.3 Hz. The closeness of these frequencies is such that a pilot would not normally be able to differentiate between a 4R vibration and a 1T vibration. In addition, various frequencies will beat together to produce a resultant that could add to the confusion, making it extremely difficult to determine the source of vibration from its frequency alone. The Operations Manual gave no guidance to help pilots in determining the source of an unusual vibration. This was intentional as it was felt that it would not be practical to do so, given the large number of variables involved. Once again, a decision made at a higher level in the system shaped the performance of the crew. In this example, the specific mistake was the use of an irrelevant checklist, but the crew response was conditioned by system design.

The Helios and Cork B-737 examples showed how the knowledge of technical systems held by the pilots informed the decisions they subsequently made in relation to information seeking and option selection. In the case of the Super Puma, the technical knowledge held by the crew was constrained by prior decisions made with regard to what information to share. In-service experience shaped the manufacturer's design of checklists and manuals, which fed through to the pilots' knowledge structures. In this example, the crew followed a checklist that had no relevance, under the circumstances, and their decision-making focused on making sense of the information available to them. The successful return to land and safe disembarkation of passengers and crew was little more than luck. In the context of a discussion of error, all three crews were actively engaged in resolving the problems they faced and initiated courses of action that made sense in relation to their, albeit limited, understanding of the world.

6.8 SENSE-MAKING, RULE-BASED AND KNOWLEDGE-BASED ACTION

Routine performance largely comprises skill-based and rule-based behaviour enacted in response to cues from the world and directed at assuring the progress of the task towards a desired goal. Occasionally, disturbance in the world or insufficient progress requires a step change in the level of effort applied to the control of the task. Rule sets have to be searched for more appropriate responses, or knowledge resources have to be mobilised to clarify situations, establish new understandings and create new rules. Sense-making, once again, is central to action at this level. Reason (1990) notes that '[i]n any situation a number of rules may compete for the right to represent the current state of the world'. The uncertainty inherent in complex situations increases the risk of triggering wrong rules. The tendency to regard the warning horn as spurious in the examples we have looked at is an example of the misapplication of a previously good rule. Some properties of the situation might have supported the selection of the rule in question, but there were other situational cues that indicated that the rule was not appropriate. The confusion of the altitude warning horn with a configuration warning is quite common. In all probability, the crew will have heard the horn in other contexts. The sound of the horn elicits one response but all the environmental cues indicate that an alternative interpretation is needed. These 'strong but wrong' rules can be compelling. Reason argues that, within a situation, there will be signs, countersigns and non-signs. Signs are those cues that trigger a response, albeit one that is not always appropriate. Countersigns are the indications that, perhaps, the situation is not what we might think. These contra-cues are often missed because of cognitive overload: individuals have difficulty processing too much information. Non-signs are irrelevant signals that may be present. Rasmussen and Svedung call them invalid cues, noise that still makes demands on processing power. Chunking of information is sometimes seen as a way to simplify the world and reduce cognitive demand. Chunking can lead to certain properties of a situation having increased significance while others will be sidelined. In part, this is the price we pay for developing heuristics, or 'rules-of-thumb', to ease decision-making. These processes all contribute to some rules appearing to be applicable when in fact they are not.

Rule-based behaviour accommodates most situations where our initial intentions are thwarted, and we have to execute an alternative plan in order to sustain operations. Knowledge-based behaviour is invoked when activity is outside of the scope of our rule set. Because either the rule we have adopted is unsuccessful or we simply lack an appropriate rule, we need to go back to the basics and construct a solution. Rasmussen and Svedung point out that one of the metacognitive skills associated with rule-based behaviour is recognising the need for situation analysis at a knowledge-based level (see Table 5.5). In discussing the cognitive competences associated with knowledge-based behaviour, they comment that 'this level is related to the extent and quality of the understanding of the relational, causal structure of the work system, that is, a correct mental model of system function'. They also observe that effective performance requires actors to possess 'knowledge about system goals, safety conditions and regulatory constraints on performance'. These represent the criteria – the constraint set – to be met to successfully manage the transitions between goal states discussed in the previous chapter. Knowledge-based behaviour represents the conscious intervention in the work process in order to overcome barriers to progress that are not amenable to rule-based resolution. The key to all of this is the ability to perform mental experiments to generate rules for action and to test hypotheses about the cause and effect of abnormal system behaviour. Knowledge about information sources such as manuals, textbooks, diagrams, etc. and the ability to use them is important at this level.

6.9 PERFORMANCE UNDER NORMAL CIRCUMSTANCES

Almost by default, the discussion, so far, in this chapter has seen error as failure. In the case of the Helios crew, examples were presented that showed that the components of the event were actually quite commonplace. It was the tragic nature of the outcome that cast their performance in negative terms. Normal work, although inherently fragile, is still fundamentally successful. The problem is that error has become a token of failure deserving of punishment when, in fact, it should be expected. It has been argued that error is a symptom of a problem rather than a cause, but this still places error outside of the process of work. By positioning error as a feedback signal, we can see that it is, instead, a property of the current status of the system. Our response to error is a reflection of the efficacy of our attempts to restore the congruence between status and state. From the examples we have looked at, especially the helicopter vibration event, we can also see how cross-scale interactions shape the context of performance.

In the rest of this chapter, I want to look at the normal variability of work reflected in the distribution of workplace error. The LOSA surveys that form the basis of this analysis used the airline's published task – the product specification – communicated in checklists, procedures or other formal texts as the benchmark, and 'errors' are departures from the prescribed activity. The fact that the job of work depends on crew being flexible and adaptive and, given that the product specification is not all-encompassing, we can say that errors represent nothing more than attempts to deal with operational challenges. I will start this discussion of normal variability with some ideas about the general distribution of error. Of course, there is no such thing as a 'correct' number of errors per flight. Leveson suggests that human error, especially when supervising complex, usually automated, systems, is not quantifiable. Error is not stochastic and, thus, cannot be estimated using probability. Given the system property of non-ergodicity, any attempt to prescribe work through procedures will encounter states of tension requiring intervention on the part of crew and, thus, be essentially error-producing.

TABLE 6.1
Errors Observed by Flight Duration

Flight Duration (minutes)	2013 (n = 53)	2017 (n = 22)
60–119	8.2	11
120–179	5.6	7.8
180–239	9.4	4.7
240–300	8.9	11.8

TABLE 6.2
Distribution of Error Types (% by Fleet)

Fleet	Slip	Lapse	Mistake
A (n = 269)	37.9	39.4	22.6
B (n = 172)	25	50.5	24.5
C (n = 156)	16.6	43.5	39.7

Accepting these caveats, it is still useful to consider event rates, if only to remind ourselves that work is fallible. Table 6.1 shows mean error rates by sector length for a sample of B-777 flights. We could argue that error rates should reflect flight time in that the longer the crew work, the more opportunities they have to commit an error. This would be the probabilistic approach in that risk is a function of exposure. Looking at the data, this does not seem to be the case.

Table 6.2 looks at error by type. The three fleets represent different manufacturers and generations of technology, but the fundamental task of flying each type was fairly generic. Fleets A and C were sampled on similar routes, and so, the crew task was comparable. The Fleet B sample was significantly different in terms of the time of day and routes flown. The data do not point to any underlying predictable pattern.

6.10 FACTORS THAT SHAPE PERFORMANCE

These snapshots suggest that normal work has an inherent level of variability and that, occasionally, performance and task demands are at variance. Because of this variability, normative models of successful performance are difficult to define. There are some common quantifiable acts that might allow us to estimate the scale of normal variability. Procedural call-outs, for example, are a standard feature of the work process. In one sample, I looked at the specific call made to confirm that the aircraft had intercepted the ILS and found a failure rate (omitted calls) of 7%. The crew are also supposed to call when approaching a level off in the climb and descent. Of course, a level off is, in part, a function of the degree to which air traffic control (ATC) intervenes in the departure and arrival process. I created a notional 'average' profile and estimated that around 4% of calls were missed.

TABLE 6.3
FO Error Rates/Sector by Experience

Experience in the Airline	Slip	Lapse	Mistake
5 Years or less (n = 6)	0.33	1.33	0.5
6 Years or more (n = 6)	1.5	2.66	1.16

All of the examples we have looked at in this section are based on normal, routine operations. None of the flights experienced a technical failure or an emergency. The crew were familiar with the routes and destinations. Unlike the events described earlier in this chapter, these flights were unexceptional. That said, clusters of events point to factors that can shape variability. For example, one study looked at crew operating a new type of aircraft. Samples were taken 12 months apart to track the aircraft's introduction into service. On both occasions, raw rates per sector were typically in double figures but if both crew had at least 12 months experience on the type, then the rate more than halved. Familiarity with new technology, then, seemed to be an issue.

The question of experience and recency possibly explains the data in Table 6.3, which shows error types committed by FOs in the pilot flying (PF) role. The specific aircraft type was used on regional flights as well as ultra-long haul routes. Those FOs with less than 5 years in the company would have recently upgraded from second officer (SO) and would be consolidating their experience flying shorter, regional sectors. However, the more experienced pilots would be used primarily as 'cruise relief pilots' on the longer routes and, as a result, be flying fewer flights in a month. The difference in error rate, which is possibly contrary to expectation, possibly reflects a recency that comes from the changed nature of their employment. This interpretation is reinforced by the fact that a common lapse was forgetting to arm the speed brake after landing gear selection on approach. This occurred on 83% of the flights with the >6-year FO as PF.

Procedure change can also induce error. After a change in the checklist sequence, the transponder was incorrectly set on 17 out of 29 occasions (58.6%). When the data were fed back to the crew, several pilots made the point that the recent procedure change was not the first. One captain joked that he had his own transponder setting procedure: 'I just turn it on when ATC tell me to'. In effect, the captain delegated an aspect of the task to ATC. The operation of the transponder on the ground allows ATC to monitor the aircraft taxying. There had been complaints from ATC that turning on the transponder was causing their screens to be cluttered when the airport was busy. As a result, by delaying selection only those aircraft in motion were visible and not those parked at the gate awaiting engine start. However, on one occasion the crew did not detect the omission until after the aircraft had started taxying and, on another, the error was detected just after take-off, during the climb. These last two examples illustrate how an error can initiate a system status that can propagate risk.

6.11 VARIABILITY

If errors represent feedback signals then it suggests that they must be capable of detection, else they would not serve a purpose. The LOSA process captures this data and Table 6.4 shows the percentage distribution of error types by mode of detection.

TABLE 6.4
How Errors Are Detected

	Slip	Lapse	Mistake
Crew	59.2	55.4	73.2
a/c systems	0.9	5.2	2.8
ATC	16.5	2.3	2.5
Other (cabin or ground crew)	0.4	1.1	0
Not detected	22.7	35.8	21.3

The first thing to notice is the high rate of undetected error. On a procedural note, during the audits, an error was deemed to be 'not detected' if there was no discernible response, either through non-verbal or verbal behaviour, on the part of the operating crew. This rate was consistent across a number of audits. The table reveals some other interesting aspects of collaborative work. A significant proportion of slips are trapped by ATC. Slips include missed radio calls or incorrect read back of clearances or instructions. Therefore, it is not surprising that the counter party in the process, the controller, affords a means of protection. Lapses include acts of forgetting to set controls. Automated systems have in-built protections and will sound an alert if the configuration is not correct. These data illustrate feedback being acted on at the system level as well as at the level of the individual participants. Mistakes, as we saw, are decision errors. The higher rate of detection of mistakes by the crew suggests that the circumstances after the active intervention failed to meet expectations, drawing the crews' attention to the error.

Amalberti (2013) offers the following modes of error detection (in order of effectiveness):

1. Direct error hypothesis. Responding to an anomaly: 'it's because you did …'
2. Suspicion: 'I haven't a clue'.
3. Affirmative checking: 'that looks about right'.
4. Routine checks: standard, independent, periodic.

We will consider the process of monitoring work in Chapter 7, but the data on undetected events seem to support the idea that 'monitoring' – routine checks – is unreliable. Amalberti makes the point that routine checking trails the other three modes in terms of effectiveness by quite some distance.

What we are seeing here is system buffering. As performance becomes marginal, either the safeguards or other actors in the system respond and afford protection. It is conventional to talk in terms of error 'trapping', but the term is misleading. The data suggest that rather than identify an aberration and somehow block its progress, buffering in the system accommodates performance, and future actions restore margins. Errors represent a trace of the work being undertaken and allow us to explore variability at the margins of performance. In the context of Safety-II thinking, things tend to go right in part because the participants in the work process have detected 'wrongness' and responded before the discrepancy became consequential. In the next section, we will look at how that happens.

6.12 KEEPING CONTROL

Buffering, then, is reflected in the way crew deal with errors and the subsequent consequence. The LOSA methodology has four ways of categorising crew responses, which can be either active or passive. Active responses include successful management (i.e., the issue was resolved) or mismanagement (i.e., the crew response created an additional error). The passive responses are to ignore the error or to fail to detect it in the first place (see explanation above). The final outcome from each action sequence will be either inconsequential or an undesired aircraft state (UAS), which is to say that, in the opinion of the observer, the margins of safety had been eroded. Of the 1107 errors captured in an audit of 177 flights, 92% were inconsequential and this rate was consistent across other audits.

The discussion that follows is based on errors observed in a sample of 86 sectors flown by one type of wide-body airliner. The rate of errors was 7.67/flight. The distribution of crew responses is given in Table 6.5. Looking at active responses first, the fact that over half of the errors were dealt with satisfactorily at the first attempt adds weight to the Safety-II argument that things, generally, go right more often than they go wrong. Active interventions rarely led to additional problems. Therefore, we can argue that the responses of the crew were, broadly speaking, adequate for the task. Active responses reflect crew working as intended, even if their actions were, at times, momentarily insufficient for the task. It is the cluster of passive responses that, I believe, throw more light on the nature of work. Passive responses suggest either a failure in sense-making or an idiosyncratic response threshold, by which I mean crew were aware of the non-conformity but chose not to act.

I said earlier that the high rate of undetected error (Table 6.4) asks questions of the validity of the concept of monitoring but we now see that crew also apparently ignore errors, too. In terms of commission, undetected errors are split more or less evenly between the captain and the FO (48.04% v 51.96%), but 65.4% of ignored errors are attributable to the captain. Given that the other crew member is the single most effective control on performance, as tracked by the error detection rate, this possibly reflects the social dynamic on the flight deck, suggesting a reluctance on the part of the FO to intervene.

Marginal performance needs to be viewed in terms of the task goal that was active at the time. The behaviours observed mainly fall into one of the three categories: incorrect sequencing of actions; communication (typically with third parties or

TABLE 6.5
Distribution of Error Responses ($n = 660$)

Response	Errors (%)
Active – Detect and manage	56.51
Active – Detect and mismanage	3.63
Passive – Detect and ignore	12.72
Passive – Not detected	27.12

procedural call-outs); wrong task actions. Sequencing issues include doing actions in the wrong order, at the wrong time or forgetting to do them, classic examples of slips and lapses.

In both passive management categories, we see that a similar percentage of errors were to do with the sequencing of activity: 13.09% of ignored and 11.17% of not detected. Examples would be:

> Landing light not switched on after gear select down
> Taxi duties commenced before exiting runway
> Checklist actioned early

Communication errors included failure to initiate, incorrect response and inaccurate or incomplete messages. These represented 13% of the ignored and 50.83% of the not-detected acts. The data suggest that undetected errors are, in fact, simply the noise in the system. They are the imperfect use of language and issues around tracking tasks.

The ignored errors illustrate the other side of doing work. During discussions with pilots about the data, a common observation about the rate of errors that were ignored was that the act that was flagged as an error in LOSA terms was probably considered to be unimportant by the actors. One simple example of this behaviour is the failure to announce approaching a specified level off. The data for lapses detected by the aircraft in Table 6.4 include this type of event. Pilots often commented that there was little point in making the call as the aircraft would level off in any case. Of course, that assumes that the automation had been correctly configured, which was not always the case. This example reflects the fact that work is socially constructed rather than a normative process. Task design and procedural guidance are construed as over-specified by the crew, and alternative, non-standard, solutions are often constructed in real time. Task modification is shaped by a variety of factors such as convenience, simplification and workload. A pilot's alignment with organisational goals and the nature of oversight also shapes performance.

6.13 'WRONG WORK' AND VIOLATIONS AS IMPROVISATIONS

The final category of performance I proposed was wrong task behaviour. Unlike the first two categories, when crew engage in 'wrong work' they are, in fact, creating their own ways to achieve the task and, at times, such behaviour can be problematic. Reason (1990) suggests that a violation is a 'deliberate error' while LOSA uses the more cumbersome formulation of 'intentional non-compliance'. Whatever the label, both capture the idea of aberrant behaviour being purposeful. Table 6.6 shows the percentage of errors deemed to be intentional in a sample of audits. Because it is difficult to be absolute about intentions (and LOSA does include some 'tests' to apply before using this category), LOSA observers can use a 'not sure' category if they are in doubt. The table only includes events where the observer was certain that it was intentional.

An example of a violation is the case of a B-777 that landed at Toronto International Airport and missed its nominated exit. The options available were to call for a tug

TABLE 6.6
Intentional Non-compliance

Fleet	Intentional Non-compliance (%)
A	8.8
B	9.8
C	26.4

to manoeuvre the aircraft on the runway or to take another exit that would require a much longer taxi route around the airport and back to the terminal. The FO was about to call for the tug when the captain instructed him to make sure his feet were clear of the toe brakes, whereupon the captain used reverse thrust to back the aircraft across the runway and make room to turn back to the planned exit. The use of reverse thrust in this context is expressly forbidden with two specific prohibitions contained in manuals. The risk is that the aircraft could tip on its tail if the brakes are touched inadvertently. The action clearly violated the published requirements. That it also deviated from social norms relating to responsibility and professionalism reflects the fact that violations are also socially constructed. The behaviour of the captain can, however, be adequately explained in terms of an individual selecting a personal goal as an expedient despite a fully understood proscription and a known risk. Violations often need to be seen in terms of the individual's sense of self and their personal interpretation of the latitude afforded to them.

A less clear-cut example of violating behaviour was seen in the study that looked at the new aircraft type entering service (Fleet C in Table 6.6). Because it was being operated as a common type rating, crew were getting used to operating both their current and the new aircraft. Differences between the two types resulted in an intermixing of procedures, often inadvertently, but with one captain going so far as to adapt the old procedures, with which he was very familiar, to the new aircraft. Some crew did not use the new head-up display as intended. Here we see individuals shaping new processes and technologies to suit their preferred approach to the task but in a way that exceeded company expectations.

The 'wrong task' category of behaviours, while it might reflect violations, captures active intervention by the crew in the work process to modify the task in a manner that was considered 'illegal' by the observer. These comprised 63.09% of ignored and 20.6% of the undetected errors. Examples of wrong task acts include:

> CN turns off flight deck recirculation fans during set up to reduce distracting noise even though he said that it was against SOP
>
> During weather avoidance, turns were made without reference to, or clearance from, ATC
>
> ATC advised of the availability of the new ATIS but crew did not check the update
>
> CN let the FO park the aircraft on a bay that only had left-seat guidance

Wrong tasks can also be interpreted as improvisation. In the sample of 86 sectors, these behaviours were seen on nine occasions (10.4% of flights). Automation was

TABLE 6.7
Crew Awareness of Violation

	No (%)	Yes (%)	Not Sure (%)
Fail to detect	73.75	2.25	24
Detect and ignore	20.3	70.2	9.5

deliberately circumvented to achieve a goal, an alternative action or expedient was implemented or a non-standard tool was used. One type of intentional error common to all fleets was PF/PM duty exchanges. For example, during ground preparation, the FO will often do tasks formally assigned to the captain in situations when the captain is dealing with, say, a problem and needs to speak to the engineer. On examination, it is clear that crew operate outside of the published procedures in order to maintain the work flow. The fact that such a high percentage of this type of event was ignored by the crew suggests that the need for flexibility and adaptability is well understood and accepted by crew, which brings us back to the nature of work. Normative models fail to take into consideration the need for work to be flexible and adaptive to cope with change. While work is undertaken within familiar frameworks, it is also a socially constructed, negotiated process.

Table 6.7 presents data in relation to passive management and violations. In the case of crew that fail to detect an error, the observer was of the opinion that the bulk of the actions were not intentional. This suggests that either crew knowledge was flawed (they were unaware of the correct action) or they were distracted by other tasks. Where acts were ignored, it was the opinion of the observer that, in most cases, crew were fully aware that they were operating outside of procedures. The data reinforce the view that crews construct solutions to problems in a deliberate manner.

I suggested in Chapter 5 that pilots engage in activity directed at achieving congruence between the desired goal state and the current status of the task. From a resilience perspective, LOSA data seem to support the idea that crew are routinely engaged in repair action to sustain progress towards the desired goal. Crew are also actively engaged in modifying and creating alternative solutions to the demands of the task.

If buffering does, indeed, describe the ability of the system to cope with crew performance variability, then the fact that most of the observed errors were inconsequential reflects the fact that, under normal circumstances, the system is able to cope. In the next section, I want to look at crews approaching the boundary of the system.

6.14 PERFORMANCE APPROACHING THE BOUNDARY

If, in the opinion of the observer, the actions of the crew place the aircraft in a state where safety margins have been reduced or eroded (itself a problematic judgement), then a UAS is recorded. In some cases, there can be a reference or recognised benchmark to refer to. For example, taxying the aircraft too fast would be considered a UAS. Some events might trigger a need for a safety report. In other cases, it falls to

TABLE 6.8
Consequential Outcomes by Error Types (as a % of all errors in class)

Outcome	Slip (%)	Lapse (%)	Mistake (%)
Additional error	2.5	3.7	4.6
UAS	4.2	1.7	7.4

TABLE 6.9
Crew Responses to Consequential Errors (as a % of all errors in class)

Management of Outcome	Slip (%)	Lapse (%)	Mistake (%)
Active	1.4	2.2	3.9
Passive	5.4	3.1	7.9

the judgement of the observer. In one study, I used a current line pilot to review 52 events classed as UASs (4.6% of total errors observed). First, events were categorised according to the pilot's view, based on their experience of operations, of the probable rate of occurrence of the event (e.g., commonplace, occasional, rare), and then they were asked to consider, given the context, what could be the next most likely outcome if the crew had not resolved the situation satisfactorily, an exercise in counter-factual thinking. Only a single UAS out of the 52 was considered noteworthy in terms of the potential exposure to hazard. What we are interested in here is how behaviour modifies residual risk: the degree to which the current task status is the precursor to subsequent failure.

Table 6.8 shows outcomes in relation to error types. We saw earlier that mistakes are associated with decisions about possible future courses of action. Active intervention, then, seems to be associated with a higher rate of adverse outcome. Table 6.9 shows a sample of 80 consequential errors (i.e., the outcome was an additional error or a UAS) analysed by type of behaviour and crew management. We saw that passive management of errors was often associated with intentional departures from expected performance. In terms of adverse outcomes, where a crew fails to act in response to a marginal performance, then the risk of encountering a UAS was doubled (RR 2.013, CI 1.05-3.859). In Table 6.5, the distribution of passive responses to undetected errors was more than double than that for ignored errors. The implication of these data seems to be that where crew actively intervene and the outcome is a mistake, the subsequent failure to detect the discrepant feedback signal is twice as likely to be at the level of sense-making (significance not perceived) rather than flawed mental models (signal significance not understood).

While these data suggest that performance variability is a product of both the type of behaviour (inputs) and the crew responses to events (feedback handling), they say nothing about how risk can be increased. What we need to understand is how margins are eroded and buffering is overwhelmed. To explore this aspect, I looked at a sample of 20 errors linked to a UAS. Three clusters of risk-shaping factors emerged:

Performance Degradation (40%)
Risk Transfer (35%)
Separation Reduction (25%)

Performance degradation arises from the configuration of the aircraft. Examples include taxying too fast and, thus, having excess energy that needed to be dissipated at some point, flying at the wrong speed for the current flap selection and applying thrust with the speed brakes extended. In one case, the crew failed to update the take-off data, thus reducing the performance margins should conditions change further when getting airborne. These examples point to the importance of efficient energy management and are all examples of mistakes. The next example is actually a commonplace lapse and is a classic illustration of the problem of habitual actions intruding in current processes. Traditionally, approaches have been flown in a full landing flap configuration. For a variety of reasons, airlines now routinely land at a lesser flap setting (Flap 25 instead of 30, for example). The crew had briefed a Flap 25 approach but, during the approach, the PF called for Flap 30. This was actioned by the PM and neither crew picked it up until they completed the landing checklist. The final example goes to the heart of modern aviation and the challenge of maintaining proficiency in manual handling. The pilot was hand-flying during the departure and, in a turn, allowed the speed to increase by 13 kts.

Risk transfer describes the condition where the actions of the crew exposed third parties to potential harm. In four cases, the crew either failed to allow adequate separation between the aircraft track and visible convective clouds or failed to turn on the seatbelt sign, having already encountered turbulence. The risk of harm as a result of being exposed to turbulence was borne by those inside the aircraft. In three other cases, crews forgot to turn on a landing light and the TCAS, thus removing potential warning to others of their presence, or forgot to turn off the weather radar after landing, which has potentially harmful effects for ground personnel. The additional risk in these cases involves others external to the aircraft. Interestingly, whereas the internal risk arose from mistakes by the crew, the external risk resulted from lapses.

Finally, airspace rules are designed to allow the maximum number of aircraft to operate in constrained airspace by maintaining a safe envelope around each aircraft. To maintain the envelope, aircraft are required to fly at specific speeds, tracks and altitudes. Any modifications to these parameters will result in reduction of the safe space around the aircraft. Separation reduction occurred on five occasions as a result of incorrect entry of data (speed and heading), manoeuvring before a nominated waypoint, taking up a wrong track and flying below a cleared altitude. The first three were examples of slips while the latter two were mistakes.

Two conclusions flow from this sample of errors. First, they were all the result of the normal behaviour of crew in relation to perturbations in the operational environment. They flowed from 'normal' work. Second, error is of interest because it is a risk multiplier. In the case of the excessive taxi speed, a change in the taxiway surface conditions could result in the aircraft losing grip and sliding. Typically, snow or standing water ought to bring about a change in pilot behaviour but, for example, a B-737, landing at an Indian airport, slid off the runway while trying to execute a 180-degree turn. The pilot had reduced speed after landing and rolled to the end of the runway, which was covered in sand and small stones. Because aircraft rarely used this part of the runway, no efforts had been made to keep it swept. The linear momentum of the aircraft, coupled with the reduced traction, resulted in the aircraft very gently sliding sideways off the end. Error, then, is of interest as much because of what might happen next as what it tells us about current performance.

6.15 CONCLUSION: ERRORS AS SIGNALS OF SYSTEM BEHAVIOUR

The study of error allows us to resurrect aspects of normal work. By looking at errors in context we can see what crews were trying to do and explore the ways in which normal work departs from organisational intentions. From a system's perspective, errors serve two purposes. On the one hand, they are the feedback loop from the environment that allows us to judge the success of our attempts to achieve our performance goals. On the other hand, they are symptoms of individual competence. Clearly, competence must contribute to error commission (allowing for violations) because it reflects inadequate task management, but an inadequate response to an error can be indicative of failures at the level of perception or processing. The challenge will always be to separate the relative involvement of each component. Errors are not predictable, they are contextual. This chapter has relied heavily on numerical analysis of a set of samples but only for the purposes of triangulating on the real question, which is how does work get done.

REFERENCES

AAIASB. (2006). Helios Airways Flight HCY522 Boeing 737-31S at Grammatiko, Greece, on 14 August 2005. Hellenic Republic Ministry of Transport and Communications.

AAIB. (1998). AAR 2/98 Report on the incident to Aerospatiale AS332L Super Puma G-PUMH over North Sea, on 27 September 1995. HMSO, London.

AAIB. (2019). Bulletin 1/2019. Report No: EW/C2017/12/03.

AAIU. (2001). Boeing 737, EI-CDB, Nr Cork Airport, 7 December 2000. Report No: 2001-014.

Amalberti, R. (2013). *Navigating Safety: Necessary Compromises and Trade-Offs - Theory and Practice*. London: Springer.

Chou, C. & Funk, K. (1990). Management of multiple tasks: Cockpit task management errors. *1990 IEEE International Conference on Systems, Man, and Cybernetics Conference Proceedings* (pp. 470–474), Los Angeles, CA, USA.

Hollnagel, E. (1983). Human error. Position Paper. *NATO Conference on Human Error*. Bellagio, Italy.

Marsh, R.L., & Hicks, J. L. (1998). Event-based prospective memory and executive control of working memory. Journal of Experimental Psychology: Learning, Memory, and Cognition, 24(2), 336–349.

Reason, J. (1990). *Human Error*. Cambridge: Cambridge University Press.

SIBF. (2006). Report No: C3/2006L Incident involving an airliner landing at Helsinki-Vantaa airport, on 20 February 2006.

7 Acting in the Public Domain – Collaboration to Achieve Operational Goals

7.1 INTRODUCTION

Level 1 in the system model represents private action. It encompasses those processes that take place largely inside our heads before they become visible activities. It would not matter if we were looking at an aircraft operated by a single pilot or a large aircraft with two pilots and a relief first officer (RFO) sitting on the centre jump seat. Each pilot will be engaged in private thought before public action. The internal world of the individual was the focus. Although the discussion of error dealt with outcomes, the actions observed were the product of internal processing. In this chapter the focus is fully on public performance. At Level 2 in my hierarchy – collaboration – our actions become known to others. Elements of my performance are accessible and are amenable to change. What we are interested in here is how two or more individuals combine their efforts to achieve a goal. Once again, the action flows from decisions but the decision process is now a shared activity: decisions happen in the space between individuals, not solely inside the heads of individuals. This example involving a crew of a business jet epitomises problems with collaboration:

> Before departure the Captain told me to order 1020 lbs of jet-A fuel. We already had approximately 1200 lbs. The flight was scheduled 1 hr 10 mins and, for the Learjet 31, the approx fuel burn is 1500 lbs. I felt that a total of 2220 lbs was not enough fuel, but was afraid to speak up because [of] his [the Captain's] attitude and that of management (I have brought what I thought were safety issues to my manager before and she told me that my job is to 'protect the ego of the captain' and not to speak up unless I was about to die. If I report safety issues or point out a captain's mistake, I fear losing my job). Once airborne, the Captain looked at the estimated fuel on arrival and became very irate. He told me to ask for a higher altitude. We were level at FL370 and we requested FL400. The controller hesitated, but after we told him the situation, he worked it out. Climbing towards FL400, the Captain instructed me to figure out why we were showing landing fuel to be so low. He asked me to look at the FMS to see if we had it programmed so that it would show low fuel. While verifying the FMS, I heard the alt warning chime. I mentioned to the Captain that we were above our assigned alt and still climbing. After using abrasive language directed towards the autopilot, he clicked the autopilot off and started a descent. We climbed to approx FL405. As we descended, I noticed that we had a high rate of descent and mentioned that we were assigned FL400.

> We descended past FL400 to approx FL396 before returning to FL400. We were later assigned FL450 for fuel flow reasons. The controllers had to move other aircraft for us to obtain the altitude. We landed with approx 25 mins of fuel on board.
>
> (ASRS. May 2005, Learjet 31).

Collaboration involves not simply immediate crew members. In the discussion of error detection, we saw that third parties can play a role. Other agencies, such as air traffic control (ATC), are involved. We can also see that collaboration is distributed across time: a past conversation with my manager will shape my actions in the present. Collaboration is shaped by individual factors, such as personality and stress, and also by group dynamics, such as rank and status.

From a systems perspective, an airline crew can be seen as a unit of delegated control. The airline allows a crew to undertake a task with a considerable degree of autonomy but control is still exercised through procedures, schedules, flight plans, etc. Coordination between crew members is facilitated by acts of communication. This chapter looks at groups as social entities doing work. Communication will be considered in more detail in the next chapter. Level 2 does contain a paradox: although we are talking about performance in a social context, the Learjet case study illustrates how people can still act autonomously within a structured group. One of the challenges of collaborative action is to manage, or constrain, individuals who have a negative effect on performance and undermine the functioning of the group. Acting in a group requires a degree of surrender: we no longer have full autonomy (even if I might think otherwise).

We will be looking at work as a coordinated action. The scope of action is bounded by rules but individuals differ in their view of rule obligation. We saw examples of this in the last chapter when we considered violations. Humans exercise agency in that they have volitional control over their behaviour and autonomy is an important component of the social construction of individual worth. We exercise control, then, and, to a degree, choose action. The implications of this need to be understood if we want to make sense of collaborative performance.

The chapter starts with an exploration of the ephemeral nature of 'collaboration' before looking at some structural aspects of groups. We will then consider how crews work together to manage tasks by monitoring the progress of work. Finally, we will attempt to apply the concept to a crew trying to cope with a dynamic, changing situation.

7.2 COLLABORATION IN A SYSTEMS CONTEXT

Just before midnight on 1 June 1999, an MD-82 with six crew and 149 passengers ran off the end of the runway at Little Rock, Arkansas (NTSB, 2001). Landing in a thunderstorm, the aircraft failed to stop, collided with structures on the airfield and slipped over an embankment. The captain and 10 passengers were killed. The flight had been delayed leaving Dallas/Fort Worth because of the bad weather. As they approached the airport at Little Rock, the pilots could see the flashes of lightning and their weather radar showed intense activity. The crew had initially planned to attempt an instrument landing system (ILS) approach to Runway 22L. In the 17 minutes they were in contact with Little Rock ATC, that plan changed three times as the crew tried

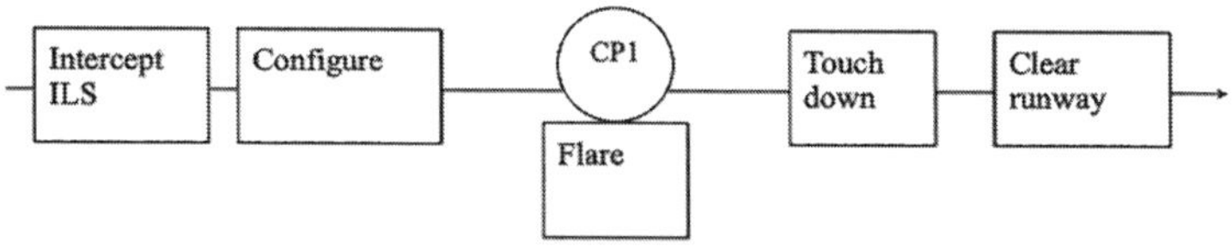

FIGURE 7.1 Little Rock initial goal structure.

to respond to the rapidly changing conditions. As the weather front approached the airfield, the shift in wind direction rendered Runway 04 more favourable and so the crew switched to an ILS to 04R. In an attempt to beat the storm and get on the ground quickly, the crew briefly attempted a visual approach but reducing visibility forced them to reposition for the ILS to 04 instead. Early in this sequence the air traffic controller suggested a visual approach to Runway 22L, but the crew declined. The omission of checklist on final approach – an undetected lapse – contributed to the captain failing to arm the ground spoilers. From a systems perspective, the aircraft's status on landing was not congruent with its required state.

Figure 7.1 represents the crew's original plan as they approached the airport. The boundary of the system constructed on that day was the point at which sufficient energy had been dissipated for the aircraft to remain on the runway surface under the control of the pilots. It was not their goal simply to land, they had to stop, too. In order to achieve a safe landing, the crew had to manoeuvre the aircraft through a set of intermediate goal states, each with a set of constraints established by the performance characteristics of the aircraft, the design of the approach profile and the prevailing environmental conditions. The crew on board, together with the company Dispatchers and Little Rock ATC, worked as a collaborative team to accomplish this task.

Crew engage in planning to meet the expected applicable constraints, changing plans as events unfold and responding to transient disturbances or momentary perturbations. A disturbance might be the dynamic influence of the environment (turbulence or reduced visibility, for example) or an unexpected change in constraints (such as tailwind limits on landing or runway visual range (RVR) requirements). The environment the crew at Little Rock were operating in comprised the complex air traffic structures codified in terms of the approach procedures to the various runways in operation as well as different classes of airspace. The meteorological situation denied access to parts of the airspace over the airport and also limited what the crew could do in other parts of the available manoeuvring space. The crew were approaching their crew duty time limit and so the legal framework within which the flight was conducted represents a further limiting factor. An array of tangible and intangible factors shaped the freedom of action of the crew but, although these might have inhibited operations, the role of the operator is to shape the task in order to achieve the goal or switch to a different goal.

7.3 COLLABORATION AS SHARED DECISION-MAKING

Level 2 in the hierarchical model is characterised by collaborative decision-making. Multiple actors agree solutions to achieving the task and then individuals act accordingly. In addition to the two crew members, various air traffic controllers provided information and support during the flight. Air traffic agencies also act to configure the legal airspace through which the aircraft can travel by approving flight plans, authorising amendments to plans and so on. ATC controllers and flight crew collaborated in real time during hazardous parts of the flight and, so can be considered a distributed crew. Other agencies operate at a remove. Dispatch, for example, had an involvement in the work process but of a different order to that of, say, ATC. A key differentiating factor between members of collaborative teams is the span of control: aircrew only have their own aircraft to deal with whereas off-board agencies are typically dealing with multiple aircraft.

The crew of AA1420 were initially faced with two problems. The first was that their intended airframe was being used elsewhere. To wait for it to arrive would have meant that the crew would not have sufficient duty time remaining to complete the flight to Little Rock. This was solved by deploying a different aircraft to the task. The second problem was the weather at their destination. By liaising with, initially, the Dispatcher, and in the final stages, with the air traffic controller, a solution was apparently found to the problem of the weather. These elements of the story – airframe switch and weather – represent 'threats' or challenges and would have induced, at the least, a sense of frustration in the crew. We also need to consider the time of day. It was close to midnight and, although the first officer (FO) said later that he did not feel that fatigue was necessarily an issue, the crew had been awake for 15 hours before they departed on this flight. We saw in Chapter 4 that hours awake before duty is a risk factor.

The original plan was to land on Runway 22L. As the aircraft descended, the crew were in visual contact with the airfield and had radio contact with ATC. The pilots and the controller were discussing the weather. The controller suggested that the aircraft's weather radar gave a better picture than his. Between them, the three actors (the two pilots and the controller) constructed a view of the weather based on actual measured meteorological parameters from the controller, the visual observations of the crew and the weather radar images. Although the weather front was approaching the airfield, the pilots estimate that it was still some 15 miles away and the captain thought that they still had 'some time' in which to make the approach.

This statement reflects the role of time in task management. Crew need to establish not just the 'here and now' but also a version of 'what will be in the future'. Human evolution has equipped us for dealing with change that progresses arithmetically (consider throwing a spear at an animal moving in a straight line. We learn to aim ahead of the animal if we want to eat), but dynamically changing situations are a challenge. This example from a line operations safety audit (LOSA) report illustrates the point. The crew had correctly identified the threat posed by a storm cell on the runway extended centreline but reckoned that, because the departure track required a turn soon after take-off, they would keep clear of the weather. Unfortunately, they had not taken the prevailing wind into consideration and the aircraft met the storm cell soon after the turn.

At this stage the crew of AA1420 have a future goal of landing on Runway 22L (Figure 7.2), but we can establish at least two intermediate goal states: the first is to fly the ILS procedure to a point at which landing can be undertaken safely and the second is to configure the aircraft for landing. Of course, these two goals are nested

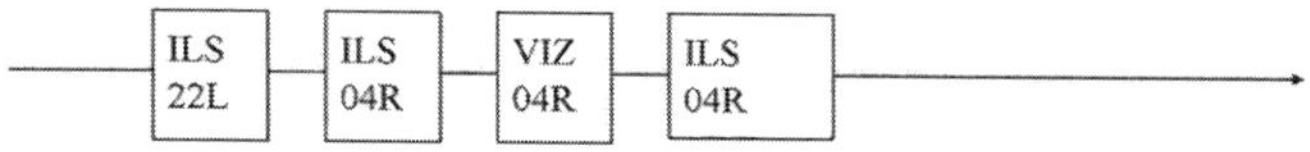

FIGURE 7.2 Little Rock intermediate goals.

in that configuring for landing can be seen as part of the broader ILS goal. However, both need to be met if the aircraft is to be safely landed. All of this must be achieved before the weather front arrived over the airfield. Although the crew have estimated the approximate distance of the weather from the airfield, they do not appear to have resolved the relative rates of the closure of either the frontal system or their aircraft with the airfield. Furthermore, the severity and implications of the weather front have not been assessed. It seems that the crew were comfortable that the problem could be avoided by landing before the storm arrived. The crews' construction of the world was approximate in that it included uncertainties that do not appear to have been recognised by the crew. Neither the current status nor the circumstances pertaining to their next desired goal state were clear and so the probability of satisfying the goal constraints was uncertain.

One such constraint was the crosswind on the runway and the crew now discuss the applicable limitations. Initially the FO stated that 30 knots was the limit, revising the figure down to 25 knots to allow for the fact that the runway was wet. The captain said that the correct figure was 20 knots and, when the FO took out his manual to check, the captain gestured to put the manual away as he was confident in his own knowledge. The captain was, indeed, correct but the episode shows how knowledge is not consistent across all team members.

The visibility was also now starting to reduce and the crew could only see the runway with some difficulty. The controller offered the crew a visual approach, presumably in an attempt to expedite the landing, but the crew declined on the grounds that the visibility was marginal and, therefore, a visual approach represented a riskier option than the ILS. Visual range is another goal constraint and the use of the ILS afforded more flexibility. The crew are now involved in trading off one option against another: an expedited, riskier arrival versus a less risky but more time-consuming one. Seconds later the crew received an updated wind report, which suggested that, at best, the crosswind limitation has been exceeded and, at worst, the aircraft would land with a tailwind (another constraint). There was no discussion of the wind situation other than the captain proposing a change to the reciprocal Runway 04R, which would now be, effectively, the into-wind runway. The change in wind direction had rendered the original plan untenable in that the wind constraints could not be satisfied. The change in goal state, rather than being the result of discussion, was based on an implicit understanding shared by both crews (a plausible conjecture, possibly, on my part).

At this point, the controller turned the aircraft away from the airfield in order to make enough space to reposition for the new approach. The FO briefed the captain of the main values associated with the new approach such as the ILS frequency, safety altitude, the minimum descent height and the missed approach procedure. However, the briefing was not complete. We have now shifted from the original intended goal to a new goal. The complete sequence of successive goals is illustrated in Figure 7.2.

During the turn away from the airfield, the crew were unable to monitor the position of the weather front as their radar was now pointing away from the storm cells and they could not see the airport environment. The manoeuvre illustrates the need to consider volumes of space and time associated with goal state accomplishment. As they turn back towards the airfield the captain was having difficulty seeing the runways because of the worsening weather. During the repositioning the weather system had, of course, moved closer to the airfield. The controller updated the crew on the surface wind conditions and the FO suggested that they attempt a visual approach. The captain agreed even though he could not, himself, clearly see the runway: 'You just point me in the right direction and I'll start slowing it down here. Give me flaps eleven'. The crew have now, seamlessly, changed to the third goal: a visual approach to Runway 04R. At the same time the captain was configuring the aircraft for landing as he called for flap and started to reduce speed. The controller, appreciating the fluctuating visibility associated with the moving frontal system, said, 'if you lose it (meaning visual contact with the runway), need some help, let me know please'. The three players in this worsening scenario appear to be working on a shared, implicit understanding of the situation.

The crew then went through a short phase with the FO providing a running commentary to the captain about the location of the runway and the approximate position of the aircraft in a normal visual traffic pattern. For the next 60 seconds the FO acted as an on-board controller providing guidance to the captain until the crew recognised that, once again, the rules applicable to visual range were being violated. This phase shows how fragile formal control structures can be. Instead of formally briefing the new plan to fly a visual approach, the crew were responding to events and creating ad hoc solutions in real time.

Unable to maintain visual contact, the crew were directed away from the airfield again in order to set up for the ILS approach to Runway 04R, the fourth new goal state. The additional time taken to position for the ILS meant that the storm front was now firmly established over the airfield. 'I hate droning around visual at night in weather without having some clue where I am' said the captain. 'Yeah, but the longer we go out here the…' replied the FO, presumably recognising that the time taken to position for the ILS was simply allowing the situation to worsen. 'See how we are going right into this crap', he said. Once the crew finally intercept the inbound course they receive an update on the weather at the airfield. The controller told them that there was heavy rain, that the visibility was less than a mile and the RVR for 04R was 3000 ft. In the next transmission, the controller updated the wind, reporting values right on the crosswind limits. The crew discuss the RVR limit and decided that they were safe to continue. Ordinarily, an approach in bad weather would be flown to a specific point and, if the landing was not possible, the aircraft would execute a go-around. The limitations and subsequent actions would be briefed by the crew. On this day, no approach briefings were given as the goals changed. The formal procedural framework unravelled, including checklist activity. Because changes were not anchored to specific points in the standard profile, the correct configuration of the aircraft became secondary to the goal of simply landing successfully. In very difficult conditions the captain managed to land. Unfortunately, because the spoilers were not armed and given the weather conditions, the aircraft did not stop in the distance available. The boundary was breached.

In this sequence we can identify a number of discrete desired goal states. Each has a set of constraints in terms of permissible wind vectors and visibility. In addition, the transitions between goals consumed time and space which, in turn, resulted in dramatic shifts in the properties of the environment. The crew had to constantly revise their plan for overcoming the constraints placed upon their intended use of the aircraft. Of course, the crew did not have to attempt to land at Little Rock. Because of the weather, a diversion to an alternate or even a hold while the weather passed through was the option. The crew did have a duty time limitation that they needed to be aware of but these were exceptional circumstances. It is customary to describe a performance such as this as 'target fixation', 'press-on-itis' or, more formally, 'plan continuation bias'. The inherent risk aversion in humans predisposes us to avoid losses and to not land at Little Rock represented just such a loss. In Chapter 4, we saw that psychological fatigue can influence risk perception. Late at night, after a long day and in worsening weather, after three previous attempts the crew appeared to be intent on landing. The cognitive load model suggests that switching tasks has a cost. The crew had repositioned the aircraft twice because constraints could not be met and so they were probably feeling the burden. We cannot know what was on the mind of the captain but there are sufficient precursors to explain the performance of the crew at an individual level.

Our main interest here is how the team functioned in its attempt to cope with the demands of the situation. What we see in the sequence is a progressive shift away from a structured team with roles and responsibilities to a situation where one pilot was trying to control the trajectory while the other pilot filled in gaps in their shared understanding. Having established that their initial plan violated crosswind limitations, the crew then lost sight of the fact that their final plan was equally 'illegal'. The volume of information from the controller about different wind conditions around the airport probably overwhelmed the crew's ability to comprehend, given the demands of the trajectory control task. As the plans changed, the structured approach briefing process broke down until, for the final two plan changes, no approach briefs were given. The formal checklist process was also degraded until, by the final approach, the checks were limited to calls for essential services: landing gear and flaps.

7.4 COLLABORATION IN NORMAL WORK

The Little Rock accident is a stark illustration of the nature of collaboration and how established processes can quickly degrade under stress. Undoubtedly, the crew were dealing with a very demanding situation but the key elements of the scenario can be seen in routine, normal operations. Based on a LOSA sample of 100 sectors, a list of 660 errors was generated of which 147 (22.27%) could be considered examples of specific task-related behaviour. We saw that Little Rock crew were constantly having to reconcile competing goals. In fact, under normal circumstances, on 9% flights, crew were required to reconcile a conflicting or confusing task goal. Under the pressure of constant change, the Little Rock crew forgot an important checklist. Task-related lapses were evident on 10% of the LOSA flights. Much of the work of the Little Rock crew was activity directed at achieving a goal state within the applicable constraints. The largest single category of observation in the LOSA sample was interventions lacking sufficient magnitude to effect change in the time and space available (21%), the result of which was that crews failed to satisfy goal stare constraints. They followed an incorrect procedure (12%), their actions were in the wrong sequence (11%), were delivered too late (8%) or were inappropriate for the specific task (5%). A fairly common event (21% of flights) that illustrates the fragility of collaboration is what I labelled task partitioning. It is a maxim that 'someone must be looking out at all times' and yet situations where both pilots were individually, simultaneously completing tasks inside the cockpit were common. The issue of roles within teams will be discussed later but 13% of observations involved unilateral interventions by the captain.

Of these events, 40.81% were actively managed and had no bearing on the subsequent conduct of the flight. Of the passively managed events (ignored or not detected), seven (4.67% of task-related events) had a potential impact on risk. For example, while the FO was off the flight deck the captain was instructed by ATC to initiate a 10-mile track offset. The captain actually chose to insert a 12-mile offset without confirming with ATC and he then failed to explain what he had done when the FO returned to the flight deck. The ingredients of the Little Rock case are, actually, quite commonplace in normal operations and risk exposure can be influenced accordingly. The collaborative process, like all activities, has inherent variability even under normal conditions.

7.5 ENGAGEMENT AS A TRANSITORY STATE

The challenge of collaboration, in part, stems from the fact that human actors can be both 'a part of' and 'apart from' the process of work. To illustrate this concept I want to look at three events with identical outcomes but each different in their development. In a 6-week period, three aircraft of the same type flew through the localiser (LOC) arriving into Hong Kong on Runway 07, one of which resulted in a loss of separation in controlled airspace. All three arrivals had one thing in common (other than the outcome): at a significant moment, one crew member became divorced from the work process (this section relies heavily on the invaluable input of five of the six crew involved, for which I am grateful. Any interpretation of their performance is my own, as are any errors in that analysis). In the case of the first event, the FO was pilot flying (PF) and was paired with a relatively new captain. Because traffic was light, given the time of day, the crew were told to maintain high speed in order to expedite the approach. In addition, the aircraft track intercepted the LOC at about 90°, requiring a tighter turn than normal. The aircraft passed through the localiser with LOC 'armed' and immediately turned back to capture from the wrong side. As the captain said in his report 'I didn't realise the jet would do such a messy job'. The second event happened 11 days later and involved an FO training flight. Because the LOC was armed late, the aircraft flew through and, again, aggressively attempted to capture from the other side. The captain took control at this point but not before separation with the aircraft following in the pattern had been lost. Finally, 4 weeks later, an aircraft, with the captain as PF, again captured the LOC late, but not to the same degree as the previous two events. In a LOSA audit of 177 flights there was one instance of the LOC 'bust', six instances of incorrect configuration (arming the system) and 11 instances of missed procedural call-outs (declaring that the system was capturing the ILS). Of the configuration errors, 66% were trapped by the other crew members but of the call-out errors only one was trapped and that was in response to a check from ATC. The survey captured the equivalent of half a day's flying for the airline and, by extrapolation, suggests that 23 crews a day omit an action while flying the ILS and two arrivals will approach the margins of accepted variability. The events we are looking at, then, are relatively commonplace.

Returning to our case studies, the first example was influenced by the intervention of ATC. In this case, arrivals were being managed by giving crew speed instructions and vectoring. The FO was not slowing the aircraft down quite the way the relatively new captain might have liked but he was reluctant to intervene, instead choosing to let the FO manage the operation. The FO was following procedures in a fashion that would work under normal circumstances but the captain felt that, perhaps, the task prioritisation was inappropriate given the speed and distance to run. However, he was not inclined to speak up. The FO flying the second arrival was trying out a new technique. Although very familiar with the VNAV automation mode, the training captain had suggested using a combination of different, commonly used, modes: FLCHG and vertical speed (V/S). The descent was going according to plan, with the occasional use of speed brake to stay on profile, but it nonetheless needed extra mental effort on the part of the FO to monitor progress. The captain, of course, was supervising his trainee as well as being the pilot monitoring (PM), which added to his own workload. In the final example, the captain, as PF, was managing the descent using flight path angle (FPA). The FO had not seen this technique before and he was watching, out of interest, to see how it worked out. Because his attention was elsewhere, he failed to keep track of the progress of the aircraft as it approached the LOC. In short, at this stage, none of the three crews were doing anything wrong but what they *were* doing influenced what happened next.

In terms of goal states, the aircraft is, first, descended to an appropriate height and position from which it can start its initial track to intercept the LOC. It is then turned onto an intercept heading before capturing the LOC. ATC gives instructions for speed and heading changes and then issues the formal clearance to intercept the ILS and begin the final approach. The final example was quite straightforward in terms of the degree to which it mapped on to the process. The captain had tuned and identified the ILS, in accordance with the standard operating procedures (SOP), but then became distracted by an indication of incorrect sensing. He then recalled someone saying that there was no need to verify correct sensing but he was not sure where he had heard that. The FO, meanwhile, reflecting on the captain's use of FPA, had not fully re-engaged with current events. He did not pick up that the captain had become distracted and that, although the aircraft had been turned onto the ILS intercept heading of 040°, the LOC had not been armed. The late arming was the cause of the slight overshoot. The captain's confusion about the current procedure says something about how pilot knowledge is recreated over time through formal and informal communication of change.

The first example was also quite straightforward in that the excess speed, coupled with the junior captain's reluctance to intervene meant that the activity was time-compressed. Of interest is the captain's belief that the FO's capability to recover the situation matched his own, which was subsequently proven incorrect. This says a lot about the validity of concepts like 'share the mental model'. It is not enough to simply have two crew with the same supposed understanding of the situation: both crew need to have the same ability to identify a corrective action and apply it within the constraints of time and space, a condition made more difficult by the simple fact that one individual's true capability is not actually known to the other crew member. On the one hand, we have to make guesses about our colleague's future performance based on little or no accurate information and then, on the other, we have to trust them to get it right.

The second of our case studies is actually quite complex. ATC issued an instruction to reduce speed to 160 kts, turn onto the intercept heading of 040° and then cleared the aircraft for the ILS. The clearance was acknowledged by the captain. On hearing the first instruction, the FO began to dial the speed back from 180 to 160 kts on the mode control panel (MCP). He then realised that this speed was below the minimum manoeuvring speed for the current flap setting. He interrupted what he was doing, selected the next stage of flap and waited for it to deploy before further reducing speed. This all took time which the crew did not have. From receipt of the ATC instruction to intercepting the LOC on this day took 40 seconds. The training captain was waiting for the FO to change heading and, much like the junior captain in the third example, was reluctant to intervene too soon. Finally, he spoke up and asked if the FO was going to turn. Having reduced speed, the FO's attention was now drawn to the track and he realised that the centreline was fast approaching. He entered a heading of 039 in the MCP – a slip – and armed LOC, which was then almost immediately captured. The management of a sequence of activities and the interplay between the time it takes for events to occur and the distance travelled in that time lay at the heart of this problem.

In two of our cases, LOC capture occurred as the aircraft was passing through it and the crew were then surprised by the aggressive response of the automation. The aircraft turned onto a heading which would bring it back as quickly as possible to the ILS LOC beam. In the case of the FO training flight, the automation turned the aircraft on to the reciprocal and so was now it was heading back the way they had just come. ATC did give the trainee FO a heading of 100° to fly to intercept but, again, the captain assumed that the automation would be a little more benign in the way it tried to capture from the wrong side. The captain in the third example did consider using HDG Select and then rearming the LOC but noticed that LOC was captured and so assumed the system would sort itself out. Awareness of system performance and capabilities was deficient here. This type of event was not something included in training and the first time you would see it was when it was happening for real.

Three almost identical events in less than 6 weeks show how crews can go from doing nothing wrong to being in the wrong place in the blink of an eye. The way the job is imagined by the designers of procedures is often simple, singular and straightforward. The way the job is enacted by crews is variable and sometimes messy. In all three cases there were small distracting factors: novel techniques were being used and crew were overseeing the work of others. ATC, through the use of vectoring, can influence the task. In a 30-day sample of arrivals onto Runway 07, ATC positioned the aircraft to intercept the LOC at distances of between 7 and 12 miles. Simple geometry means that a 7-mile intercept will be rushed whereas a 12-mile intercept affords ample time. The procedure is designed for a 30° interception angle but, again, vectoring results in a variety of angles up to and beyond 90°. The interception angle places different demands on automation. The sources of variability, then, are more than simply the performance of the crew. The collaborative nature of aviation creates variation. These three examples illustrate buffering. The system was able to cope with a variety of solutions to the task of intercepting the inbound course to the runway. The first two events were similar in terms of the extent to which the crew flew through the LOC but, because of the time of day, the heavier traffic during the FO training flight's arrival meant that margins were much reduced. The volume of space and time available to rectify the situation while maintaining separation from other aircraft was constrained. Time of day and, therefore, traffic density can turn an event from an embarrassment into a hazardous condition.

These examples bridge Levels 1 and 2 in the hierarchical model. All of the individuals involved in these three events privately engage with the world and conduct activity that was additional to the immediate needs of the task. This internal dialogue interfered with the normal conduct of work. In the case of the use of FPA the 'additional activity' on the part of the FO was no more than curiosity: he was interested in seeing how things worked out. The junior captain was distracted by the challenge of fulfilling his new role. In the case of the training flight, the supervising pilot was dealing with managing his intervention without adversely affecting the mental state of the trainee. What we would call 'collaboration' or 'teamwork' often involves the resolution of these internal mental states. These are the 'interfaces' we have to negotiate when acting in teams.

7.6 CONTROL IN WORK GROUPS – MONITORING AS COLLABORATIVE TASK MANAGEMENT

Collaboration on the flight deck requires individuals to reliably fulfil their part of the joint task. The basic framework of activity is provided by procedures but collaboration also involves activity at a metacognitive level in that individuals' need to independently exercise oversight. Crew members coordinate action with others but also validate progress in order to decide whether to intervene or modify activity as required. We call this process 'monitoring'. Failure to 'monitor' is now well-established in the lexicon of accident report writing. The UK Civil Aviation Authority (2013), the Flight Safety Foundation (2014) and IATA (2016) have all produced guidance material on improving PM skills and it has even been suggested that 'training in monitoring' be included in the revised European airline recurrent training guidance (EASA, 2018). Dismukes and Berman (2010) offer a cautionary note: whatever you do to improve monitoring, the process will fail. Monitoring, or routine scanning, was shown to be the least effective method of error detection in Chapter 6. This section positions 'monitoring' as a social process at the heart of collaboration.

Historically, the operation of an aircraft was considered to be the job of the aircraft commander who sat in the left-hand seat. With the advent of two-pilot cockpits on commercial passenger aircraft, the right seat was occupied by an assisting, albeit qualified, pilot. This division of responsibilities is manifested in the layout of early generation cockpits where the left seat had a full array of instrumentation while the right seat pilot typically had a sub-set of displays but also had many of the controls for aircraft systems. The body in the right seat was clearly intended to be in the supporting role.

The change in operating philosophy, such that both pilots took turns to operate the controls, required a change in nomenclature. Various terms were used to differentiate the pilot actively in control of the aircraft from the pilot acting in support. Federal Aviation Administration (USA) (FAA) Advisory Circular 120-71A (FAA, 2003) formally proposed the adoption of PF and PM to distinguish between the two roles. The Circular also introduced an appendix on crew monitoring and cross –checking but addressed the issue through the development of SOPs rather than explaining what monitoring is and how it should be undertaken. A common understanding of the PM (or 'pilot non-flying' (PNF)) role is to 'monitor the flight management and aircraft control actions of the PF and carry out support duties such as communications and checklist reading. The Operations Manual will specify fully the roles for the PF and PM/PNF, but one of the most important aspects of the duties of any PM/PNF is the cross-check of the actions of PF (Skybrary, 2018)'. In order to clarify how pilots view the concept of monitoring, pilots attending an annual crew resource management refresher course were asked to discuss 'monitoring'. Nearly 2000 pilots participated over a 12-month period. Pilots were divided into two groups, captains and first officer/second officers (FO/SOs), and set a number of tasks designed to clarify and elaborate on the concept.

One interesting insight was that 'one size does not fit all', the implication being that monitoring, like flying, is an adaptive skill and needs to be modified to suit the monitoring load represented by the nature of the operation and the skill set of the PF. This is further supported by a cluster of comments relating to differences in experience levels within and between the two groups. For example, it was observed that training captains often lacked the routine familiarity with line operations and some domiciled (based) pilots were unfamiliar with clusters of destinations they rarely operate to. Interestingly, trainers themselves freely admitted that they were aware of a lack of 'sharpness' in their skills and many suggested that they needed breaks from the training programme just to keep their line currency up to standard. This increased the monitoring load on the FO. These observations reinforce other comments about differences in competence and experience as being limiting factors in collaboration.

It was clear from the outset that the groups did not see the PF/PM roles as being mutually exclusive. As PF you are also monitoring the activity level of the PM to ensure that you get the support you need. As PM you are actively involved in the flying task simply because of procedure design and task allocation. This can interfere with monitoring the bigger picture because it is a cause of distraction. With this proviso, the comments of the groups fall into three main clusters: describing what is monitoring, describing the activity associated with monitoring; elaboration of the role of the monitoring pilot.

Turning, first, to monitoring as a concept, both the captains and the FO/SO groups described monitoring in terms of creating and sharing mental models, keeping the big picture, maintaining awareness and mentally flying the aircraft. So, monitoring is underpinned by accurate mental representations of the intended progress of the aircraft. As such, then, effective monitoring is predicated on sound Level 1 behaviours.

However, clear differences emerged between the two groups that reflect the social dynamic. These differences relate to the way in which the two groups differ in terms of deciding on how progress was to be gauged: in effect, who decides if progress is acceptable. When captains are in the monitoring role, the FO is, of course, the flying pilot. This means that the captain has delegated, to a degree, responsibility for the conduct of flight to the FO. Comments by captains included:

- 'Ensure that what you think should happen does happen, knowing what you are trying to achieve'
- 'Ensuring the aircraft is doing what it is supposed to be doing'
- 'Looking for abnormalities; how well is it going?'

These comments from captains reflect the importance of having a clear plan for the progress of the flight. The final comment also captures the idea of sense-making involving pattern matching between the expected and observed states of the world. The FO/SO group, on the other hand, talked in terms of:

- Balancing your expectations with what is happening and comparing to your comfort zone
- Anticipation/expectation; comparing expectations
- Mentally challenging yourself and asking yourself questions

The captain groups' comments are active, they relate to comparing actual progress against the intended plan. The FO/SO group, on the other hand, used passive terms that reflected a need to verify their own understanding of what they are observing. So, although the group have a common understanding of how monitoring functions (by establishing an accurate overview of the task) they differ in the extent to which they have confidence in the accuracy of their world view. This will degrade an FO/SO's ability to undertake monitoring. The workshops did not explore differences between FOs in terms of experience but one conclusion that flows from these comments was that competence was based on knowledge structures and, at some point, pilots shifted from an internal view of the work process ('am I sure I understand what is happening?') to an external view ('is what is supposed to happen actually what I'm seeing?'). This first group of comments, then, is an elaboration of the discussion of sense-making in Chapter 5 but in a more applied context. The differences between captain and FO comments reflect how an experience changes the nature of sense-making and immediately points to a potential for breakdown between the two roles.

The second cluster of comments related to the actual task of 'monitoring': what is going on when you are monitoring. The views of the captains suggest that monitoring is a set of fixed and variable tasks. The significant fixed tasks were:

- Ensure compliance with SOPs
- Confirm performance of tasks
- Task management (prioritise – monitor rather than get distracted by minor tasks; reduce PF overload; knowing what you have to monitor more closely (pay attention to))

In the context of competence, captains stressed the importance of 'having an idea of what to expect before checking', which links to the earlier comment about knowledge structures. Importantly, captains also said that the task involved 'have a comfort zone and know when it has been exceeded'. This comment points to the social dynamic at work. The implication is that impinging on the 'comfort zone' will trigger a change in the dynamic: captains will intervene. This is the variable component of monitoring. The FO/SO groups' view of the task of monitoring offered a much richer description. Initially, they offered a conventional description of the task:

- To carry out non-flying pilot duties
- To support flying in all aspects of the operation
- Backing up and assisting PF

Of interest here is that the comments reflect the subordinate role of 'assisting' whereas the captains' comments were more global in outlook. The next cluster of comments elaborates on the captains' observation about SOP compliance:

- SOP adherence
- Call-out deviations and omissions
- Question non-standard operations/unexpected operations
- Offer suggestions/reminders
- Challenging the PF
- Picking up errors

One syndicate referred to monitoring as 'quality control' and this function was borne out by several comments that reflect the idea of process control:

- Cross-checking and challenging
- Checking configurations
- Verification
- Comparing to plan
- Checking task completion
- Keeping track of automation, level awareness and PF actions

Finally, a repeated theme was the need to actively observe the actions of the PF:

- Watching other guy; watching the PF
- Watching everything that is going on
- Watching/scanning

Of course, there was no evidence that captains did *not* do the things on the FO/SO list when they were in the PM role. In all probability, they undertook these actions but did not see them as noteworthy. In fact, one interesting comment was that monitoring 'happens subconsciously'. This second cluster of comments supports the importance of fundamental procedural knowledge. Collaboration is underpinned by a shared procedural framework which has been sufficiently internalised such that performance is almost automatic. Of course, the challenge is to have mastery of procedures while keeping track of any changes. Maintaining currency is effortful and one reported mechanism for updating procedural knowledge is simply relying on being corrected during recurrent training sessions rather than arriving for training having reviewed the current procedural status. This behaviour suggests that the procedural framework held by some crew is imperfect, drifting towards a generic status until being periodically recalibrated. The gap between the intended framework and that held by individuals might only be small but it is an incremental addition to the broader gap between crew members that reflects differences in competence and experience. I proposed in Chapter 6 that homogenisation was a category of drift at the level of the individual. This discussion of the problem of maintaining procedural currency suggests how homogenisation might start.

These first two categories of comments build on the previous discussions of sense-making and underpinning knowledge. Although both are essential to individual action (Level 1 activity), they create the context for collaboration. The final cluster of comments from the workshops touch on social control of performance. I mentioned earlier that, when the captain is monitoring, the implication is that the task of flying has been delegated to the FO but the social construction of that act of delegation is implicit in the observations by some captains who said that monitoring was:

- Babysitting
- Making sure other pilot is doing what s(he) is supposed to do
- Making sure FO does what (s)he is told
- Ensuring what we expect is what we get

These comments indicate that 'delegation' is an implicitly negotiated transaction. The FO in the PF role has full authority to operate the aircraft, but only up to a point. The view that the FO operates at the discretion of the captain was further revealed by the second cluster of comments:

- Making sure other person's selections are correct
- Ensuring outcome correct
- Object to deviations
- Must stop errors; pointing out errors

The tone of the comments is interventionist. Not only does the captain PM exercise oversight but they do so from the perspective of what they consider to be the 'correct' method of working. On the other hand, the FO/SO groups were quite blunt in their assessment of the monitoring role:

- Captain management (what is (s)he doing, what does (s)he want me to do)
- Being the ultimate diplomat
- Make the captain feel happy
- Saving the captain's bacon

These views echo the comment of the Learjet FO in our opening case study: 'my job is to "protect the ego of the captain" and not to speak up unless I was about to die'. These role-related comments flow from the manner in which roles are interpreted, a reflection of role drift, and the way in which power is exercised in the context of the crew. And, of course, they are not unique to aviation. Rather, it is the cost of being human.

All the groups made frequent reference to the need for a 'shared mental model'. It seems that monitoring is the epitome of collaborative action and to be an effective collaborator requires a clear understanding of:

- What is the task?
- How do we achieve the task goals?
- What is the team doing?
- What is the technology doing?
- What is the other person capable of?

Monitoring, then, occurs at the intersection of fundamental competence, prescribed work, social dynamics and internal dialogues. Rather than consider 'monitoring' to be a metacognitive control process somehow separate from, or additional to, other activity, it seems that it is simply the task of doing work in a team. It suggests that to understanding normal work we need to be cognisant of these multiple dimensions.

The groups were also asked why they felt monitoring fails. There were six key groups of factors that lead to a breakdown. Three were to do with aspects of the task (with some physiological aspects linked to the theme). The other three relate to discretionary behaviour. In the task-related group, familiarity and associated complacency featured strongly in the lists of reasons why monitoring failed and this is borne out by some airlines' LOSA findings that show that home base is often the location for most errors by the crew. Task overload or under-load, leading to boredom and distraction, was the next most frequently cited causal factor. Stress and fatigue, especially, were also included in this group. Finally, operational issues such as unfamiliar airports, procedural changes, language (strong accents or pace of delivery) and the unexpected (e.g. technical failures, go-arounds) were operational factors that cause problems. This last cluster captures the idea of threats and perturbations.

In terms of discretionary behaviour, the crew dynamic (flight deck gradient, captain's style), inconsistent application of SOPs and style of communication were the main causes of a monitoring breakdown. In discussions, examples were given of failures to extend the crew concept to augmented crews, that is, when additional crew members were carried. So, for those phases of flight when three or even four pilots were on the flight deck, there was often little or no attempt to formally include the additional crew in the process of flight management. This can also be seen, for example, in smaller aircraft operations. Some types, such as the King Air or early model Cessna Citation jets, are approved for single-pilot operations but the terms of a commercial contract or an insurance requirement might mandate two-crew operations. I once worked with a helicopter operation that used a single-pilot aircraft but required an additional pilot for operations beyond a certain range offshore. No attempt was made to adapt procedures to accommodate the additional pilot. Where there were no clear multi-crew procedures, accidents can, and have, occurred.

I suggest that an emergent property of Level 2 in my hierarchy is efficiency. Effective collaboration will deliver an efficient performance. Sense-making and competence are building blocks of collaboration and will support efficiency but, clearly, the social dynamic is the third pillar.

7.7 WITHIN-GROUP SOCIAL DYNAMICS

Joining a group requires an individual to subordinate their own self-interest in order to benefit the group. It is this sacrifice that supports the efficient achievement of the task goal. Self-interest is more than simply personal gain or material well-being. It reflects demands for recognition, esteem and other emotional enhancement. For example, I once ran some workshops for UK Police Helicopter units. Typically, only one or two pilots were on the unit payroll and leave or sickness was covered by a contracted pilot from an agency, often recently retired from the military. Police helicopters are fitted with a range of sensors and aids for supporting ground operations. A crewman sitting in one of the front seats directs the sensors. The first problem often encountered with a short-term contract pilot was that they were used to being the 'captain' and sitting in the left-hand seat, as is the convention in almost all multi-seat aircraft. However, depending on the equipment installation, the 'pilot' might sit on the right. This could bruise some egos. Furthermore, when arriving on task, the pilot might prefer to arrive with the target area on their side of the aircraft so that they can evaluate the approach in case of hazards. Unfortunately, the helicopter needed to be positioned so that the crewman can assess the scene and begin to deploy the sensors. Again, a source of dissatisfaction for some. My examples might seem quite trivial. A case of rampant ego, maybe. But it illustrates that, for some, subordinating ego and prior experience to better meet the needs of the team and the task can be emotionally difficult. The issue here is whether these factors lead to disruptive behaviour. In critical teams, even an adverse comment can undermine confidence in peers.

Teams are convened to achieve goals that cannot be accomplished reliably by individuals. The task, by default, shapes the performance of the team. The discussion of violations in the last chapter showed how teams adapt to circumstances and modify tasks. Fatigue-related performance impairment or a lack of competence might require individuals to compensate for others. All of these factors will shape the performance of the team. Compensating for others requires individuals to be willing to invest effort. The police helicopter example illustrates how team members have specific tasks assigned to them, have a nominated position within the team but they also play a role in the group. So, the helicopter pilot is the nominated captain of the aircraft, their task is to fly the aircraft but their role is to act in support of the other crew at the appropriate time. A role is more than just the specific job an individual has been assigned in a group. A role captures 'all the other stuff you have to do'. The nature of roles creates two problems: role conflict and role ambiguity. Role conflict reflects the fact that, sometimes, there might be too much to do at a specific moment or that actions might seem contradictory. Therefore, something has to be shed or traded off. Role conflict might be actual or simply perceived. So, changes in procedures might increase workload at specific phases of flight but, if we go back to the discussion of stress, my perception of demand might create tension but the actual demand is within my capacity. Role ambiguity reflects the fact that, in the absence of clear guidance, individuals create a version of what their role demands, which might differ from what others expect. In particular, role ambiguity can give rise to different understandings of an individual's freedom of action and the extent of their authority. The ambiguity that surrounds role construction supports drift at the individual level.

Opportunities for idiosyncratic role construction also flow from an individual's status. Status can be described as group stratification in response to evolutionary selection pressure. High-status individuals have access to rewards (food, reproduction) denied to lower status group members and high status usually grants greater latitude in terms of conduct. Within formally constituted teams, an appointed position within the immediate (the crew) or extended (supporting agencies) group is a source of status. The legitimate power associated with status is usually derived from job descriptions, terms of reference or, in some cases, even a legal mandate. The effective exercise of power, and use of authority, is associated with accepting responsibility and being held accountable. Role ambiguity, in particular, can lead to modes of behaviour that undermine group cohesion.

The social dynamic that flows from the existence of roles and status in the group is, in part, engineered (the organisation formally appoints individuals to positions and defines responsibilities) but also negotiated (individuals are tolerated, or compensated for, until behaviour exceeds a threshold). These forces, although ephemeral, are tangible. One way we try to constrain variability in group behaviour is to appoint a gatekeeper, or 'leader'.

7.8 LEADERSHIP – A SPECIAL CASE IN SELF-DIRECTED TEAMS?

The formal appointment of a leader is an attempt to exert social control over the team process and ensure that performance meets organisational goals. Although the captain of an aircraft is considered the 'leader', the highly proceduralised nature of the aviation task coupled with high levels of training and standardisation mean that, under normal circumstances, work flows with little need for direction. The job of the captain is not so much to give direction but to resolve issues arising from normal perturbations. In aviation, captains act as the company agent with delegated authority to take certain action once the crew have left the main base. Here, again, we see opportunities for both role conflict and ambiguity in that making a decision in the best interests of the company might not necessarily be in the best interests of the crew at that moment.

The nature of leadership in aviation is apparent in this next case study involving the crew of a B-777 en-route to the USA. The captain and FO were at the controls with the RFO and SO taking their rest in the crew bunks. The first sign of a possible problem was when the captain detected a smell of burning. Burning smells can come and go on aircraft and are commonly associated with the galleys and food preparation, so the captain turned on the door camera to see what was happening in the forward galley, immediately behind the flight deck. A check was also made of the list of dangerous goods on the cargo manifest to see if there was anything flammable in the forward hold. Nothing untoward was found.

The crew had already updated the navigation information and the nearest en-route alternate airfield was a small military runway at Shemya, in the Aleutian Islands. The captain altered heading, wanting to at least be pointing in a sensible direction, and put out a precautionary PAN call. The status could always be downgraded if there proved to be no problem and the track alteration did not add significantly to the flight time. At this point the captain donned an oxygen mask. The FO, meantime, was unable to detect any smell but, seeing the captain's concern and subsequent donning of a mask, he put his on, too. Full-face masks are clumsy and many pilots who wear glasses find that they cannot get the mask to fit properly, as was the case here with the captain. With no glasses, the captain admitted to having difficulty reading the instruments properly. The resting crew were called back to the flight deck and the cabin manager, together with a flight purser, were asked to check for smoke.

When the relief crew returned they were surprised to find the operating crew wearing oxygen masks. They queried the reason given that there were no discernible fumes or smoke. The masks were then removed and the new crew were briefed on the situation. The airport charts for Shemya were reviewed and the weather was updated. The diversion airfield had an automated weather reporting facility and, at every update so far, the conditions were below minima. On the latest update, they were above minima. This change in information created an additional stress in that, given the consistent trend during all the previous updates, how much confidence could be placed on this one report? And what if the weather deteriorated again? In order to land at the alternate destination the aircraft would have to dump fuel, and once that was completed the crew would be committed. There was no second chance. The various tasks of checking approach charts and weather, communicating with the company operations centre needed the effort of all four pilots and one later said that it would have been difficult if there had just been two pilots on the flight deck. While all this work was going on, the smell returned and, so, it was agreed that a diversion was definitely needed. The RFO and the SO were busy with the work associated with the diversion. The FO was working with the data link and ATC. The captain had decided to delay the fuel dumping until they started the descent so as to not commit too early. They planned to land on the least favourable runway because, despite a slight tailwind, it had an ILS: an example of trading off relative benefits. The captain commented that it was difficult to create the space to think. It was important to review the port page information, to understand what to expect on the ground and to make sure that nothing was missed. In the midst of all of this, one of the crew noticed a status message indicating that the Equipment Bay cooling fan had failed. The caption had been missed by all and the RFO said that he had to tap the captain on the shoulder to draw attention to the message. The captain expressed relief: they now knew what the problem was. The FO said that he was relieved because the problem was now over. Neither of these two positions was technically correct. Interference between systems can trigger unrelated warnings and so the caption might not necessarily have been linked to the actual problem. Captions do not necessarily convey information about ongoing status and illumination is simply a means to attract attention. The RFO then proposed going into the Equipment Bay to check for himself, an action proscribed in procedures. He returned, stating that the problem had, indeed, been resolved with the failure of the cooling fan. A safe landing was made at Shemya and flight was resumed a few hours later.

This episode says much about the reality of leadership and collaboration. Despite the subsequent 'disappearance' of the smell, the need to divert assumed an importance in the mind of the captain because, if it did return, they 'might not be able to control the situation'. The initial solution developed a momentum. After the RFO's return from the Equipment Bay, the captain said that 'I couldn't clearly see him – the FO – (because of the mask) but he seemed to be suggesting that there was no need, now, to divert. It was his tone of voice'. But once fuel dumping had taken place there was no alternative. Crew commented that they 'wanted to slow things down. Create breathing space'. Thought process 'kept getting interrupted, trying to remember what you are supposed to do'. It was hard to focus on just one task. The need to communicate with multiple off-board agencies – operations, ATC – with variable quality communications links meant that the task was simplified: 'the Operations Centre was a distraction so we dumped them and dealt just with ATC'. 'People did stuff without being told, just jumped in and got on with it'.

The nature of the situation was understood differently by different team members. Work was undertaken not in a coherent, structured manner but in an ad hoc sense. Tasks were taken on by whoever was to hand. The process was kept safe by activity achieving a critical mass. Nothing was missed. Individuals created space to think and activity that was not deemed to be adding value – speaking to the main base – was abandoned. A collaborative solution was created in real time. The role of the captain was to interpret the current status of the task and create a narrative that guided the action of the team. The captain made decisions about key aspects of the situation – whether to divert, when to dump fuel, which runway to use – but the work of the team was a sequence of spontaneous interventions directed at preparing the aircraft for the goal of landing at an alternate airfield.

It is an implicit assumption that aircraft captains exercise leadership. Leadership is a formal embodiment of power in groups. Despite a high degree of self-direction, aviation teams still occasionally need an intervention by an individual and, of course, in an emergency effective leadership will be essential. An important function of a leader is to make sense of a situation such that the team can establish the best course of action. Leaders, then, create a version of the world that can then be used to direct future action. Figure 7.3 proposes a model of leadership in aviation and attempts to capture the implications of work in self-organised, self-directed teams. The model starts with some external event that triggers a need for an intervention that we might call 'leadership'. Triggers would include a situation that becomes highly ambiguous or uncertain; an emerging need for coordination; competition or contradictions between task elements or goal and, finally, danger. The next step calls for the discrepancy – the disconnect between the planned goal and the current state – to trigger behaviours aimed at clarifying the problem. So, at this stage we are problem-solving.

The next step is to identify actions that satisfice. Satisficers are those criteria that, once met, remove the need for a 'leadership' intervention. Again, this idea links to the earlier discussion about sense-making. It reflects the fact that the task, at this

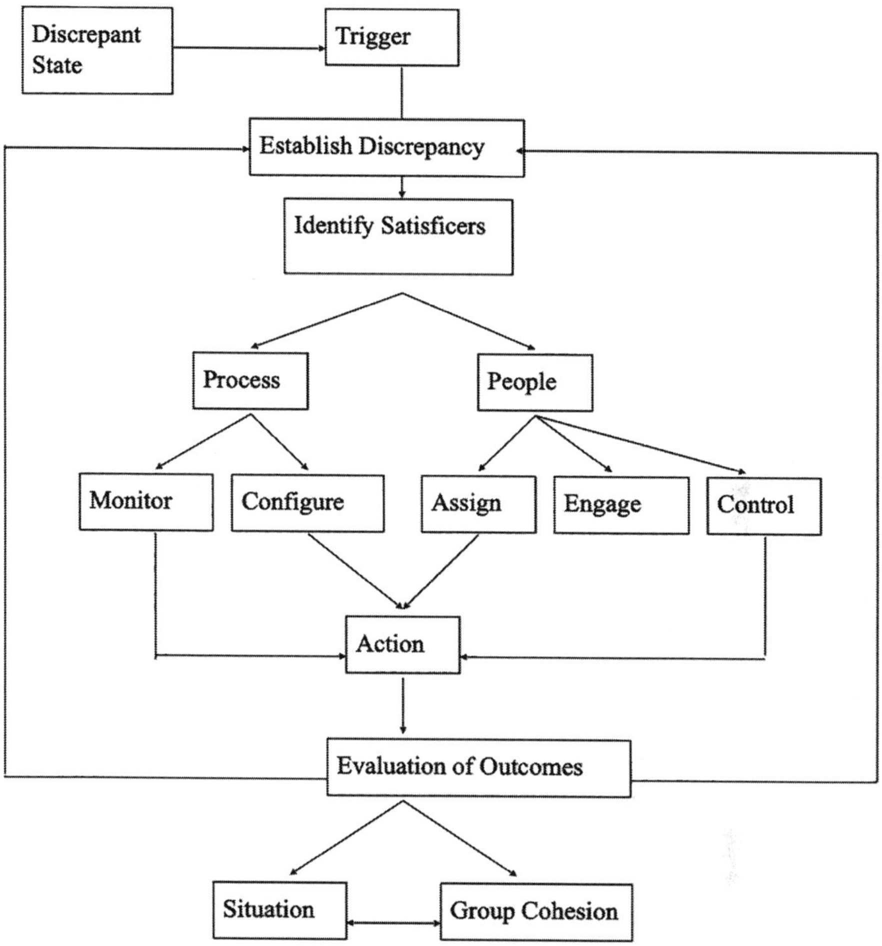

FIGURE 7.3 A model of leadership dynamics.

stage, is to bring stability to both the process and the team in order to return the situation to a steady state. Monitoring and configuring are actions that deal with the task and technology. Assigning and controlling relate to the team. Assigning work involves directing and delegating while controlling involves checking against the plan. In a study of captaincy skills, Senko (2010) identified mentoring as an important function. Mentoring is a form of transaction where one team member invests in the development of another, but can also be seen as a form of control. Where the leader choses to mentor rather than control through the exercise of authority, their social power will be enhanced. Engagement recognises that individuals have a responsibility to fulfil their own roles in the team but the leader needs to promote engagement. Having acted, the leader needs to consider the outcomes and assess if further action is required. The leader also needs to consider the implications for the group, its identity and cohesion.

In the case of the Shemya crew, the initial interpretation of the problem was framed by prior knowledge. Aircraft have been lost because of crew incapacitation resulting from smoke and fumes and crew now have protective equipment for dealing with such situations. The initial response was driven by a desire to reduce risk by activating sources of support (the PAN call) and manoeuvring closer to a suitable runway. Once the full team reconvened, the activity was driven by a need to communicate the situation and build an understanding of how the potential plan would unfold. Once the decision to commit to the diversion was made events built a momentum of their own, even when the scale of the problem was better understood. Leadership, in this case, can be seen, first in terms of initiating action. But, second, the captain was also considering future risk. The captain of the aircraft bears a responsibility not shared by the rest of the crew and this can cause a disconnect between the leader's and the team's view about best courses of action.

7.9 CULTURE

Once final, intangible, dynamic that shapes group behaviours is something we label 'culture' and which is probably best characterised as the group's shared identity. Culture, broadly speaking, can be categorised as being a top-down or bottom-up property of a group. Top-down models are exemplified by the work of Hofstede (1993), who looked at managers across international divisions of a single company. His analysis identified a number of dimensions such as power distance, individualism, masculinity, uncertainty avoidance and long-term orientation. The work of Hofstede is highly problematic. The fact that the original study was on managers within a single, albeit large, organisation suggests that the sample might not be representative. If you looked at the questionnaire data, whereas means between countries might differ, the range of scores showed considerable overlap. Hofstede's work was, nonetheless, the first to be widely adopted and underpinned early attempts to consider culture in aviation (Helmreich & Merritt, 2001). Helmreich identified three 'cultures': national, professional and organisational.

An opposing view is that culture is a bottom-up process of group identity creation. A study of safety culture in Japanese rail track maintenance teams (Itoh et al., 2004) demonstrated that there were as many 'cultures' as there were teams. In Table 7.1 I present some data for an airline's command upgrade course pass rate by nationality. On the

TABLE 7.1
Command Course Pass Rates by Nationality

Nationality	% of Pilot Population	% Failure Rate	% of Trainers
A	26.16	12.8	37.7
B	16.87	10.7	29
C	15.22	19.4	12
D	11.49	60.8	5
E	7	33.3	1.2

face of it, the data might suggest either significant differences in ability between pilots of different nationalities or, possibly, discrimination in the workplace. However, significantly, there was no published graduation standard for the course. In the absence of an objective output standard, a 'norm' had been created that reflected the background of the dominant group, which could be described as 'Male, British, Commonwealth and ex-Military was a bonus'. The 'norm' was socially transmitted through trainers having themselves been trainees. The oversight mechanism – management review – was the prerogative of senior members of the same culture who were also trainers. To be successful on the command upgrade, individuals not only had to demonstrate proficiency, they also had to conform to the prevailing 'type'. Social 'norms' represent emergent culture within groups. In short, culture is the tacit set of rules that allow the group to cohere, the rules of group membership. Cultures are transmitted through assimilation and are reinforced through endorsement and preferment. Unfortunately, 'cultures' also allow risky practices to emerge and be condoned or, at least, ignored.

7.10 CONFLICT RESOLUTION

Given that collaboration has ample opportunity to create tension between group members there needs to be a process for resolving conflict within a group. Sources of conflict can be overt, usually a disagreement over a course of action, or covert. Covert conflict might flow from a snub or insult.

Unfortunately, status and power can militate against effective conflict resolution simply because the position in the group affords control over the agenda. Lack of self-awareness, equally, can inhibit the emergence of a need for intra-group conflict management. One well-established concept of interpersonal relationships is that of 'assertiveness'. Rooted in the concepts of civility and personal rights, assertiveness proposes that both parties have a position worthy of recognition. An assertive position sits between 'aggression' and 'non-assertion'. An aggressive position is one-sided: 'my views are more important than yours'. A non-assertive position is submissive: 'do what you want, I just want a quiet life'. The assertiveness paradigm points to concepts already discussed: self-awareness and metacognition.

The significance of conflict in terms of group cohesion is underlined by recent research into workplace incivility (Porath et al., 2015). Incivility describes aggressive or abusive acts that demean the recipient. Often committed between seniors and subordinates in hierarchies they comprise a range of workplace behaviours from inappropriate jokes, public acts of humiliation, threats and intimidation. Research has shown that not only does incivility impair the performance of the recipients, it can also affect witnesses to those acts. The behaviour of the Learjet captain we saw at the opening of this chapter represents classic incivility.

7.11 BEHAVIOUR BETWEEN GROUPS

The theme of this chapter has been within-group collaboration but work also occurs across organisational boundaries. We have seen examples of this with the ATC controller supporting the Little Rock crew, the refuellers working with the ATR-42 captain and the challenges of working with different agencies encountered by the B-777 crew diverting to Shemya. We can identify four dimensions of inter-group collaboration:

Domain agreement
Solution agreement
Positive evaluation
Work coordination

Domain agreement relates to ownership of the parts of the task. In effect, who is responsible for what. Role ambiguity can sometimes inhibit domain agreement. For example, where one actor believes that their requirements should take priority agreement can be difficult. Solution agreement requires that all parties have a clear understanding of how work will proceed. This creates a requirement for a formal approach to planning. Positive evaluation requires parties to endorse and encourage one another. I once did an analysis of ramp safety events over a 3 month period. The two main parties involved were the cargo representatives and the ramp management staff. Approximately 50% of reports started with the comment ‘it was their fault’, meaning the other party, or ‘they admitted it was their fault’. This would not count as a positive evaluation. Finally, and hardly surprisingly, work needs to be coordinated. This usually involves active tracking of progress. Active support of other teams is facilitated by consideration of others’ needs, appreciating requirements for information, communicating in a timely manner and in a format that is understood by the other team.

7.12 CONCLUSION

This chapter has explored how work is undertaken in the public space between crew members. It has mapped out some of the formal, organisational structures that shape teams and looked at the social dynamics of humans at work. A number of themes have emerged. First, there is an overlap between processes at the individual level and at the group level but these processes are qualitatively different at Level 2. What is new at Level 2 is the emergence of a social dynamic that shapes performance. We have also seen that system-level performance involves both processes within and between groups. This chapter has looked at structures and forces that shape behaviour but collaboration is fundamentally powered by communication. That is the topic of the next chapter.

REFERENCES

CAA. (2013). Monitoring matters. Guidance on the development of pilot monitoring skills. CAA Paper 2013/02.

Dismukes, R.K. & Berman, B. (2010). *Checklists and Monitoring in the Cockpit: Why Crucial Defences Sometimes Fail.* Moffat Field, CA: NASA Ames.

EASA. (2018). Notice of Proposed Amendment 2018-07(B). Update of ORO.FC — evidence-based training subtask RMT.0599.

FAA Advisory Circular 120-71A dated 27 February 2003.

Flight Safety Foundation. (2014). A practical guide for improving flight path monitoring. Final Report of The Active Pilot Monitoring Working Group. Alexandria, VA.

Hofstede, G. (1993). *Cultures and Organisations: Software of the Mind*. New York: McGraw-Hill.

Helmreich, R.L. & Merritt, A.C. (2001). *Culture at Work in Aviation and Medicine National, Organisational and Professional Influences*. Aldershot: Ashgate.

IATA. (2016). Guidance Material for improving flight crew monitoring. Montreal.

Itoh, K., Andersen, H.B., & Seki, M. (2004). Track maintenance train operators? Attitudes to job, organisation and management, and their correlation with accident/incident rate. *Cognition Technology and Work*, 6(2), 63–78.

NTSB. (2001). Report No: AAR-01-02: Runway overrun during landing, American Airlines Flight 1420, McDonnell Douglas MD-82, N215AA, Little Rock, Arkansas, 1 June 1999.

Porath, C.L., Foulk, T., & Erez, A. (2015). How incivility hijacks performance: It robs cognitive resources, increases dysfunctional behaviour, and infects team dynamics and functioning. *Organisational Dynamics*, 44(4), 258–265.

Senko, K.A. (2010). Qualitative Phenomenological Study of Leadership Perspectives in Commercial Airlines. Unpublished PhD thesis. University of Phoenix.

Skybrary. (2018). Pilot Flying (PF) and Pilot Monitoring (PM). www.skybrary.aero

8 Communication

8.1 INTRODUCTION

The hierarchical model I have outlined functions through communication and feedback loops. Communication includes all the signals present in the operational world that relate to the status of that world. At Level 1, communication is represented primarily by inputs made to devices and other actions while environmental responses received by the individual represented feedback. At Level 2, communication comprises verbal and non-verbal acts that facilitate the building of working relationships and coordination of activity, both within and between collaborative groups. At higher levels, actors communicate mainly through documents and other artefacts. Communication is the driving force of the system. This chapter builds on some of the concepts already developed around sense-making and task management and looks at how these things get done. We will primarily look at the structure and function of communication in an operational context. For convenience, in addition to communication between individuals and teams at the level of collaboration (Level 2), I will also look at the implications of managerial communication as forms of control.

I start with a brief review of the evolutionary evidence for the development of communication. A consistent theme of this book is that we are the product of our inheritance and the echoes of this development are still present in how we communicate today. I take a functional view of communication in that the ability evolved to meet a need. We will look at some examples of communication both within a crew and also between an aircraft and air traffic control (ATC). We will also look at aircraft checklists as tools of communication. What I hope emerges from the discussion is the fact that communication is a remarkably plastic, flexible process with communicators patching and repairing incomplete signals and, yet, still managing to arrive at the required outcome.

8.2 THE EVOLUTION OF COMMUNICATION

It is now widely understood that a range of different species is capable of quite complex communication. Bees can share information about the location of food sources. The distress calls of young chimpanzees can differentiate between different predators. Dolphins coordinate their behaviour when attacking schools of fish. In hominids, it seems that the structures necessary for speech evolved around half a million years ago. Being capable of speech, though, does not mean that language was actually used; it just means that the structural adaptations necessary for forming words were present. In all probability, our ancestors were using gestures long before they started speaking to one another. You only have to watch someone on their mobile phone to appreciate just how deep rooted gestures are in our communication toolkit and we have all witnessed how, when struggling for a specific word, the hands come into play to try to get meaning across.

From an evolutionary perspective, we need to ask why language evolved. It makes no sense to invest in the anatomical development of bones, muscles, nerves and brain capacity unless it serves some purpose in terms of survival. There are two main candidates for why we communicate: to exchange information and to promote social bonding. As Dunbar (2014) observed, there is no compelling reason to prefer either option, but the evidence suggests that language developed in a social context. We talk to facilitate living in groups. Experiments that look at recall of information from stories have shown that social content is more memorable than technical – factual – story elements. We are inveterate gossips. Another finding that has a parallel for our discussion is that communication about scarce resources is greater between kin than between non-related individuals. In a study of radio transmissions by lobster fishermen off the coast of Maine (Palmer, 1991), in situations where there were many strangers on the frequency, fishermen were less likely to broadcast the locations of good fishing compared with when it was just fishermen from the same community. The case can be made that communication for social bonding preceded the transmission of facts. I do not want to draw too many conclusions from this brief discussion other than to suggest that we underestimate the evolutionary nature of communication and the implications at our peril. Ironically, while the evidence suggests that communication evolved in a social context, in aviation, we use constrained, codified language to achieve a primarily technical function. It seems, then, that we have almost designed communication to fail. I start this chapter with a case study that illustrates the nature of that fragility.

8.3 SAN JUAN, PUERTO RICO

On 9 May 2004, an ATR 72 bounced twice and then crashed on landing at San Juan, Puerto Rico (NTSB, 2005). At 1849:46, the cockpit voice recorder (CVR) captured the sound of a thump as the aircraft landed heavily followed by one of the pilots whispering the single word – 'God'. The aircraft had struck the runway 1600 ft beyond the runway threshold after a challenging approach flown by the first officer (FO). The aircraft skipped to a height of 4 ft. At this point, the captain seems to have taken control. Six seconds later, there was a second, very loud, bang. The aircraft touched down for the second time 2200 ft beyond the threshold, bounced, and climbed to 37 ft with a nose-up attitude of 9°. The engine torque increased from 10% to 43%. The aircraft then pitched down to −3°, and the torque was reduced to 20%. The aircraft pitched further down to −10° and struck the runway about 3300 ft beyond the threshold with a bank angle of 7° left-wing down. A vertical load of 5 g was recorded. The aircraft bounced to an altitude of 24 ft before, 5 seconds later, making its final contact with the runway 4000 ft beyond the threshold. The aircraft ran off the side of the runway. Both engine mountings had broken and, among other damage, the captain's seat had broken free from the flight deck floor.

The crew had departed from Mayagüez, Puerto Rico, climbing to a cruising altitude of 7000 ft for the short 25-minute sector. The crew were handed over to San Juan approach controller about 15 minutes out from the airfield. They were told to expect an instrument landing system (ILS) to Runway 10, the anticipated approach and the one that was briefed before departure, and to descend to 6000 ft. They were further cleared down to 5000 ft and then given vectors to the ILS. At this stage, the aircraft was directed around bad weather. The crew were instructed to reduce speed to 210 kts and, about 8 minutes from touchdown, were told to reduce speed further to 180 kts and to intercept the localiser. Just before handover to tower, the crew were informed of a Boeing 727 that was 4 miles ahead of them and warned of possible wake turbulence. The crew were then told to reduce speed to 160 kts and to contact the tower controller.

At this point, the captain instructed the FO to reduce speed further, to 120 knots, to maintain separation between themselves and the Boeing 727. He described the effects of being caught in wake turbulence in an ATR 72. The FO questioned the fact that this was contrary to the ATC instruction. The captain's next radio call was to tower during which he requested a change to the visual approach to Runway 08. With 3 minutes of the flight remaining this was approved by ATC. At this point, the aircraft was approaching 1000 ft and was flying too fast. The FO suggested slowing down, but the captain commented that there was an aircraft behind them and they had already been going slower than the other traffic, having arbitrarily reduced the speed from the instructed 160 knots. The FO confirmed his approach profile using the VASIs but, at this stage, the configuration of the aircraft, its speed, power setting and attitude became confused. The aircraft drifted low on the glideslope, and the FO pulled the nose up to compensate with the effect that the aircraft ballooned and floated for some distance before making its first impact with the runway.

The bare bones of the event would indicate poor aircraft handling. However, the short duration of the flight means that all of the conversations between the two pilots, from just after starting the engines on the ground at Mayagüez to the moment when power is shut off after the crash, was captured on the CVR. It is very rare to have access to the crew's conversation for almost an entire flight, and this is the evidence we will use in our examination of communication.

8.4 WHAT IS 'COMMUNICATION'?

Traditional models of communication portray a closed-loop process. On one side, we find the sender of a signal or transmitter. The sender encodes a message and transmits using a variety of communication media. The media available include written and spoken words, formal hand signals, the whole gamut of non-verbal signals and so on. Communication often comprises assemblages of media combining to transmit the message. On the other side, there is the receiver, the intended target for the communication. The receiver has to receive, decode and understand the message for communication to occur. The loop is closed when the receiver provides feedback to the sender to establish that communication has been successful. Failure can occur at each stage of the encode–decode cycle. Unfortunately, the apparent simplicity of the closed-loop model does not necessarily fully capture what happens in the real world. According to one view, 'communication is one of those sprawling, complex topics that has many facets and many uses because it is fundamental to all social processes' (Kanki & Palmer, 1993). In fact, so pervasive is the topic that, when developing a set of behavioural markers to assess crew resource management, one group of researchers decided not to include a specific marker for communication. They argued that communication underpinned other observable crew behaviours and general performance was predicated on effective communication (Flin & Martin, 1998). While these two positions may reflect the fuzziness of the topic, it is generally accepted that communication has identifiable properties and structure, fulfils specific functions in social settings and is exercised with varying degrees of effectiveness. These are the aspects that we will explore.

8.5 COMMUNICATION IN AN AVIATION CONTEXT – WHAT IT DOES?

In an analysis of CVR transcripts (Gougen & Linde, 1983), the authors identified four basic speech act types: requests, reports, declarations and acknowledgements. Requests are all those speech acts that require the addressee to perform some actions. Reports are indications of the state of the world. They can be further subdivided into supporting statements, challenges and expressions of belief or feelings. Declarations are specific statements that bring about the correspondence between the propositional content of the communication and reality. Acknowledgements are reciprocating responses from the addressee. In the excerpt (Excerpt 8.1) from the San Juan transcript, I have used this structure to analyse an interaction between the captain and the FO. The captain is questioning the FO's technique, pointing out the difference between the two models of aircraft – ATR 42 and ATR 72 – in terms of power

TABLE 8.1
Bales' Small Group Analysis Categories

Socio-emotional Factors in Interaction: Positive
- Showing solidarity: raise others' status, gives help, rewards
- Showing tension release: jokes, laughs, shows satisfaction
- Agreeing: shows passive acceptance, understands, concurs, complies

Task Factors: Information
- Giving suggestions, direction, implying autonomy for others
- Giving opinions, evaluation, analysis, clarifies, confirms
- Giving orientation, information, repeats, clarifies, confirms

Task Factors: Questions
- Asking for orientation, information, repetition, confirmation
- Asking for opinion, evaluation, analysis, expression of feelings
- Asking for suggestion, direction, possible ways of action

Socio-emotional Factors: Negative
- Disagreeing, showing passive rejection, formality, withholding help
- Showing tension, asking for help, withdrawing out of field
- Showing antagonism, deflating others' status, defending or asserting
- Self-oriented behaviours

Source: Bales (1950).

setting. The FO had done his initial training on the ATR 42 and, having only recently joined the company, had been given a short course to introduce the differences on the ATR 72.

Excerpt 8.1

C:	The 72, it stays in the notch.	[Declaration]
FO:	Oh that's right, sorry.	[Acknowledgement]
C:	Yeah. You're familiar with the notch, right?	[Request]
FO:	Yeah, no yeah, I'm just-	[Report (Supporting)]
C:	Yeah.	[Request]
FO:	Used to having- my last captain....	[Report (Supporting)]
C:	What? he want you to have-	[Report (Challenge)]
FO:	Nuh. whu- he's like –	I'm more used to it
	when we were hand flying it, I was having you know like kinda doin a lot of this with the with the power, when I was when I was flying on final and approaches,	[Report (Supporting)]
	so I just naturally just had my hand here, cause (when) we're comin' back in and we're getting ready to s- slow down so I just had, I always just had it resting here... just so to remind me.	[Report (Supporting)]
	Its, its fine.	[Report (Expression)]

The first thing to notice is the messy, incomplete nature of communication. Few sentences were finished and meaning is often implicit. Bales (1950) identified two clusters of communication acts that distinguish between events that are associated with task activity and those related to the socio-emotional state of the group. The task cluster was further subdivided by the direction of the intended action, while the socio-emotional acts were identified as either positive or negative.

Bales' framework (Table 8.1) was an attempt to provide a generic template for exploring small group interactions in a social science research setting. Kanki and Palmer (1993) looked specifically at an aviation context and identified five main communication functions:

- Providing information;
- Establishing interpersonal relationships;
- Establishing predictable behaviour patterns;
- Maintaining attention to task and monitoring;
- Providing a tool for management (implementing plans, solving problems, making decisions).

There are some points of commonality between Bales' framework and the functions listed here. Orasanu (1994) proposed sharing information, directing actions and reflecting thoughts as the key aspects of pilot communication. There seems to be little to differentiate between these various taxonomies of communication acts with task direction and relationship building emerging as key themes. Kanki and Palmer include both specific functions, such as 'providing information', and some desired outcomes, such as 'establishing predictable behaviour'. Similarly, Orasanu's categories overlap, such as with 'sharing' and 'directing'. Her final category, reflecting, suggests behaviour aimed at sense-making. Interestingly, these various attempts to describe communication events all reflect the social/informational evolutionary dichotomy discussed earlier. From a systems perspective, communication serves to clarify goal-directed activity by affording a means to bring some structure to the task and also to manage relations across interfaces. Next, I want to explore how communication works.

8.6 HOW SPEECH WORKS?

Gougen and Linde (1983) used concepts from linguistics in order to identify ineffective speech patterns associated with crews involved in aviation accidents. They describe three main aspects of speech: first, communication has propositional content, that is to say, an utterance contains information. Second, an utterance can be characterised by its linguistic form. The linguistic form is the particular verbal formulation used to deliver the utterance. Third, an utterance has social force, which is the intended outcome of the utterance. In Excerpt 8.2, the captain states that 'you'll pro[bab]ly [...] circle to eight'. The propositional content of this statement, clear in hindsight, is that the captain expected to land on Runway 08. From the subsequent outcome, we know that the social force of the speech act was inadequate: the FO did not understand that the filed flight plan would be altered at some stage. The linguistic form of the speech act can be characterised as indirect; in effect, it took the form of advice and guidance rather than a formal instruction. If the linguistic form and the social force are not aligned, then communication will fail.

The authors discuss aggravation and mitigation as two major modifiers of the linguistic form of a speech act. Both modifiers can be seen as being related to the concept of politeness or 'face'. Face can be thought of as the public self-image of an individual and links to the concept of self-esteem discussed in Chapter 4. Face-threatening acts limit actors' perceived freedom of action: they impose upon or inhibit individuals in a similar way to incivility. The opposite, polite acts, bring the goals of two actors into alignment and sustain a consistent self-image of the individual concerned. In effect, communication is enhanced when the self-esteem of the other is taken into account. Aggravated speech is often of a direct nature, which, in turn, can often seem impolite. Studies by Fischer and Orasanu (1999) and Sexton and Helmreich (2000) have found that direct orders or commands and the use of the first person ('I want…', 'give me…') are perceived to be less-effective forms of communication. Mitigated speech tends to be of an indirect nature and is considered more polite. It tends to be phrased in the first-person plural ('I think we should…'). It works by making an implicit appeal to the sense of a shared crew responsibility for task accomplishment. Mitigated speech often comprises suggestions about what the crew should do next. Gougen and Linde observed that the operational context affected communication. They examined two situations, crew-recognised emergencies and crew-recognised problems, and in both, mitigation was reduced as the need to consider causing offence was less relevant.

Nevile and Walker (2005) noted two characteristics of poor flight deck communication: overlap and pauses. In aviation, the particular constraints of using radiotelephony have resulted in protocols and standard phraseology for communication. So, one speaker cannot start until the previous speaker has finished broadcasting. Human communication has some similar characteristics in that people are quite good at detecting when one speaker's turn is about to end and the second speaker can begin. Overlap occurs when one speaker starts before the previous speaker has finished, and often before important information has been conveyed. Pauses, in this context, refer to a failure of the second speaker to respond when a response would normally be expected. The authors also identified acts of repair, in which one speaker corrected the performance of the other, even where no intervention was necessary. Excerpt 8.2 illustrates overlapping. Note how the FO never finishes a sentence:

Excerpt 8.2

FO: Alright, sorry, so I...again we...brief whatever approach they gave us, and then expect the visual?..so...so you're sayin'
C: Ninety nine percent of the time approaches you do in San Juan are visual approaches
FO: Ok so you...
C: They give you an ILS...
C: They give you an ILS to get you inbound, an' then they prolly gonna ask you 'you got the airport runway eight in sight?' 'roger', 'clear Lagoon Visual Eight'. You're gonna circle to eight.
C: That's the most common approach they do here
FO: Ok
C: The only time they do approaches here, to minimums, is when it's raining.

The effectiveness of interpersonal communication is reflected in the degree of congruence between the speaker's intentions and the addressee's response. Consider this next excerpt (Excerpt 8.3) from the San Juan CVR transcript. It represents the formal briefing for the approach, conducted with 12 minutes to run.

Excerpt 8.3

C: Approach briefing
FO: Uhhh, we're doin' the ILS ten
C: Yup

This excerpt comprises the totality of the technical content of the brief. The use of structured communication protocols is intended to bring a degree of reliability to the process, and the approach brief is a recognised component of the preparation for arrival. However, the quantity of information in an exchange is not necessarily important if both actors possess a common understanding of the task (Hutchins, 1995). However, as we have seen, this was not true of our crew on the day in question. The excerpt is an example of how repetitive tasks easily become abbreviated with the result that the risk of a communication failure increases. A form of consolidation drift, crews undertaking multisector days are known to resort to the use of 'standard briefing' as a shorthand for the required brief. This use of linguistic analysis has allowed us to examine what gives communication, its force and allows it to achieve a goal. The process is disrupted by mismatches between communication acts or inefficient structure.

8.7 A FUNCTIONAL MODEL OF COMMUNICATION

Rasmussen and Svedung (2000) identified communication acts that support performance within their skill–rule–knowledge (SRK) hierarchy (column (a) in Table 8.2). Their framework includes communication at both Levels 1 and 2 in my hierarchy. For example, the movement of an index or marker on a navigation display will trigger a response at the skill level in order to keep the aircraft on the correct track. The failure of an undercarriage indicator lamp to turn green will trigger a response at the rule level. A discrepant system value will trigger processing at the knowledge level in order to resolve the information on the display. While these signals are all processed at the individual level, it is often these environmental cues that are the starting point for interpersonal communication at the collaboration level.

In column (b) of Table 8.2, I offer some specific functions that flow from their analysis. Unlike the other taxonomies discussed earlier, the SRK structure allows us to map performance onto underlying knowledge structures and to link communication to the conduct of flight.

Flight deck communication is largely constrained by design. To promote the predictability mentioned by Kanki and Palmer and increase the efficient use of limited bandwidth, crews follow checklists and use formal, prescribed language, often described as ICAO standard phraseology. These frameworks exist to coordinate aviation in a multilingual world with crews often comprising foreign nationals working with no shared mother tongue. However, this technocratic view of the world ignores the fact that humans are social animals and, as we saw, language first evolved to facilitate bonding. It is important to remember that social communication – by which I mean not task focussed – facilitates cooperation and, so, I need to add an additional category to the three listed in column (b) and that is developing social bonds. Communication in aviation, then, has three functions directed at task management and one directed at social bonding.

TABLE 8.2
Task-related Functions of Communication

Level	(a) Type of Communication Act (Rasmussen & Svedung)	(b) Purpose
Knowledge	Interpreting message content in relation to mental models. Talking in 'shorthand'	Clarify future goals and intentions
Rule	Controls sequence of actions, identifies discrepancies or unexpected status change	Control (option selection)
Skill	Time/space signals that update and synchronise, the body language of actors	Control (current action) and verification of status

The task management functions are, first, those communication acts designed to control activity and to verify progress. At Level 1, this could be simply the internal dialogue that accompanies work. At Level 2, it includes such things as the challenge and response management of checklists, requests for activity in support of aircraft control and statements that verify the current status of the aircraft. An example would be the procedural calls associated with the aircraft reaching a cleared altitude. Acts in this category would also include observations of discrepant signals suggesting that goal constraints might not be satisfied. The next group of functions is designed to control the near-term management of the operation. Here we see decision-making being verbalised. Acts in this cluster will include proposed plan modifications and discussions of probabilities of meeting goals. The final category of task-related acts deals with the future state of the aircraft and includes the statement of future plans and intentions, anticipated responses, expected next steps, etc. The discussion among the crew relating to the diversion to Shemya and the dumping of fuel we saw in the last chapter fall into this category. Speech acts in this category often attempt to bring clarity and, so, are linked to sense-making.

On the aircraft, social speech presents a problem. We try to minimise non-task-related conversation in order to maintain a task focus and to reduce the risk of missing radio calls. This is codified in the 'sterile cockpit' rule. Unfortunately, we need to recognise the importance of social interaction at the appropriate time. The social category of communication events includes utterances intended to reduce stress, moderate the tone or atmosphere on the flight deck or establish social bonds. These four core clusters of communication apply equally to activity within teams and also to all collaborative action across organisational boundaries. I want to return to the crew of the ATR 72 and consider their interaction in the light of this analysis before I apply the framework to an example of collaborative action with multiple actors.

8.8 EXPLORING COMMUNICATION DYNAMICS: CONTROL AND VERIFICATION

This next excerpt drawn from the San Juan accident (Excerpt 8.4) illustrates the use of the 'challenge and response' structure in formal communication and is an example of action control and verification. Three items in the workflow, numbered in parenthesis, are actions or events that must be verified so that the crew can agree that the flight is proceeding as planned.

Excerpt 8.4

FO: Flight attendant's chimed (1) flight controls are checked (2).
C: Check.
FO: Rudder cam (3)?
C: Centred.

We can also see here the nature of aviation communication as a prescriptive code. Hutchins (1995) noted that task-related communication in complex social settings, especially with external agencies, is usually conducted via limited bandwidth media. In an attempt to reduce the risk of failure, communication takes place using defined protocols. This is apparent in Excerpt 8.5. The meaning of this exchange is embedded in a shorthand description of the current status of the aircraft (current altitude and next waypoint) and requires both parties to understand what 'Juliet' refers to – the current airport information – and that it is a formal requirement to be in possession of the latest information before commencing the approach.

Excerpt 8.5

C:	San Juan Approach, Eagle 401 seven thousand direct Dorado	[Current status]
ATC:	Eagle flight 401 San Juan Approach expect the ILS	[Future goal]
	runway one zero, descend and maintain six thousand.	[Control]
C:	Seven for six, Eagle 401	[Verification]
ATC:	Eagle 401 you said that you have Juliet?	[Verification]
C:	Affirm.	
ATC:	401.	

The bulk of this interaction relates to verification of status, but the first transmission for ATC does include information about a future goal: 'expect the ILS'. In this next excerpt (Excerpt 8.6), we can see how social modes of interaction can substitute for a prescriptive code. It occurs after the FO has just taken off. As the aircraft leaves the ground, it starts to drift to one side of the runway. Detecting the discrepancy through visual cues, the captain intervenes to keep the aircraft on the correct track. Instead of directing the FO to maintain the runway centre line, he phrases his intervention in a softer, social tone:

Excerpt 8.6

FO:	Positive rate, gear up.	[Control]
C:	In transit	[Verification]
C:	Be careful of drifting like that…'s a bad habit.	[Control option]
FO:	Yup.	[Verification]

8.9 COMMUNICATION, OPTION SELECTION AND DECISION-MAKING

In Excerpt 8.7, the aircraft is on late finals and has been cleared to land on Runway 08. Earlier, having been instructed to fly at 160 kts until the outer marker, the captain had arbitrarily told the FO to reduce speed to 120 kts and then increase to 140 kts. This was because the captain wanted to maintain separation behind the Boeing 727 ahead of them. Each action was, of course, a decision, but it also illustrates role ambiguity. Despite the instruction from ATC, the captain's idiosyncratic interpretation of his role, and any associated freedom of action that flowed from it, resulted in a unilateral change that had consequences. The target speed for the approach was 104 kts. The checklist that contained the step that established the required approach speed had been missed earlier in the flight and so no benchmark had been agreed by the crew in advance. The intervention by the captain had left the FO in a non-standard configuration, and he was seeking clarification about how best to reconfigure the aircraft.

Excerpt 8.7

FO:	You still want me to keep it up … slow down to one twenty?	[Option]
C:	Nah... just do whatever you need. Just ya ya got traffic on a five-mile final behind you anyway. You get too slow because it screws their sequence up	[Option] [Current status]
FO:	Ok	[Verification]
C:	Cause I've I slowed you up more than he wanted, so. …. you were you were about four miles behind 'em [the B-727] and there's not very much wind	[Current status]
FO:	Ok	[Verification]
C:	And below his glide slope, which is a bad combination	[Current status]

The FO's utterance is an example of control action selection. Although too fast for the final approach, their option to reduce speed has been removed because they had been flying slower than the planned speed according to ATC, allowing the traffic behind to catch them up. The captain is rationalising his action in relation to the Boeing 727 and their relative position in the traffic pattern. He is engaged in knowledge-based behaviour in order to rapidly update the current status. The excerpt illustrates how different types of behaviour interleave. This episode is representative of action as a stream of events, each of which has a prospective effect on the ability to implement alternative future courses of action. By not following the instructions of ATC, future options have been constrained. The action now feeds forward to shape future actions. Similarly, sense-making can be retrospective. Having reached a point where a decision has to be made in relation to speed, the current status of the aircraft in relation to the desired goal state is being reviewed in the light of the past action. A conventional situational awareness analysis would argue that the captain failed to project ahead. In fact, he made a decision that made sense in the current context and in relation to his prior experience. To continue at the requested ATC speed, he would, in his opinion, have put the aircraft at risk of exposure to wake turbulence. His action made sense to him at the time. This whole episode is actually quite complex. For example, if we look at the propositional content of each utterance and the linguistic forms being used by the two speakers, we can question the effectiveness of the communication. However, I have limited the discussion to how decisions are embedded in communication.

8.10 CREATING FUTURE PLANS: SHARED UNDERSTANDING

Mental models and the manner in which they represent a shared understanding of the world were discussed in a previous chapter. In Excerpt 8.5, San Juan ATC told the crew to expect the ILS approach to Runway 10. This was in accordance with the flight plan filed by the crew before departure. The statement made by the controller was confirming that the expected plan of action still held (i.e. the runway in use had not changed since the crew departed Mayagüez). Excerpt 8.8

continues the interaction we have already seen in Excerpt 8.3, where the FO gave his approach briefing. The captain interrupts with a comment about the more likely runway.

Excerpt 8.8

C:	Approach briefing	[Control]
FO:	Uhhh, we're doin' the ILS ten	[Explicit future goal]
C:	Yup. You'll probably circle to eight	[Implicit goal]
FO:	Ok.	[Verification]
C:	Probably be Lagoon	[Restated implicit goal]

The Lagoon Visual approach is a published arrival procedure that, initially, brings the aircraft inbound on the ILS Runway 10 heading and then, passing over the western edge of San Jose Lagoon, the aircraft is turned left onto the approach to Runway 08. Later, in Excerpt 8.9, as they begin their approach, the captain effectively negotiates a change of runways.

Excerpt 8.9

C:	San Juan Tower Eagle 401 ILS runway one zero with eight in sight.	[Explicit goal] [Implicit goal]
ATC:	Eagle flight 401 San Juan Tower cleared to land runway one zero {...}	[Explicit goal]
C:	'kay. One zero and eight is not available?	[Implicit goal]
ATC:	Eagle flight 401, uh, confirm you have eight in sight?	[Verification of implicit goal]
C:	Affirm	[Verification]
ATC:	Eagle flight 401 change runways, cleared lagoon visual runway eight approach, runway eight cleared to land.	[Goal state change]

The company gates at this airport are adjacent to Runway 08 and, so, a landing on this runway, as opposed to Runway 10 saves a long taxi. It is highly probable that the captain always intended for the FO to land on Runway 08 as he had even mentioned that this was common practice before they had even taken off from Mayagüez. He failed to make his intentions clear to the FO.

8.11 SOCIAL CHAT

The study by Gougen and Linde commented on the role of non-operational utterances in sustaining alertness during low-workload phases of flight. Given that communication seems to have evolved to support group functioning, it is hardly surprising that social chat is a normal part of human interaction. Communication can reduce tension and build a positive group atmosphere, but unfocussed chat at the wrong time can be a distraction. The 'sterile cockpit' concept, which tries to limit non-essential chat at critical stages of flight, is an example of a bureaucratic response to what is considered to be inappropriate behaviour.

The crew of the ATR 72 engaged in extensive social chat, most of it from the captain and directed at the FO. We see the crew discussing points of technique (the FO had trained on the ATR 42 and only had 20 hours in total on the ATR 72), differences between training and line practice (work as imagined and work as enacted?), and conveniences such as the use of flight deck power outlets to recharge mobile phones and to power DVD players to watch movies in flight.

8.12 WHAT 'NORMAL' COMMUNICATION LOOKS LIKE?

I mentioned that the San Juan accident was one of those rare occasions where we have an almost complete transcript of the crew's interaction and, in Table 8.3, I present an analysis of the communication acts using my four-factor taxonomy. Of course, just as there is no such thing as a 'correct' error rate, there is, equally, no preferred distribution of speech acts. Although context-dependent, it is rather assumed that an efficient crew will use communication as a tool in an appropriate manner, as dictated by procedures and circumstances. However, I think we can draw two significant observations about the San Juan data. First, given that we saw earlier that the captain's implicit assumption about the probable runway at San Juan was a plausible contributory factor in the accident, the frequency of directed speech acts intended to clarify goals and intentions was the lowest of all. Sexton and Helmreich (2000), analysing transcripts from a simulator study, suggest that increased communication between crew will reduce error because of the increased predictability of behaviour induced. Predictability can only flow from a shared understanding of the goal. The second observation concerns the high frequency of social speech acts. In particular, 29% of speech acts during the descent and final approach phases fall into this undirected category. While social interaction serves a purpose, the operational focus at the appropriate time is essential.

What can line operations safety audit (LOSA) tell us about 'normal' communication? In the sample of 100 sectors discussed in the previous chapter, 42.57% of

TABLE 8.3
Distribution of Speech Acts

	Ground	T/O & Climb	Cruise	Descent	App/land	Total
Time (%)	14.9	16.8	12.4	43.7	12.04	
External (ATC/cabin/Ops)	14	17	7	46	9	93 (20.08%)
Control and verification	53	14	7	35	33	142 (30.66%)
Option selection	4	8	4	8	40	64 (13.82%)
Clarify goals and intentions	0	2	9	6	8	25 (5.3%)
Social	22	20	21	51	25	139 (30.02%)
Total	93 (20.08%)	61 (13.17%)	48 (10.36%)	146 (31.5%)	115 (24.8%)	463

'errors' related to communication. This high rate simply reflects the fact that communication forms a significant part of the work of the crew and, importantly, it is also accessible to observation and, thus, audit. Unfortunately, LOSA gives us an incomplete understanding because it only captures departures from proceduralised communication and does not capture the quality of communication in the sense that it has been discussed so far in this chapter.

Given these caveats, communication was either within the crew (endogenous), such as procedural call-outs, or between the crew and external agencies (exogenous), such as ATC calls and acknowledgements. The ratio of endogenous to exogenous communication was 2:1. This simply reflects the density of the ATC environment the flights were operating in. Shorter sectors increase the time spent dealing with ATC. Looking at endogenous behaviour first, of the four factors in my model, control and verification were easily seen, as were formal structures associated with clarifying future goals. Controlling action (calling for a checklist) and verifying progress (calls associated with configurations and status) make up the largest cluster of interactions. Control can be immediate (a call) or delayed (actions arising from a prior briefing). Briefings, of course, represent opportunities to clarify future goals. In the sample, there were two points in each flight where a briefing was mandated: before departure and at the top of decent, prior to the arrival. In the sample, we observed 200 briefings. In 7% of briefings, either the structure of the briefing was incomplete or it was delivered at the wrong time (typically in the descent because of an intervention by ATC). Errors do reveal something of the fragility of performance that flows from individual variability (incorrect phraseology, incomplete or incorrect messages) and from the challenge of balancing competing demands. Dealing with ATC while completing other tasks being a classic example. Missed procedural call-outs were the largest category of endogenous errors. Interestingly, observations of 'social' communication acts comprised just 1.7% of all episodes, but I am inclined to think that this value is depressed simply because the crew knew they were being audited.

When dealing with external agencies (exogenous), typically ATC, we see the impact of contextual facts such as local accents, available technology, traffic densities and, hence, controller workload. In this instance, the notional 'error rate' is as much a measure of regional infrastructure as it is of crew performance. Incorrect read-back was the largest cluster of exogenous errors.

We saw in Chapter 5 that LOSA allows you to tag errors in terms of crew responses. Did they actively respond or did they passively ignore the error, for example. The discussion so far has only looked at errors where the crew responded in an active sense, which comprised 62% of all communications errors. If I turn now to those episodes where the crew response was passive, we see that the endogenous to exogenous ratio changes to 7:1. In both cases, missed procedural call-outs were the largest cluster (44.5% of active and 59.8% of passive errors). Interestingly, missed calls have been cited as evidence of poor 'monitoring' in some studies, but it seems to me that crew elect to expend effort based on a number of possible factors: alignment with the intent behind the requirement (why call if the automation will do the job?); workload (trading off a verbalisation against a more important use of attention); crew synergy (we both know what's going on so why chatter needlessly?).

Normal work, then, affords space for the crew to modify responses – the extent of compliance with prescribed communication protocols – in accordance with local circumstances.

8.13 COMMUNICATION IN A SYSTEMS MODEL

The discussion, so far, has looked at the nature of communication between individuals. Ultimately, though, collaboration is goal-directed and, so, we need to consider how communication supports goal accomplishment. We have seen that the San Juan crew did not appear to have a shared understanding of the approach to be flown on arrival. The explicit, filed flight plan approach differed from the custom and practice solution, which was a manifestation of local knowledge. The 'normal' trajectory was summarised by the captain when he said 'they give you an ILS to get you inbound, [...] then [...] ask you "you got the airport runway eight in sight?" [...] "clear Lagoon Visual Eight"'. The first goal is to establish an inbound track, and the second goal is to meet the requirements necessary to switch to a different runway. The transition from the ILS Runway 10 to a visual approach to Runway 08 has a critical point at which the two approach paths intersect. The correct approach speed for the approach must also be achieved. The system safety boundary is the point by which surplus energy must be dissipated. The margin associated with safety boundary extends from the point at which the crew are cleared to reduce speed to the target for the approach, which would have been the outer marker on the inbound course to Runway 10, to the point on the runway where sufficient energy has been lost for the aircraft to stop on the paved surface. In a previous chapter, I suggested that the margin was also the zone in which discrepant task-related signals become apparent.

Given the time of day and duration of the flight, fatigue was unlikely to have been a factor. There was a significant experience differential between the captain and the FO. The airline's induction course included just four supervised sectors before a pilot was cleared to the line (personal communication). The captain was not a trainer, and the FO was considered competent to undertake line duties. The subsequent investigations did uncover the fact that the FO was receiving counselling and was on medication for anxiety. The flight on that day, then, needs to be seen in the context of the crew dynamic.

The captain's construction of a wake turbulence risk did not take into consideration the actual separation of the two aircraft and the prevailing wind. The risk of turbulence was less that the captain seemed to believe. While the buffering capacity at this stage could cope with his intervention, the margin was reduced. The failure to maintain the speed requested by ATC was a signal suggesting that the plan modification would have implications that would need to be evaluated. This was subsequently recognised, in retrospect, by the captain: 'Just ya ya got traffic on a five-mile final behind you anyway. You get too slow because it screws their sequence up'. The segment of the profile covered by the CVR transcript includes seven points at which the crew are required to complete a checklist. Four checklists are completed, two are partially completed and one is missed out altogether. One of the partially completed checklists includes the step that establishes the target speed for the approach. The failure to complete checklists and to establish key parameters, such as the desired approach speed, were also signals. The relatively high rate of social chatter between the crew suggests that the workload was not high.

The FO's struggle to reduce speed in the final approach phase while maintaining the correct path to the runway suggests that the system was approaching its boundary. In effect, the FO was fighting the laws of physics. The aircraft's potential and kinetic energy, coupled with the FO's skill set, was beginning to exceed the buffering capacity. Ironically, the subsequent intervention of the captain after the initial touchdown was possibly the act that caused the system to transition to unsafe.

Hutchins describes communication as a mechanism by which teams construct a shared memory of the progress of a distributed task. In part, this is reflected in Excerpt 8.7. By constructing a representation of the movements of aircraft in the traffic pattern behind them, what has happened in the past, the captain can reflect on the options remaining for the crew. Unfortunately, at this point, he delegates decision-making to the FO.

8.13 DISTRIBUTED COMMUNICATION

In more complex, distributed systems of collaboration, communication is often mediated through tools and artefacts. Consider the following example, which involves aircraft ground movement at an airport (TSB, 2004). On 29 October 2004, a BN Islander nearly collided with a DHC Dash 8 on take-off from Vancouver International Airport (see Figure 8.1). The Islander, call sign Navair 612, was holding at Taxiway A at the threshold to Runway 08R on the south side of the runway. The DHC Dash 8, Jazz 8191, was positioned on Taxiway L2, to the north and further along the runaway. It was early morning and still dark at the time of the event. The ground controller issued the taxi instructions for the Dash 8 crew and then completed the flight progress strip. This 'strip', a piece of perspex with sections for different types of information, is an example of an artefact used to communicate between controllers. The design of the strip is such that space for information is limited and the information

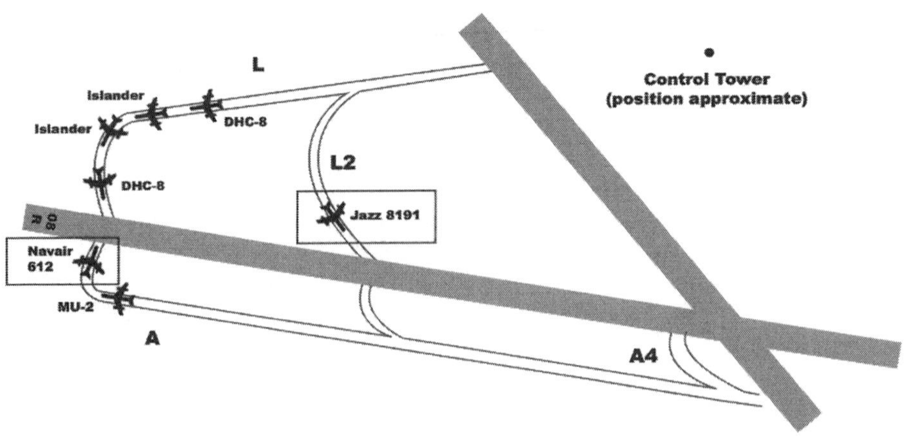

FIGURE 8.1 Vancouver International Airport.

about the cleared taxiway – in this case, the Dash 8 was cleared to enter the runway at L2 – was obscured. The ground controller passed the flight progress strip to the tower controller who was sitting beside him.

The tower controller cleared the Islander to line up, asking the crew to move forward to allow the Dash 8 to line up behind it. The tower controller clearly assumed that the two aircraft were on the same taxiway where there were several other aircraft waiting for take-off at the time. The crew of Jazz 8191 was then cleared to line up. They could not see the Islander nor, given their angle of approach to the runway, could they see the threshold area. Because it was still dark, the Islander could not see the Dash 8.

Controllers are required to name the intersection in the take-off clearance if it differs from the threshold. Unaware of the actual entry point, the tower controller omitted that piece of information. Pilots are not required to repeat the intersection in their read-back of the instruction and so the error was not picked up. The formal rules of communication, designed to validate differences but obviate the need for unnecessary use of a limited medium – the radio frequency – were easily circumvented through the omission of a single item. The design of the flight progress strip represented a risk in that its completion by the ground controller was a catalyst for failure.

Vancouver had area surface detection equipment (ASDE) installed for detecting aircraft moving on the ground. Issues of reliability and dependability meant that it was used as an aid only, and controllers had little faith in it. The Dash 8 was shown as a target on the ASDE at the L2 position, but this was disregarded as a spurious return by the tower controller despite the fact that the return moved after the Dash 8 was cleared onto the runway.

As the Navair 612 started its take-off roll, Jazz 8191 was moving towards the runway, the crew looking ahead and to their left for the Islander they were supposed to be following. At the last moment, they decided to check the up-wind area, to their right, only to see the lights of the Islander coming towards them. They stopped the aircraft and turned on all external lights as the Islander lifted off in front of them.

Communication between two co-located workers, the ground and tower controllers, was mediated through the flight progress strip, but the design of the strip hampered effective communication. The tower controller was nonetheless able to render the communication sensible as he held a view that all aircraft were in a common location. Interestingly, the ground radar showing the location of the Dash 8 was noticed but dismissed. The crews of the two aircraft constructed an understanding of the situation through direct instructions from ATC and also calls to other aircraft. In the case of Jazz 8191, their actions were linked to instructions assigned to another aircraft, Navair 612. While obeying their own instruction, they were also looking out for the other aircraft upon whom their own progress was dependent. If nothing else, they knew that their own take-off could not commence until Navair 612 had departed. Although they had started to move, there was clearly a doubt in their mind. We do not have details of the conversation between the crew of Jazz 8191, but the evidence suggests that they were uncertain of the situation. An instruction from ATC, set in the context of an instruction to another aircraft, triggered a discussion by a crew directed at resolving ambiguity. The indication of Jazz 8191's position on the ASDE is an example of a signal that lacked sufficient salience to intrude on the controller's understanding of the situation. It was a busy time of day with many aircraft preparing to depart, so workload might have been a factor. It was also still dark and so any visual cues would have been hard to detect. These events point to how, in collaborative systems, messages propagate across networks. In this example, the network was small, but the next case study explores propagation across both time and space.

8.14 COMMUNICATION AS INFORMATION PROPAGATION ACROSS A NETWORK

On 10 September 1991, an Embraer 120, N33701, was pulled into the hangar for scheduled maintenance (NTSB, 1992). The time was about 2130 and the second shift was at work. One of the tasks for the night included the removal and replacement of the left and right horizontal stabiliser deice boots. The job required the leading-edge assembly to be removed, the old deice boot to be separated from the stabiliser, the new boot to be bonded to the structure and the leading edge to be reattached to the stabiliser by 47 retaining screws on both the top and bottom surfaces. The job originally was assigned to the third shift, but the second shift supervisor started the job to help with the overall workload.

The hangar supervisor, who I will call S1, was busy with a C check on another aircraft, so he assigned two mechanics (M1 and M2) to assist a second supervisor (S2) in the task of replacing the boots. This second supervisor was normally responsible for the line (routine rectification and replenishment of aircraft) but, on this occasion, had been asked to oversee work on N33701 in the hangar. A quality inspector (QI1), overhearing the mechanics being tasked, and still with an hour to run on his shift, volunteered to lend a hand. Being the smallest of the three, he went up onto the upper surface of the tailplane, 20 ft above ground level, to remove the screws along the upper surface of the leading-edge assemblies. He placed all the screws, apart from a few damaged ones, in a bag, which he then placed on the 'manlift', a hydraulically operated platform used for working at heights. He wrote 'assisted mechanics with removal of deice boots' in the Inspection Department Written Report to account for his time.

The second supervisor (S2), in the meantime, was observing the progress on the aircraft and saw QI1 on the top of the right-hand stabiliser. The mechanics had come to the end of their shift and the third shift was arriving for work. The oncoming supervisor (S3) checked the supervisors' shift handover briefing sheet but saw no comments relating to N33701: this is because S1 had not yet written up the task. He then inquired directly of S1 what work has been done on the aircraft. He asked if work has started on the left stabiliser. S1 looked at the aircraft and said 'No'. S3 then expressed some concerns about the time remaining to finish the job and decided to defer the replacement of the left deicing boot. S3 told S1 that he was putting the left boot back into stock. S3 also checked the job card, which gave no indication of any work done on the left stabiliser.

S3 assigned a new mechanic to the task (M3) and told him to check with S1 and the two mechanics (M1 and M2) working on the aircraft as to the state of progress.

They reported that they had managed to remove the leading-edge assembly on the right stabiliser and remove most of the boots, but their progress had been hampered by some stripped screws that needed to be drilled out.

As he went off shift at about 2230, M1 remembered briefing S1 on progress but could not later remember if he had mentioned the work on the upper left leading edge. M1 also briefed the fourth mechanic (M4) on the job but failed to complete any documentation as the job cards had not been given to him. The job cards were still in the documentation package awaiting the third shift supervisor (S3). M1 remembered telling M4 about the removal of the screws along the left-hand upper leading edge. However, M4 was reassigned before he could start work, but before he left, he then briefed another mechanic (M5) that the screws had been removed and that the boots needed replacing. He saw a bag of screws on the 'manlift' and gave them to M5.

The third shift mechanics finish the removal of the right-hand boot and bonded the new one into place. A second quality inspector (QI2) was lending a hand at this stage. QI2 had come on shift early and had checked the department's written report before QI1 had entered his comment about helping the mechanics. There was no direct verbal communication between QI1 and QI2.

At this point, N33701 was pushed out of the hangar to make room for another aircraft. M5 completed the fitting of the right leading-edge assembly outside, making use of the screws in the bag together with some new ones drawn from stores. When the right leading edge was refitted, M5 noted a few screws left over but paid these no attention as he understood that there had been some replacements used. S3 climbed up to the right-hand stabiliser to check the reconnection of the deicing fluid lines. He did not check the left-hand stabiliser as he was not aware of any work having been done. It was, of course, dark and the top surface of the stabiliser was not, in any case, visible from the ground. There was no direct light on the aircraft. Finally, M3, QI2 and M5 went off to complete the paperwork. M3 returned to clean up around the right leading edge, telling M5 that the replacement of the left assembly had been deferred. The aircraft was then cleared for flight and departed at 0700 from Houston to Laredo. On the return flight, as the aircraft began its descent into the Houston area, the aerodynamic forces on the tail structure tore off the left-hand leading-edge assembly causing the major structural failure of the tail. All 14 passengers and crew were killed.

Three supervisors, two quality inspectors and five mechanics were involved in this scenario with work spread across three shifts and distributed in space between the inside of the hangar and outside on the ramp. Maintenance work is typically controlled through documentation. In this case, the various shifts had access to logs for recording information and job cards for controlling specific actions in relation to the aircraft. These are the primary artefacts associated with communication, but we also have the physical presence of the replacement for the left stabiliser boot and the bag of screws left on the 'manlift'. These items, too, allowed individuals to construct meaning in the scenario. Although the engineering staff were assigned to specific shifts, the transition between shifts was fluid as individuals reported for duty and began to support the work of others or assume control. As a result, communication was asynchronous. Individuals briefed themselves on the status of work before records had been written up. Verbal updates were incomplete or incorrect.

Finally, personnel did not have equal access to information: mechanics were not issued with work cards as the documents had yet to be issued to the appropriate supervisor.

The bag of screws left over after the leading edge had been refitted shows how meaning is constructed in situ. The screws came, of course, from the left-hand stabiliser. Because the status of the task was not fully understood and, thus, accurately communicated, the screws became 'leftovers after drawing new screws from stores' rather than 'screws not replaced after the job was changed'. The meaning of the 'bag of screws' was constructed based on comments made by others. This example neatly illustrates work as the part-processing of tasks that are then handed off to other workers for additional input. Three groups of workers – supervisors, mechanics and quality inspectors – all participated in an asynchronous and distributed sense during the time the aircraft was being worked on. Because of the way people communicated, the actual state of the aircraft was forgotten and, instead, an imagined state was constructed and transmitted between actors. The propositional content of messages was flawed and, as a result, the job was not finished correctly. On 24 October 2002 (TSB, 2005), a DHC-8 lost a three-foot section of the wing leading edge because 14 screws that fasten the section to the bottom of the wing were missing. The aircraft managed to get airborne and land again safely. The events that led up to the wing failing were very similar to those we have just seen.

8.15 COMMUNICATION AS HIERARCHICAL CONTROL

Much of this chapter has looked at communication between individuals, but the maintenance example introduced the idea of logs, diaries and job cards as control mechanisms. As well as facilitating action, communication is probably the primary control mechanism in my hierarchical system. Figure 8.2 captures some of the hierarchical communication aspects of the Embraer tragedy. In this section, I want to look at communication as a broader control process at the system level, partly because it is convenient to do so here and also because formal communication structures and protocols shape interpersonal communication. In this section, we will be looking at policies and procedures, orders and instructions as communications artefacts. In the maintenance case study, quality inspectors' written reports, shift handover sheets, maintenance document packages and job cards represent the range of artefacts used to exert control over subordinate levels in the system. These communication artefacts are created to both satisfy a regulatory requirement and also to communicate to the end-user of the technology. Thus, they serve two purposes and this can sometimes present problems. We also need to consider not only how artefacts are constructed but also how they are interpreted and applied. The risk of communication failure is as prevalent in this domain as it was when we considered speech.

Control is typically exercised through the creation of rule sets. We can identify three types of rules: goal, process and action (Grote, 2008). Goal rules are about broader conditions and are represented by policies. A policy is a description of a state or top-level constraint. For example, the 'clean wing' concept which forbids dispatch with contamination on the aircraft is a 'policy' and, therefore a goal rule. Not all goal rules are communicated as a formal 'policy' but are usually written in such a way that their force is implicit. Process rules guide decisions. For example, rules are written

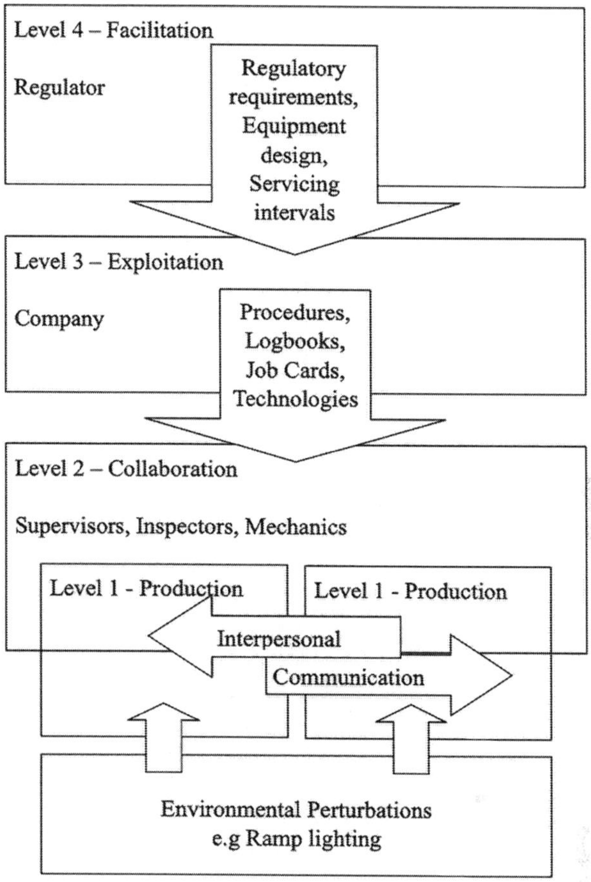

FIGURE 8.2 Communication in a system.

to direct pilots to deice aircraft before departure if certain environmental conditions apply. Action rules are about task completion: how to do the job. In a cold-weather operations briefing, there might be specific instructions about how to conduct checks of the airframe to verify that it has been properly deiced. As we move from the action, through the process to goal rules, descriptions tend to become more generic. Policy manuals tend to be collections of different types of rules, usually formulated at the goal and process levels. Action rules are usually delegated to checklists for use on the job. Checklists, however, are subsets of larger rule sets designed to communicate specific actions. They are abbreviated process rules that describe key steps in a specific element of the task of operating the aircraft. Non-normal and emergency checklists also contain decision rules to allow for contingencies. This next case study looks at checklists in action.

On 6 April 2006, a Bombardier DHC8-Q400 (SHK, 2007) experienced a propeller overspeed. A combination of the inputs of the captain, who was pilot flying, and the performance of the aircraft automation resulted in the aircraft rapidly descending from 2000 to 1200 ft before the crew could regain control. The case study is complex. Both pilots were very experienced, having both flown the B-747, and had accumulated considerable flight time on turboprop aircraft. The technical failure occurred during the descent, prior to the aircraft levelling off but before then turning left to begin the final approach. The initial indication of a problem was via a white propeller electronic control unit warning lamp. The lamp is simply an attention-getter. The crew then need to diagnose the problem and, in this case, the right-hand propeller was found to be over-speeding. This was correctly diagnosed by the crew. The overspeed checklist was not consulted, instead the initial 'memory items' were completed. These call for the following actions:

Power Lever ....Retard Toward Flight Idle
Airspeed .... Reduce
IF unable to control propeller RPM:

Because the aircraft was in the descent, the power levers were already at flight idle and the aircraft speed was reducing towards the final target speed for the approach. Therefore, it was felt that these conditions had been satisfied. As the aircraft levelled off at 2000 ft, the captain increased power on the left-hand engine in order to maintain level flight. The aircraft's own automatic systems increased power to the left-hand engine such that there was now greater than 100% torque applied to the left side. The right-hand propeller, still at flight idle, presented a negative angle of attack to the airflow and, as such, was an effective speed brake. The resultant drag caused the temporary loss of control.

Before we look at the formal communication represented by the checklist, we should just look at the communication between the crew. On two occasions, the FO suggested that they 'secure' the right engine. Both times the captain rejected the idea. In his evidence, the captain said that he thought that was the engine was 'dead'. He had constructed the situation as the loss of the engine but, given that they were commencing the final approach, he did not want to start the complex process of shutting down the engine at that moment. The FO conceived of the situation as a propeller problem and the action of 'securing' the engine would automatically cause the propeller to auto-feather and, thus, reduce the drag. By referring to the engine while actually meaning reducing the drag from the propeller, the FO failed to communicate her intent to the captain. The propositional content of the speech act was obscured. It is interesting that the pilots were, nonetheless, communicating in formal, professional terms.

Returning to the checklist, documentation can be evaluated in terms of its formulation, intent and scope. To a degree, these categories overlap those we saw earlier when we looked at the properties of speech. The formulation is similar to the linguistic form in that it captures the use of language – in this case, written – to communicate information. The intent is similar to propositional content in that it relates to the information contained in the checklist items. Social force is implicit in that the checklist is an official document. I have added a third dimension, scope, that possibly does not apply as widely to communication at the interpersonal level. It relates to the range of circumstances under which the communication applies. The context may change the appropriateness of a specific rule.

The intent of the first two actions in the Q400 checklist is to reduce the load on the propeller to stop the situation getting any worse: we are stabilising the problem. In the context of the event, the crew had simply identified an issue and taken the initial steps. The checklist actions neither diagnosed the problem nor represent an appropriate response.

Moving on to scope, there are three phases of flight during which a crew might need to cope with a propeller issue: during take-off and climb, in the cruise and during a descent. During take-off, the pilots activate an automatic feathering system designed to cope with a failure at a critical stage of flight. The captain had completed his periodic company training the week before the event during which he rehearsed a propeller overspeed after take-off. It seems that the checklist was designed to cope with an overspeed in the cruise phase of flight, hence the need to reduce the load. This is not an issue in the descent. Once the aircraft levelled off, the propeller now became an issue because of the drag it induced. Ironically, the correct action at this point would have been to advance the power level to a position of about 10% torque, which equates to the minimum drag position. The appropriate action was the opposite of the action required by the checklist. While automation can deal with the take-off, the checklist can deal with the cruise, it seems the designers forgot about the descent.

Formulation describes the use of language and is the textual equivalent of linguistic form. The propeller overspeed checklist includes the following:

> Note: If the engine is not shutdown immediately after feather the propeller with the Alternate Feather system, the propeller may unfeather [...]

This statement is not correct in its use of English. Its meaning, mirage-like, flickers and fades with every reading. The text would be correct if 'after feather' was changed to 'after feathering'. The meaning would be clear, however, to a pilot already experienced on aircraft fitted with feathering propellers. Could such a simple change be difficult to make? We will return to this checklist in Chapter 10 but, for now, we can see how control fails at a systemic level because of poorly framed documents.

8.16 COMMUNICATION, SAFETY DRIFT AND SCALE LAW AT A SYSTEMS LEVEL

Most of the time, the world works according to plan. Work routines contain low levels of variety or a small number of exceptions over time (Kluge, 2014). Occasionally, we encounter inconsistent or unusual situations. Inconsistency will trigger speech acts to seek clarification or may result in a state of confusion in the individual. Problems arise when acts of clarification are not initiated either because the individual can rationalise signals or can tolerate the state of confusion (and, therefore, does not feel a need for clarification). By finding ways to accommodate the current situation, we can defer taking appropriate action until it is too late. We also see, at an individual level, that messages can be incomplete or delivered in a non-standard format. Of interest, LOSA observations reveal that crew engage in repair behaviour when communicating with off-board agencies. They fill in gaps in understanding by guessing what a speaker said or fall back on previous experience to interpret statements. Despite the prevalence of uncertainty and repair, the system still functions.

Moving on to consider the implications of the Embraer 120 and the Dash 8-400 examples given above, what we are really looking at is communication across networks. The network comprises both individuals engaged in activity and artefacts designed to inform and constrain those individuals. In addition, as we move up the hierarchy, the distance between the point of design and the point of application grows both physically and temporally. I think that this presents a challenge. It is a cliche in aviation that the manufacturer's documents are designed more to limit liability in the event of failure that to guarantee successful workplace performance. Whether this is true or not, the issue of information ergonomics and system scale effect is clearly a problem. The risk seems to be a function of the number of nodes in a communication network and the distance between initiators of communication and actors responding to that communication.

8.17 CONCLUSION

Communication is both the glue that binds teams and the mechanism that enables work to be done. It is a skill we all take for granted. This chapter has outlined the functions served by communication in the work process and also the properties of communication that allow it to fulfil its purpose. Within a systems context, communication serves as both control and feedback. Orders and instructions, commands and tasks are all direct action, whereas responses, acknowledgements and queries are forms of feedback. Unfortunately, the process is inherently fragile and requires effort if it is to function effectively. That said, humans are adept at filling in gaps in conversation based on the context of speech and deducing meaning. We repair flaws in communication. We will continue our examination in the next chapter where we will look at the role of the organisation in my hierarchical model.

REFERENCES

Bales, R. (1950). A set of categories for the analysis of small group interaction. *American Sociological Review*, 15(2), 257–263.

Dunbar, R. (2014). *Human Evolution*. Harmodsworth: Penguin.

Fischer, U. & Orasanu, J. (1999). Say it again Sam! Effective communication strategies to mitigate pilot error. *Proceedings of the 10th International Symposium on Aviation Psychology*. Columbus, OH.

Flin, R. & Martin, L. (1998). Behavioural markers for crew resource management. UK Civil Aviation Authority Paper 98005. London: Civil Aviation Authority.

Gougen, J.A. & Linde, C. (1983). Linguistic methodology for the analysis of aviation accidents. NASA-AMES.

Grote, G. (2008) Rules management as source for loose coupling in high-risk systems. In Hollnagel, E., Nemeth, C.P., & Dekker, S. (Eds.), *Resilience Engineering Perspectives* (pp. 99–100), *Volume 1: Remaining Sensitive to the Possibility of Failure*. Aldershot: Ashgate.

Hutchins, E. (1995). *Cognition in the Wild*. Cambridge, MA: The MIT Press.

Kanki, B.G. & Palmer, M.T. (1993). Communication and crew resource management. In Wiener, L., Kanki, B.G., & Helmreich, R.L. (Eds.), *Cockpit Resource Management* (pp. 99–136). San Diego: Academic Press.

Kluge, A. (2014). *The Acquisition of Knowledge and Skills for Taskwork and Teamwork to Control Complex Technical Systems: A Cognitive and Macro-ergonomic Perspective.* London: Springer.

NTSB. (1992). Accident Report AAR-92/04.Continental Express Flight 2574 In-Flight Structural Breakup EMB12ORT, N3370 Eagle Lake, TX. 11 September 1991.

NTSB. (2005). AAR/05-02. San Juan, Puerto Rico. Crash during landing executive airlines (doing business as American Eagle) Flight 5401, Avions de Transport Regional 72-212, N438AT. 9 May 2004.

Nevile, M. & Walker, M. (2005). Using conversation analysis to represent and analyse recorded voice data. ATSB Aviation Research Report B2005/0108.

Orasanu, J.M. (1994). Shared problem models and flight crew performance. In Johnston, N., McDonald, N., & Fuller, R. (Eds.), *Aviation Psychology in Practice.* Aldershot, Hants, England and Brookfield, VT: Avebury Technical and Ashgate.

Palmer, C.T. (1991). Kin-selection, reciprocal altruism, and information sharing among Maine lobstermen. *Ethology & Sociobiology*, 12(3), 221–235.

Rasmussen, J. & Svedung, I. (2000). *Proactive Risk Management in a Dynamic Society.* Karlstad, Sweden: Swedish Rescue Services Agency.

Sexton, J.B. & Helmreich, R.L. (2000). Analysing cockpit communications: The links between language, performance, error, and workload. Human performance in extreme environments. *The Journal of the Society for Human Performance in Extreme Environments*, 5(1), 63–68.

SHK. (2007). Report No: RL 2007:12e. Aircraft accident to LN-RDA at Kalmar Airport, H county, Sweden, on 6 April 2006. Case L-08/06.

TSB. (2004). Report No: A04P0397. Risk of collision. Nav Canada. Vancouver International Airport, British Columbia. 29 October 2004.

TSB. (2005). Report No: A02O0349. Aircraft difficult to control. Jazz Air Inc. de Havilland DHC8-311, C-FACF, Toronto/Lester B. Pearson International Airport, Toronto, Ontario. 24 October 2002.

9 Organisational Factors – Level 3

9.1 INTRODUCTION

The discussion so far has looked at individuals and groups engaged in work. Even though some of the activities had been conducted at a distance (with air traffic control (ATC), for example) or across time (in the case of maintenance shifts), it still involved people engaged in a specific task. The work of those individuals took place in a context shaped by external factors. Organisational 'pressure' is occasionally cited in incident reports and often commented on by participants in crew resource management (CRM) classes. Indeed, the presence of external factors can be identified in some adverse events, even if their exact contribution to outcomes is more difficult to determine. For example, on 29 March 2001, a Gulfstream III business jet crashed on approach into Aspen, Colorado, killing the three crew and 15 passengers (NTSB, 2002). The airport at Aspen had a curfew in place, and the weather was marginal. Because of the late arrival of the passengers, who were being flown to a dinner party, the crew calculated that they would have enough time for just one approach, and if that was unsuccessful, they would have to divert. While they were waiting for the passengers the crew did consider landing at a different airfield and providing vehicles for the client and his guests. This was relayed to the customer's PA and, in response, the pilots were told to keep their comments to themselves. He, the client, had spent a substantial amount of money on dinner. None of them made it. The captain and the dispatcher had discussed the attitude of 'the boss' over the radio en route, so it was clear that the experience was in the minds of the crew. What constitutes a 'pressure' and how it affects outcomes are difficult to determine, but it seems fair to suggest that, in this case, the captain's decision-making was influenced by his experience with the customer.

In this chapter, I want to look at the role of the organisation in shaping the conduct of work. I have called this level in the hierarchy that of 'exploitation' because, in general terms, the role of an organisation is to exploit assets in order to generate a return. The fundamental point we need to make at the outset is that airlines exist to make money. The economics of the aviation industry is precarious, as the 2020 Covid pandemic revealed. Many small operators survive on slender margins, and a single unwanted event can wipe out any surplus generated if risks were not covered by insurance. It is perhaps an irony that insurance cover and a large pool of disposable, low-cost labour has probably impeded efforts to raise safety standards in some sectors of the aviation system such as ramp operations.

Organisations comprise individuals working in teams and so what we have already discussed in terms of Levels 1 and 2 applies equally within the corporate entity. The difference is one of outputs. So far, we have been looking at how crews function in order to deliver the service being offered by the company. At Level 3, the output is of a different nature. Rather than a task or a flight, the output is the return on the capital invested. Control over the conduct of business flows from the company business plan, industry audit requirements, regulatory requirements and social expectation. Feedback comprises measures such as financial performance, staff turnover, inspection and audit findings, adverse events and accidents. Financial success and the long-term survival of the company might be measures of performance from a shareholder's perspective, but the emergent properties we are more interested in are those of morale and organisational citizenship – or its corollary, resistance. In terms of safety, organisations create the circumstances within which individuals act safely.

9.2 WHAT IS AN AIRLINE?

In very simple terms, we can view an airline, like any commercial enterprise, as a mechanism designed to generate a return on the capital invested. The return takes the form of a surplus of income from various activities once costs have been covered. In order to make money, an airline attempts to optimise its offering within the marketplace. The type of aircraft flown needs to match capacity to demand in terms of seats available and economy of operation. The frequency of services and sector lengths will have implications for aircraft utilisation and crewing requirements. The routes flown will reflect markets, and thus, specific passenger groups and the premium that can be charged. This, in turn, will influence the service standards offered. The infrastructure supporting the operation and the mode of ownership (i.e. the extent to which functions are subcontracted) affect costs.

The rise of so-called low-cost operators reflects an apparent polarisation of airline models with, on the one hand, the full-service 'legacy' carriers being contrasted with the 'no-frills' operators on the other hand. In fact, 'low-cost' is a misnomer. The ticket price may be lower but the business model is based on high utilisation of assets and, therefore, is in line with other industries attempting to maximise profits through efficiency. This is clearly demonstrated in the adoption of minimum turn-round times on the ground. Time on the ground is, in effect, lost revenue and so the process of aircraft turn-round, in the low-cost model, has been refined such that it can be completed in a notional 25 minutes. Southwest, the originator of the low-cost model, calculated that extending the turn-round time to 35 minutes would require additional aircraft if existing schedules were to be maintained (Calder, 2002). Of course, maximum profits are obtained from a full aircraft but another low-cost operator found that high passenger load factors increased the probability of delayed departures simply because of the time it takes to disembark and board a full aircraft. By manipulating ticket prices, the airline was able to control load factors at a level that allowed the fast turn-round of aircraft but still generated the same revenue as a full aircraft (Westjet, personal communication).

The design of the operation, the range of income-generating activities and the markets served all have implications for the nature of the tasks undertaken by the workforce. For example, quick turn-rounds place a burden on the operational staff to complete tasks against the clock, with an increased probability of injury. In the United Kingdom, the Health and Safety Executive did consider mandating minimum turn-round times in order to mitigate injuries to baggage handlers and other ground staff (personal communication). Strangely, the idea did not take hold in the industry. The seasonal nature of some operations means that operators in these markets have a continual and significant outflow of experience (see discussion of Table 9.1). The business model adopted by a carrier sets up contradictions that have safety implications.

9.3 HOW THINGS GET DONE IN AVIATION

An airline, then, is an assemblage of functions configured to deliver a service. In so doing, it has an inherent level of risk associated with the manner in which operations are conducted. Risk needs to be seen as a property of the design of the operation as much as it is a product of the way entities operate. The following two events have the same end result but get there by different routes. They illustrate the point already made that work is distributed and show how risk emerges from the way an entity is organised.

On 27 October 2002, an Airbus A320 was positioning from Manchester, England, to collect holidaymakers on the Greek island of Kefallinia (AAIB, 2004). En route, the captain discussed passenger loading with the cabin supervisor (CS). The plan was to board 69 passengers at the first destination and to collect more passengers from the nearby island of Zakinthos. The company used a standard passenger loading plan which required one-third of the passengers to be seated in each of the three 'bays' in the cabin, each bay comprising ten rows of seats. At Kefallinia, the company handling agent (HA) passed the load form to the captain indicating that the passengers had been seated in accordance with the procedures. In fact, the HA had assigned half the passengers to the front bay and the remainder at the rear, assuming that the middle rows of seats would be taken by the passengers boarding at Zakinthos. The HA's local solution to the problem of managing the aircraft's weight and balance requirements was 'legal' in terms of aircraft performance but not compliant with the company procedures. Equally, custom and practice were not communicated across all participants in the process of loading the aircraft. After boarding had started, the captain asked the CS about passenger locations and was told that most of the passengers appeared to be filling the front rows of seats. The captain then instructed the CS to move passengers to the rear of the aircraft. The HA questioned this action but was told that it was at the captain's request. At this stage only half the passengers had boarded: those assigned to the front rows. The captain's sampling of the boarding process was premature and resulted in an inaccurate understanding of the final status of the task. By the time boarding was complete the second batch of passengers had taken their assigned seats at the rear of the cabin. With all the passengers on board, the aircraft prepared to depart, and as it began to accelerate along the runway, the nose reared up striking the tail on the runway. In his attempt to avoid a nose-heavy condition, the actions of the captain resulted in the aircraft attempting to take-off in the opposite, tail-heavy condition. Despite the captain's intervention, the HA did not challenge the modification to his plan.

In this example, the aircraft was deployed to collect two-part loads from different destinations. The loading procedure was locally modified to cope with the task. The role of communication between the various actors was, of course, of fundamental significance in this event and elaborates nicely on some of the key themes of the previous chapter. It was the way the aircraft was deployed to exploit a business opportunity that created the opportunity for normally adequate processes to fail. From an organisational perspective, we are interested in not just how the company designed the boarding procedure but also tested its rigour; what training was given to cabin crew around the topic of aircraft weight and balance; what quality controls were in place to ensure correct loading and accuracy of paperwork. In fact, the local working practice we have just explored was not unique, and I have spoken to many captains who fell foul of the opposite condition. Because a part load of passengers was seated at the front of the cabin, the aircraft could not rotate and so a high-speed rejection of the take-off was required, a much riskier manoeuvre.

This next example contains many of the same ingredients but with the addition of more sophisticated technology and what we described in Chapter 2 as drift. On 7 December 2003, a Boeing 737-800 tipped on its tail as it started its take-off roll from Göteborg Airport, Sweden (SHK, 2004) because all of the passengers were seated at the rear of the cabin. How this scenario unfolded is almost the opposite of the previous excursion to the Greek islands. The aircraft had left Salzburg, Austria, with a full load of 180 charter passengers, 59 of whom disembarked at Göteborg, the remainder destined for Stockholm. No new passengers boarded at Göteborg. Because the flight was being conducted as a charter, the responsibility for passenger seating fell to the travel agent who booked the charter, not the airline operating the flight. To ease the disembarkation, the Göteborg passengers had been seated at the front of the cabin. The airline used an electronic system to produce its load sheets. As passengers checked in, the IT system generated a 'seat occupied' message, which was then used to determine that passenger loading was within limits. In this instance, two load sheets were required: the first for the sector from Salzburg to Göteborg and the second for the Göteborg to Stockholm sector. As the check-in desk at Salzburg is not connected to the networked system, a hard copy passenger load sheet was faxed to the central passenger and load control office. In the meantime, a default value of 'passengers evenly distributed' is used to produce the load sheet. The system also sends a warning message to the next destination to the effect that a default value has been used and that the passenger distribution must be verified. Unfortunately, the faxed passenger details from Salzburg were sent to an unmanned machine because the Salzburg office had an out-of-date telephone directory. Furthermore, because no passengers were boarding at Göteborg, the check-in desk was not opened, and so the default message warning was not seen by gate staff. The load sheet prepared for the second sector was based on the 'passengers evenly distributed' default message whereas, in reality, the aircraft was tail-heavy.

In contrast to the previous example, the task was distributed with networked systems and automated processes being employed in order to get the job done. In this case, work had been delegated to a third party – the tour operator – who then had to integrate activity with the airline's processes. Change, represented by the office relocation, was not communicated, resulting in essential information not getting to the appropriate human agent. We see processes designed for a scheduled operation being applied in a changed context without the implications being considered. So, it was assumed that passengers would both disembark and then board each time the aircraft was used. The process and supporting technology was predicated on that assumption. If you remove a component – boarding in this case – then the process fails. The use of default messages in IT systems is a common design element, but one that carries with it an inherent risk unless all possible use cases have been explored.

These examples cast some light on the complexity of airlines as an organisation and underscore the fact that we cannot look at safety as simply failure at the level of the individual. The airline, as an entity, is a business model embodied as a set of processes coupled with technology. A deficiency in any part of the jigsaw will induce a risk, and the manifestation of risk might be in ways that were unexpected.

Table 9.1 shows data from a small European regional airline operating a fleet of turboprop and narrow-body jets. Because the operations were seasonal, there was a core of permanent cabin crew with temporary staff being recruited for the summer. Recruitment and training are two functions that are part of any organisation's human resources function, even if they are subcontracted. Staff numbers must be matched to need. Too many staff are wasteful whereas insufficient staff represents a potential production limiting factor. Recruitment and selection involve finding a pool of applicants and then choosing the most suitable. Each part of the process has an inherent lead time and a cost. Having selected recruits, they need to be trained. Again, this requires a period of time determined by the training needed to meet the qualification standards defined by state regulations. Recruitment, selection and training need to be sequenced so that qualified staff enter service as the demand increases. New recruits need a period of on-the-job training and experience before they can be considered fully proficient.

The time series in Table 9.1 illustrates these processes at work. The shift from winter to summer starts in April as the hours flown by airline starts to increase. Recruitment has started and staff numbers show that the cadre of 'summer casuals' has been recruited. The new staff are on the payroll but not yet productive: they will be in training or undergoing line training. From May to August the flying task is at its maximum, and the tasking per crew member reaches its peak. As the Autumn approaches, the task starts to ease, and the summer casuals are released. If we look at the data for sick days, we see that there is a lag between effort peaking and staff falling ill. Furthermore, the winter data suggest that the burden of sickness is borne by the permanent staff: the summer casuals have gone, but the sick rate remains above average. Of course, some of the winter data reflect seasonal effect, such as 'winter flu', but the rate rises in the late summer and, so, possibly reflects workplace stress. But there is one other indicator not shown in the data. On the day that I visited

TABLE 9.1
Cabin Crew Productivity and Sick Leave

	J	F	M	A	M	J	J	A	S	O	N	D
Block hrs (1000)	1.2	1.1	1.3	1.7	2.2	2.2	2.2	2.3	2.1	1.9	1.1	1.1
Staff	134	134	134	176	176	176	172	172	172	170	143	143
Block hours/staff	9.1	8.5	10.2	9.9	12.6	12.6	13.1	13.4	12.6	11.7	7.8	7.7
Sick days/staff	3.5	3	2.4	2	2.45	2.35	3.05	3.01	3.25	3.59	3.42	3.18

this airline, I was in the office of the head of cabin crew when a report came in that an aircraft escape slide had been accidentally triggered. From the next-door office, I heard someone call out 'Summer's here!'. The lag between the increase in the flying task and the arrival of competent new staff resulted in additional strain on the existing crew that could be measured in the rate of accidental escape slide deployment during door operation.

Organisations, then, exhibit 'behaviour' in much the same way as humans do. In part, that behaviour is a product of how organisations set about satisfying the demands of production. The hierarchical model proposes that levels exercise control over subordinate levels, and a large part of the work of 'management' comprises deploying a suite of control methods. Because management processes are dynamic, it is possible to identify inputs, on the part of management, and corresponding outputs manifested by the workforce. In a perfect world, all outputs should contribute to the safe and efficient delivery of the desired outcome. Alas, the world is not perfect, and often, we see unintended consequences arising from what management might consider being logical inputs. Equally, management sets up feedback mechanisms, incident reporting being a classical example, and yet these processes, too, do not always work as intended. This chapter will examine the methods of control and the resulting unintended consequences. In the next section, I want to look at some fundamental economic problems airlines have to address before, then elaborating a framework for understanding organisational behaviour.

9.4 FUNDAMENTAL CHALLENGES AT THE HEART OF ORGANISATIONS

In order to achieve its goal of generating a return on the capital invested, a company has to resolve two specific tensions: risk and control (Garud & Shapira, 1997). In an economic sense, as opposed to safety, the risk is the anticipated difference between aggregated revenue inflows and promised payments to agents: in effect, it is the probability of generating the expected return. Control, again in an economic sense, refers to the right to make decisions about asset use not explicitly defined in law or contracts. From an organisational perspective, risk, in the sense it is used here, will shape decisions about asset configuration. Control can be seen as a negotiated space between the management and the workforce, and it is in this space that the safety risk is amplified if the economic risk and control are not aligned with system performance.

Although the term 'risk' has been used in two different contexts, the underlying concept is the same: will what we expect to happen actually come to pass? In a safety sense, we offset risks by putting mitigations in place. In a financial sense, we offset risks through interest rates and financial instruments, such as fuel hedging and insurance. Mitigation is an additional cost, but the excessive intolerance of risk (extreme caution) can represent foregone opportunity with, as a result, reduced reward. This interplay of the two modes of risk can be seen in this report of a Southwest Airlines 737-700 aircraft that ran off the end of the runway at Chicago-Midway Airport on 8 December 2005 (NTSB, 2006). The airline used three models of Boeing 737, the -200 and -500 have performance characteristics that allow a landing at Midway without the use of thrust reversers. The -700 series aircraft can only land at this airport without the use of reversers if the runway braking action is good. At the time of the accident, snow was falling and so the braking action was reduced. The crew had planned to use the thrust reversers, as was expected, but their delayed deployment resulted in the over-run. Although all the passengers and crew were unhurt, the aircraft could not stop on the runway. It passed through the perimeter fence and struck cars on a public road, killing a child.

The three models of aircraft in use all have different capabilities in terms of passenger capacity and range. The use of thrust reversers for landing performance planning allows the company greater flexibility in terms of routes and capacity it can offer into Chicago-Midway. In effect, the airline can carry more passengers from more airports. The financial risk, in this case, was the probability that the use of the -700 would be viable: could the airline fill the aircraft? The safety risk associated with this combination of aircraft and runway condition was managed through a performance calculation (runway braking action and the use of reverse thrust) and was dependent upon the crew's correct execution of the landing technique. This event is a manifestation of the safety/production paradox. Rather than being mutually exclusive, safety and production are two facets of the same problem that must be resolved in part by the organisation through its configuration of systems and, in part, by the crew through competent performance.

At an organisational level, financial risk is managed through the configuration of assets. The selection of routes, acquisition of aircraft, pricing of tickets, quality of service and so on reflect decisions made by management. The conduct of operations is controlled in such a way that crew performance will support the generation of anticipated returns. The problem we face is that the concept of 'control' as outlined in this section is fundamentally about the right to decide on the use of assets. So, implicit in this view is that it is the right of management to decide on how the crew will perform in relation to the company's business goals.

Just as the concept of risk has two different manifestations, control also has two meanings. These different perspectives create tension in a system. Control, in the sense that I have previously used it in my hierarchical model, invariably involves delegation. Each level in the model has, to some degree, to delegate authority for the actual conduct of work to the layer below for the system to function with any degree of efficiency. Where delegation occurs, performance is still expected to be delivered such that the observed outcomes fall within a tolerable range. The contradiction here is that the delegated authority contains a degree of discretion, or freedom of action, resulting from the gap between the formal controls and the flexibility needed to complete the task, given the vagaries of everyday life. The delegation also comprises both authority and responsibility. Captains, for example, are required to exercise authority in order to conduct business in such a way as rewards (profit per flight) are maximised but to act responsibly in order to minimise risk (and here we can also include financial risk in that crew behaviour can affect actual returns from a flight). Responsibility is bounded, on the one hand, by legal and procedural constraints and, on the other, a mutual understanding of what constitutes acceptable behaviour in the absence of controls. We saw in Chapter 5 that risky decisions are typically made under conditions of ambiguity, uncertainty and information overload. In order for the outcome of the decision process to be optimal, different levels in a hierarchy must have shared perceptions and understanding and a convergent set of expectations. The degree of convergence between the hierarchical tiers is reflected in the level of trust existing between the levels in an organisation.

This discussion rather assumes that an organisation exists as a coherent entity sharing a common set of goals. We know that organisations exist independently of the people filling specific positions within that organisation, but it is also true that organisations comprise disparate collections of groups, each with their own set of goals (Hutter, 2005). Members of an organisation take on obligations and expectations and contribute to corporate goals. However, at the same time, different levels within my model, and different functional groups within each hierarchical level, have independent goals and processes that are often in conflict. The problem of organisational silos was addressed by Tett (2015). She identified three main properties of silos: competition, ignorance and misplaced trust. Competition occurs when organisations have different performance goals or reward structures between the departments. In effect, it is not in my interest to cooperate with you because my own performance measures might be impaired. Ignorance is created when bits of an organisation fail to communicate. Data about a problem may exist but are buried in a department. Misplaced trust occurs when one bit of an organisation assumes that the other bit knows what is going on, which seems like a plausible precursor to organisational drift.

The issues discussed in this section represent a strategic view of an organisation but, in order to understand how strategy is transformed into action, we now need to look more closely at processes at the operational level. It is here that strategy becomes tactics. In effect, decisions about strategic goals are translated into desired work processes.

9.5 MANAGEMENT CONTROL AND WORKER RESPONSES

One view of a company is that it is a system for generating returns for shareholders. In fact, company directors, in most jurisdictions, have a fiduciary duty in law to look after the best interest of shareholders. Another model of a company is that it is a collective of stakeholders, all with competing interests, and that is the view that guides the analysis in this chapter. Earlier, I proposed that organisational citizenship and resistance were emergent properties at Level 3. Organisational citizenship is a concept that captures behaviour that we could describe as an employee 'going the extra mile'. Citizenship is the voluntary investment of effort for the benefit of the organisation. Organisational resistance is the opposite. Hodson (2001) identified categories of legitimate management activity that, if implemented poorly, generate resistance. I have recast Hodson's analysis in terms of management intentions and resultant outcomes. By resistance, Hodson means behaviour that usually falls short of direct action, such as going on strike, but still results in inefficiency and reflects a loss of goodwill. It is not necessarily sabotaging or vandalism (although that is not unknown in labour disputes), but it is an unwillingness to support, or a lack of alignment with, company goals. Resistance could be characterised as intentional suboptimal performance. The balance between citizenship and resistance might be seen as the practical manifestation of morale. The four key management functions and their associated actions derived from Hodson's work are:

Control v Management Chaos and Abuse
Demand v Overwork
Delegation v Challenges to Autonomy
Continuous Improvement v Contradictions in Employee Involvement

9.6 CONTROL, MANAGEMENT LEGITIMACY OR CHAOS AND ABUSE?

I want to start this section by restating the concept of 'control' as a legitimate right to make decisions about the employment of assets as it goes to the heart of many problems between the management and workforce. The important part of the definition is that control also relates to power 'not explicitly defined in law or contract'. Control affords management the space to act unilaterally in terms of asset use for which no specific authority or precedent exists. It is the tension between the management and the workforce that flows from this unregulated state that gives rise to issues of citizenship and resistance. Anderson (2017) has coined the phrase 'Private Government' to characterise the nature of post-industrial employment. She cites the ways companies exercise almost tyrannical control over employees and makes the point that if actual governments were as abusive, they would not be elected to power in democratic countries. One example she uses, which is becoming increasingly widespread in aviation, is random testing for substance and alcohol consumption. A rational argument can be made for random testing given that pilots have been caught attempting to board aircraft when above the legal alcohol limit and the effects of alcohol on performance are well known. However, Anderson's point is that to demand the right to test based *on no suspicion* could be construed as an abuse of management's power. In effect, management (albeit supported by law or delegated to a third party in some states) has drawn to itself the right to intrude into the *prior* life of the employee. By prior life, Anderson means life preceding the contractually obligated duty period. Although alcohol is voided through the system in a fairly short period, some recreational drugs might still be detectable weeks after use. Drug use during the employee's own time might leave a detectable trace that, although possibly so slight as to have no effect on current performance, remains grounds for action. Furthermore, management typically mandates punishment should an employee choose not to participate, even if innocent of any possible charge. Anderson's point, of course, is not to defend drug users or drunks. She is suggesting that the distribution of power between the workers and management has shifted significantly in favour of management such that we are now blind to the imbalance. On the one hand, this can encourage cavalier action on the part of management while, on the other hand, increase the likelihood of acts of resistance on the part of workers.

Another example, which will be picked up in the next section, is that of job demands. We saw in Chapter 4 that work-induced fatigue can impact a pilot's personal life and can have long-term health effects. In a study of nearly 400 pilots in 25 different airlines, 45% agreed or strongly agreed with the statement that they would not report fatigue 'for fear of investigation of off-duty time', and 34.8% responded similarly to the statement 'fear of investigation of commuting arrangements' (Greig, 2019). These responses reflect, in part, a fear of self-incrimination because pilots do have a duty to report for work in a fit state. In a significant number of cases, respondents are rejecting the company's right to intrude into their personal lives. Garud and Shapira observe that those who stand to lose their investment, in an economic sense, should have control over risk-taking activities. However, if that investment is the health and well-being of employees then the implication of this position suggests that current approaches to, say, setting flight time limitations need a radical review.

The unintended consequence of this imbalance between the management and workers lies at the heart of Hodson's analysis. In an attempt to demonstrate some semblance of independence, workers develop ways to express their frustration while stopping short of industrial action or resignation. Workers find ways to withdraw emotionally, and sometimes physically, from the workplace without provoking management sanction.

Hodson's framing of the reciprocal, management chaos and abuse, describes the way in which management's attempts to direct and control the workforce on a daily basis can often result in turbulence and confusion. A good example would be the stability of the published crew roster. The roster, of course, defines the life of a crew member over the period for which it is valid. Changes to the roster often disrupt personal plans, and yet the roster production process can be fragile and unstable. One airline I worked with managed to produce the pilot roster 2 weeks in arrears. Another airline employed an individual to run rostering, equipped with no more than a pencil and graph paper. The rosters were always incorrect. Pencil and paper methods have probably seen their day, but even modern computerised systems can still fail. Of course, the reasons for disruption do not solely rest with the management side of the equation: individuals fail to report significant personal changes to the rostering team; short-notice company tasks occur requiring individuals to be taken offline, and people fall sick. In a study of roster stability, 3% of changes were due to operational disruption, either weather or technical faults. Crew recognise that such disruptions occur and are tolerant of the subsequent need to change rosters. The biggest cause of roster instability was crew 'no-shows' for duties or training events. In fact, on an annual basis, the airline involved had to plan 33% more simulator events than was needed to meet the training plan because of crew no-shows (personal communication). However, notwithstanding the cause of the disruption and the challenges of staffing aircraft, the solutions generated by management to deal with disruption are often seen by the workforce as adding to the problem. For example, opaqueness around the proper establishment of the crew for the number of aircraft operated can be an issue. Having insufficient reserve pilots available, such that rosters were guaranteed to be changed, can be seen as simply pushing the problem downstream. Managements have to act in real time to resolve issues around staffing but the nature of their solutions sometimes causes friction.

The interpersonal styles of individual managers, together with perceptions of competence, company responses to the grievance and the silo effect mentioned earlier all represent chaotic and abusive management. In terms of how employees respond to it is worth just pointing out one way in which pilots and cabin crew differ from many other types of employee. Airline crew spend their working days *away* from the office. They spend little time in a central location and so more general communication can be difficult. One common form of resistance is choosing to ignore management altogether or finding ways to work around management. Individuals find ways to generate their motivation to work and create social groups within the workplace that provide a sense of belonging that does not flow from the company. These are forms of psychological protection that are, nonetheless, responses to the perceived management inadequacy. More active forms of resistance include not keeping up-to-date with company communications, using unauthorised workarounds and shortcuts to make the job simpler or reconfiguring the job to your satisfaction. In this latter case, role ambiguity, which was one of the sources of safety drift at the level of the individual, can be triggered by the management action.

9.7 DEMAND AND OVERWORK – EMPLOYEE SICKNESS/ABSENCE AS RESISTANCE

Flight crew salaries make up between 5.9% and 10.5% of airline direct operating costs (IATA, 2017). In my hierarchical model, the current gross availability of pilots is influenced by environmental factors. Forces such as the reduction in the scale of military aviation (and, hence, the supply of trained pilots), increasing costs of pilot training, reduced access to funding from banks for training after the financial crash of 2008 and the rapid expansion of commercial aviation in Asia (with the subsequent increase in demand for pilots) are all beyond the control of individual airlines. What airlines can control is the utilisation of crew: exploitation of the asset.

The expectation of effort imposed by management on the workforce is reflected in the design of the work schedules that sometimes rely on overtime, duty period durations and constraining time off or controlling access to contractual vacation allowances. We saw earlier that recruitment and training can lag behind the operational requirement thus creating a potential for overwork. Not advertising for a replacement until the present incumbent has left can also create a temporary vacancy, thereby representing a saving on staff costs. In aviation, flight time limitations represent a legal constraint on pilot utilisation and, thus, available production capacity. Defining crew productivity requirements represents an archetypical example of 'control' as framed earlier. While the contractual obligation may be defined in units of time, the consequences of how that contract is met place a significant burden on worker health and well-being. Where contractual demands are perceived to be either excessive or unreasonable, such that employees feel overworked, the classic response is absenteeism.

Company policies on eligibility for paid sick leave and absence management derive from state regulations, which differ considerably between the states. Companies can choose to offer more generous paid sick leave and more flexible absence policies than the required minimum, but the interplay between the regulatory frameworks and individual behaviour is actually quite complex. Establishing policies to deal with both sickness and absenteeism is a legitimate management function.

There are a variety of reasons why the crew absent themselves from the workplace. One is, of course, genuine ill health or a significant medical condition and is what we could call involuntary absence. The other main category of absence is voluntary: the employee chooses to not report for duty. Voluntary absence may be legal, an example being allocated vacation, or it may be illegal. An example of an illegal voluntary absence would be a decision to prioritise personal requirements, such as a family commitment, over the obligation owed under the conditions of employment. Another cause of illegal voluntary absence is disaffection. For whatever reason, the employee holds the employer in such low regard that absence is a form of private industrial action, not to be confused with more organised attempts to disrupt operations by mass 'sickness' events, for example.

Unlike the broader working population, there are several reasons why a pilot might be unfit while another worker might still be available, a head cold is one example. Pilots are also required to hold a valid medical certificate, and various rules relating to the issuing of a certificate also render the pilot unavailable for duty (minimum recovery times for specific conditions, restrictions on who can certify fitness). Pilots who are off work for more than a certain period are required to seek recertification. As a consequence, pilots might be fit for duty but still unavailable. Such pilots are deemed long-term sick. It is difficult, if not impossible, to make direct comparisons of lost productivity between the airline pilots and the general working population. Factors such as state rules on access to paid sick leave and the general job market affect absence rates.

Attendance represents a contested space between the workers and management where tensions over demand are visible. My first encounter with this issue was when I looked at daily operations in a UK low-cost carrier in its first year in business. Relations between the management and the 99 cabin crew employed by the airline were poor but the company did not recognise unions, and so the cabin crew felt that they had little representation. In response to perceived increased rates of absenteeism, management issued instructions that annual leave would be awarded on a point system based on attendance records. The new policy came into effect at the end of the summer season, just as the weather started to change and a common time for colds and flu. The crew now started reporting for duty with head colds, and several had issues with clearing their ears in flight, with their balance, and some were so ill that the captains refused to let them fly on the return sector. Over a 2-week period, nine cabin crew were offloaded at destinations, causing serious disruption, including passenger load restriction on return flights because of cabin crew to passenger ratio limitations and flight re-routing to collect stranded passengers. Another airline I worked with actually issued a policy that said 'colds and flu was not a sufficient reason for failing to report for duty' while another stated that 'stress' was not an acceptable cause for absence, just as the head of cabin crew departed for 6 months of stress-related sick leave.

The interplay between policy and behaviour can be quite complex. Figure 9.1 shows pilot absence rates and pilots with unbroken annual attendance records (not absent) over a 10-year period for a medium-sized airline. Both sets of data suggest a change over time with absence rate increasing as more pilots took days off. While interesting, the data do not reveal the cause.

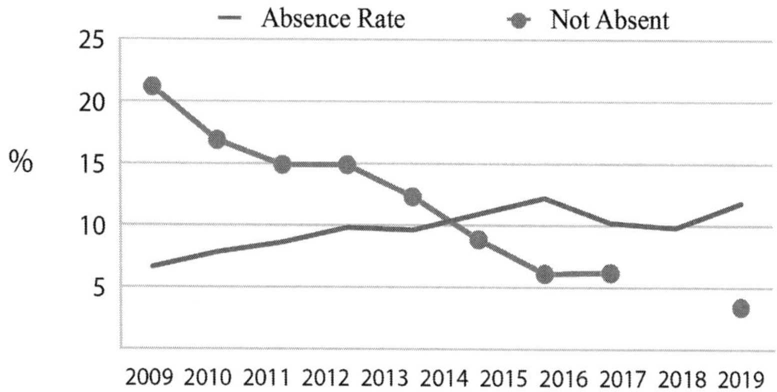

FIGURE 9.1 Pilot absence.

In order to understand the data, we need to look at the distribution of absence events. Patterned absence is a common feature of employee behaviour. For example, people are often ill on a Friday or a Monday. Bank holidays often result in spikes in additional time off. These patterns are common across all areas of employment. Figure 9.2 shows the distribution of absence events per pilot for the same airline over an 11-year period. The number of pilots having no absences declined over the full range of the data set. (Figure 9.1 shows this as a rate for all pilots while Figure 9.2 shows the rate as a percentage of all absence episodes). The rate for one or two episodes/pilot remained fairly constant, which is to be expected given that humans do fall ill. There is a change in the number of crew taking three or four episodes and again for five or six episodes a year with the trend generally upwards for each category.

Figure 9.3 shows the distribution of episodes by duration. Again, the data show that taking 1 or 2 days off at a time remained fairly constant while absences of 4 days duration show a gradual decline. The episodes of 3-day duration show a remarkable

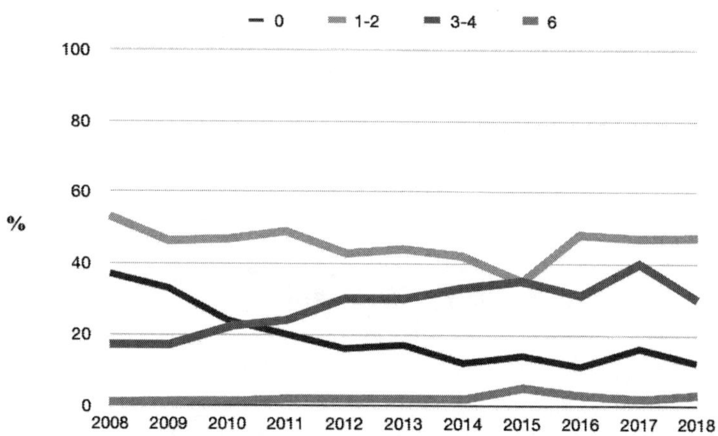

FIGURE 9.2 Absence events per pilot.

increase in 2016, coinciding with a decline in the rate for episodes of 6-day duration. While the pattern for events shows an increase over time, the data for episode durations shows a change. Broadly speaking, a couple of days off a couple of times a year might be considered normal wastage, but crew appear to then taking more episodes of 3- or 4-day duration.

For the period from 2008 to 2015, the absence management rules in force at the airline were six events or 30 days total in a rolling 12-month period. The rate for six episodes in Figure 9.2 probably reflects the difference between a calendar year and a 'rolling 12-month' period. It is possible to be absent every 2 months and not trigger the threshold. This is possibly the first clue that pilots are able to game the system. In 2016, the rules were tightened up to try to reduce the rate of absence. Trigger events were set at five episodes in 12 months and a total of 21 days. In addition, whereas 7 days of self-certified absence per episode was permitted under the old rules, this was changed to 3 days, after which a doctor's certificate was required. This change in self-certification policy seems to be reflected in Figure 9.3. Absence durations reflect the rules for submission of a sickness certificate. To explore the impact of this policy change in more detail, Figure 9.4 shows the data for a specific fleet for the year in which the self-certification limits changed. The new policy was introduced in April. There is a clear shift away from episodes of 7-day duration to the new 3-day limit.

Absence is a legitimate concern for management. No-notice absence is disruptive and unwarranted absence represents moral hazard: a theft of time and wages. Looking at episodes and durations together, patterns could be detected of which the most common was the '4 and 20' rule: 5 days off every 3 months. The fact that there were clearly discernible patterns of absence across the year can only suggest intentional behaviour. The willingness to fulfil contractual obligations is a barometer of the state of the relationship between the management and workforce. It could be seen as an indicator of morale. The data suggest that pilots respond to management interventions, but they still do not address the cause. In these patterns, we see pilots coping with genuine illness but also, in all probability, some are using absence to recuperate from the demands of the task or manage the competing demands of family life. Others may be expressing their level of disaffection with the organisation or choosing to place self-interest ahead of contractual obligations.

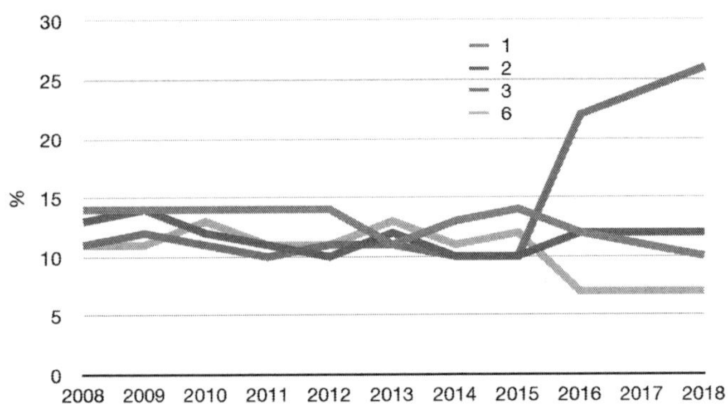

FIGURE 9.3 Absence episode duration (days).

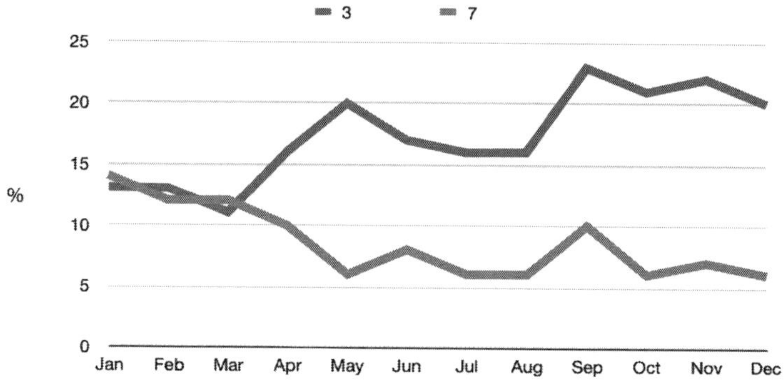

FIGURE 9.4 Number of absence episodes (% of all episodes).

Policies and procedures are tools that can be used by management to exercise control, but this example suggests that rather than control an issue, policies shape the nature of the response. There is a tension between the management of legitimate action and abuse. In 2018, the UK confidential human factors reporting programme (CHIRP) received several reports from the crew who felt that absence management rules were being used to apply undue pressure in order to coerce crew to operate when unfit. At the time of writing, a study is underway to consider the implications of absence policies from a safety perspective.

As we saw in Chapter 4, aviation makes significant demands on the individual, and ill health is often the result of succumbing to those demands. Absence from the workplace is usually legitimate but, where it is not, it is often unavoidable. The issue we have been considering here is that the space between the management and workforce where individuals exploit rules out of personal interest but also as an expression of disaffection. The next area of tension between the management and pilots is what Hodson called 'challenges to autonomy'.

9.8 DELEGATING CONTROL: CHALLENGES TO AUTONOMY – WHY FUEL EFFICIENCY MEASURES ARE RESISTED

Occupational groups deemed 'professions' are characterised as being of high status, having significant responsibilities associated with their work and having high levels of autonomy in how they do their work. It is widely understood that the loss of autonomy undermines self-esteem and, hence, is a stressor. In this section, I want to consider autonomy from Hodson's perspective. Delegation of authority, as discussed earlier, is an essential requirement in all organisations. What Hodson suggested was that having delegated responsibility to an individual, management must be careful when questioning the subsequent judgement of that individual. Delegation implies trust whereas a constant questioning of decisions suggests the opposite. A classic example would be the use of the commander's discretion to extend the duty period in order to finish a task. Company procedures state that, as the name implies, discretion is for the commander of the aircraft to apply based on an assessment of operational need and crew capability. It is not unknown for a commander to choose not to operate into discretion for perfectly legitimate reasons only to have the decision challenged. Equally, the industry has emphasised the importance of discontinuing the approach and initiating a 'go around' if the criteria for a safe landing are not met. A 'go around' and the subsequent additional flight time to repeat the approach is a cost, in terms of fuel, and possibly a source of inconvenience to passengers. Although the situation is changing, it is still possible to find situations where a company may advocate the safest approach but are clearly unhappy if a pilot complies with the requirement. In the previous section, we looked at employee absenteeism, and I said that one reason for being absent is ill health. In Chapter 4, we looked at fatigue, and most airlines have a procedure for reporting unfit for duty if the pilot considers themselves to be too fatigued. In a 2018 study, the Australian Transportation Safety Board (ATSB, 2019) surveyed 625 pilots about their experiences of fatigue and perceptions of the culture in their organisations. Of this sample, 277 pilots reported that they had removed themselves from duty in the previous 12 months because of fatigue. Even though they had used the extant company process, 57.1% claimed that their action had left a negative impression with the management. These contradictions between the espoused positions and enacted realities generate resistance behaviours. Hodson observed that workers create independent meaning systems and develop their views of processes and how they should be conducted. In short, people start to 'do their own thing'. They take pride in finding solutions to problems despite what the company wants. Fuel efficiency is an area where we can see these forces at work.

Fuel costs comprised 23.5% of airline operating costs in 2018 (IATA). Of course, fuel costs are linked to the price of oil, which reflects both production output and global security issues. Fuel efficiency addresses the issue of economic reward, but one component of efficiency is the performance of the crew, a potential risk. In terms of challenges to autonomy, the commander is responsible for the final fuel uplift decision. Airlines use sophisticated programmes to calculate fuel requirements and policies usually state that if a commander takes the fuel suggested by the flight plan then the company accepts responsibility for any fuel-related outcomes. However, commanders retain the right to adjust the fuel taken. Most airlines have tools for monitoring fuel uplifts by individual pilots, and many have mechanisms for feeding back to pilots where uplift is considered excessive. Pilots can affect fuel efficiency by adopting specific procedures and through the general operation of the aircraft. In part, the willingness to adopt fuel efficiency measures reflects aspects of pilots' morale and attitudes towards the company.

Vogel (2014) looked at a range of fuel efficiency measures and investigated the rate of uptake and also the perception of workload as measured by the NASA task load index. Table 9.2 shows the adoption of a range of measures while Figure 9.5 shows the subjective assessment of workload for the two pilots in the different roles of pilot flying and pilot monitoring. The perceived workload by the phase of flight shows a relationship with fuel efficiency measures. Single-engine taxi out is the most obvious measure that appears to be discarded because of the high workload. More than one-third of pilots reported performance impairments resulting from the application of fuel-saving techniques, and 2% ($n = 19$) reported involvement in an accident or serious incident. Trying to satisfy a management policy creates an apparent operational risk.

TABLE 9.2
Adoption of Fuel Efficiency Measures

Fuel-Efficient Policy	Never	Seldom	Sometimes	Often	Always
Single-engine taxi out		x			
Single-engine taxi in			x		
Adherence to optimum flight levels				X	
Flaps 3/25 landing			x		
Idle reverse				x	

Source: After Vogel (2014).

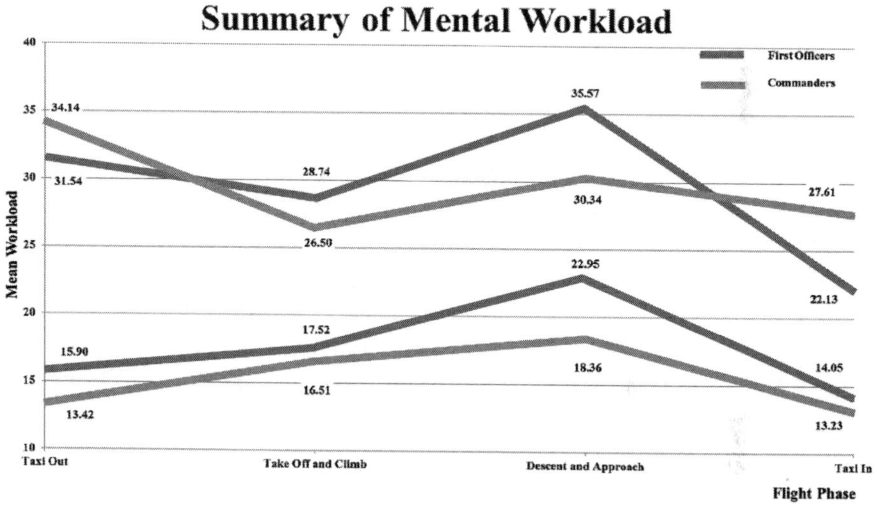

FIGURE 9.5 Workload (From Vogel, 2014).

Fuel efficiency sits at the intersection of management activity aimed at increasing returns and pilot perceptions of effort, safety and net benefit. The increased workload is a safety risk but also represents a discretionary commitment to the task. To meet the requirement to be more fuel-efficient, the crew could be thought of as having to sacrifice operational margins. The question becomes one of who benefits the most? In the context of this section, although the responsibility has been delegated to the commander of the aircraft for decisions about fuel, the requirement to satisfy an efficiency target is now a constraint: the two policies are contradictory. Having delegated the responsibility for safety to the operating crew, their perception of priorities – safety first – is undermined by demand for efficiency. The problem is exacerbated by the management responses to excessive fuel uplift. In some airlines, the judgement of pilots who take more than the flight plan fuel is sometimes questioned.

Gosnell et al. (2019) addressed the pilot's adoption of fuel efficiency policies from a behavioural economics perspective. All 335 captains in an airline were informed of a study being implemented to explore fuel efficiency, but only some of them were to be assigned to specific intervention groups. Interestingly, all captains improved their performance during the study period irrespective of the treatment group they had been assigned to. Giving feedback on recent performance had a moderate effect on efficiency, but setting personal targets for captains was more beneficial. After the study concluded, captains said that they would value more information about their individual performance. In the broader context of competence development, these comments from pilots point to opportunities to shape behaviour.

Hodson's concept of denial of autonomy highlights the tension identified by Anderson when she discussed the asymmetries of power. The company has delegated authority for decision-making but retained the right to chastise if the operational goal is not always compatible with the commercial goal. If we stay with fuel management, unforeseen circumstances have resulted in fuel-related accidents and incidents, and here, we only have to think back to the Portland, Oregon, accident that led up to CRM in the first place. The probability of fuel becoming critical is often a function of sector length and airspace characteristics. The need to avoid weather or cope with arbitrary ATC interventions can affect the track miles flown and, thus, the fuel consumed. Whereas a fuel policy is often a fixed point, actual fuel consumption, and therefore the reserves required, is a variable. Ergodicity suggests that a failure to recognise this fact will result in failure.

9.9 STRIVING FOR EFFICIENCY: CONTRADICTIONS IN EMPLOYEE INVOLVEMENT – WHY SAFETY REPORTING FAILS

The final area of interest in our discussion of management behaviour is the apparent tension between employee involvement in the broader design of work and the implications for future operations. Management actions that seem to be inclusive of the workforce, such as quality initiatives and safety reporting, appear to represent a rational approach to issues but can often result in additional work and added restrictions and constraints on workers.

Safety reporting is a fundamental tool in safety management but is also a good example of this idea of seeking employee commitment while not always rewarding that apparent investment in safety. Having a 'strong reporting culture' is usually seen as a sign of a healthy safety culture in an organisation, and yet, under-reporting of events is common in any safety-critical domain that relies on worker feedback. Little research has been done on rates of under-reporting in aviation. In a study of a German carrier (Helsabeck, Schmidt-Moll & Schubert 2015) a questionnaire was administered to 86 pilots. The authors found that under-reporting rates differed according to the nature of the event. For significant events, such as landing with less than 30 minutes of fuel, reporting was 100%. Events that could be attributed to outside agencies, such as time pressure induced by organisational deficits, were reported on 94% of occasions. Adverse weather event reporting achieved an 82% reporting rate while events involving near-misses on the ground were reported on 33.3% of occasions.

In an attempt to quantify under-reporting, I looked at 1107 errors observed on a line operations safety audit (LOSA) in 2013. Three were deemed, by the observer, to be reportable events. Because of confidentiality rules, it was not possible to verify whether a safety report was ever submitted for the relevant flights. However, as no reports were submitted for routes between the relevant city pairs during the study period, I think it is safe to assume that the events went unreported. I then looked at the safety reporting for the audit period and identified comparable events. For example, floated or hard landings or deviations off track to avoid weather without ATC clearance were seen in both the safety data and the LOSA. By comparing the number of reports submitted over the study period with the occurrence rate in the LOSA data, I calculated the rate of under-reporting. For those events where there was a chance of being caught by a third party (i.e. ATC might file a report or the flight data analysis programme might pick up the event) there was a 12.5% reporting rate but where there was no chance of being observed by a third party then reporting was closer to 2%. In another study, I looked specifically at arrivals and departures into the airlines' home port. This destination generated most data in both modes of sampling: all aircraft have to come home at some point. The reporting rates for three different fleets are given in Table 9.3.

There is quite an extensive literature across a range of domains on why safety reports are not submitted although, again, the specific evidence from aviation is limited. In the German study, the reasons for not writing reports differed between the short and long-haul pilots. For short-haul pilots, the effort required was the most frequently cited reason for not reporting. In the context of this chapter, negative feedback by their superiors was the next most frequent reason followed by the lack of any observed change as a result of reporting. For the long-haul pilots, negative response from managers was the biggest disincentive followed by the lack of any change as a result. Long-haul pilots commented that they generally felt reporting was meaninglessness and insignificant. Hutter and Lloyd-Bostock (2015) interviewed a sample of British pilots about reasons for not reporting. Once again, the effort needed and wariness about management responses were core themes. I fed my data on under-reporting to groups of pilots and asked why they, specifically, might not submit a report. The effort, mistrust of management and the lack of any observable response were core themes. This idea of reporting being potentially harmful probably explains some of the differences shown in Table 9.3, reflecting cultural issues within fleets as reflected in management responses.

Of particular interest was the fact that many pilots talked about reporting as a variable, not a constant. As one pilot said, 'the first time I encounter a problem

TABLE 9.3
Reporting Rates

	Arrivals (%)	Departures (%)
Fleet A	27	29.4
Fleet B	74.1	43.4
Fleet C	25	14.7

I might report it but the next time I come across it I know what to do so why report it?'. Another training captain said that 'If I'm doing a training flight with a new FO I might report something but if it was a normal line flight for me I probably wouldn't'. The normalisation of operational challenges was commonplace. Some destinations were renowned for the unreliability of services and facilities, and crew simply stopped reporting issues 'because there is nothing that can be done'. The idea that reporting is socially constructed also emerged from Hutter and Lloyd-Bostock's work. Events, such as fatigue, became normalised and, so, the crew stopped reporting it. The formal 'rules' around reporting were of less importance than perceptions around how reports will be used. The authors found that younger pilots learn about reporting by participating in discussions with more senior captains. What we are seeing here is another form of culture. Social norms around reporting are communicated to the next generation of pilots.

The paradox of reporting is neatly captured in this encounter I had while running a reporting system for a small Scandinavian airline. Talking to one of the cabin crew about the gap between the reports submitted and the fact that I knew of events happening on the line that did not get reported her response was, 'My job is to fix problems. When I'm just doing my job, why should I write reports about it?'. Reporting of events is required under rules issued by aviation authorities and is elaborated in company policies and procedures. The evidence, though, suggests that reporting is a negotiated process. The effort required to report is balanced against crew perceptions of benefit and personal risk. The resolution of this equation is seen in terms of the rate, coverage and quality of reports submitted. The reporting of hazards and safety-related events is considered to be the cornerstone of safety management which, in turn, is seen as part of a broader initiative aimed at incremental improvement. The benefit to the organisation is seen in terms of reduced risk through better defence against hazards and possible loss prevention. Unfortunately, the participating crew do not always see any advantage in reporting. Crew expend effort while running the risk of sanction yet rarely see any positive change that might make their jobs less demanding. Interestingly, commercial air carriers are required to have an approved safety management system and, therefore, the establishment of the scheme is a sunk cost. The actual effectiveness of the system is rarely measured and, so, there is little incentive to address the issues raised by the crew as barriers to reporting. This behaviour is suggestive of self-organised criticality, discussed in Chapter 3. In this view, protection is seen as a component of the cost of production. Organisations will accept a level of protection that is deemed adequate but does not impede production. Because protection represents an impediment, organisations migrate to a position in a notional 'safety space' where they are 'safe enough' at least cost. Addressing issues around the effectiveness of reporting represents an unnecessary additional cost when the condition imposed by the regulator – having a reporting system in the first place – has been met.

Workers respond to these apparent contradictions by building a stronger sense of identity within the workgroup rather than within the company as a whole ('it is against them') and reducing the emotional commitment to the company ('screw them!').

9.10 DO ORGANISATIONS LEARN?

I want to end this chapter with a brief discussion of organisational learning. Organisations create risk through their routine operations, and they implement controls to minimise or mitigate those risks. Controls include rules and procedures, practices and surveillance (audit and reporting systems being forms of surveillance). As we have seen, all of these controls are fallible. Normal operations are subject to routine non-conformities which occasionally become anomalies. In the worst case, an anomaly becomes a public catastrophe: failure is now visible. There will always be a tension between control and chaos and an effective learning organisation actively seeks evidence of potential failure. All organisations need to be flexible and adaptive and to be able to respond to changes in their operating environment. Resilience flows from effective oversight of the operation and meaningful engagement with practitioners to generate learning. Hollnagel (2009) offers the following four cornerstones of a resilient organisation:

1. Respond to events
2. Monitor ongoing developments
3. Anticipate future threats and opportunities
4. Learn from past failures and successes

From these, we can see that monitoring ongoing developments will be impaired if reporting systems are inadequate.

Learning can be challenging for two reasons. The first is that success is the expected outcome of work, and it can be hard to identify why success occurred other than the crew just did their job as planned. The second is that organisations respond badly to adverse events. The short-lived low-cost carrier 'Go' had two B-737s take-offs from London Stansted Airport in the space of 2 weeks with the nose gear pin still in place (personal communication). As a result, the aircraft had to land in order to have the pin removed. It was made known to the crew that the next person to take-off without removing the gear pin would lose their job. A few weeks later, a crew was about to depart. It was a wet day and visibility was poor. The captain looked out to see what he thought was the ground engineer holding the nose gear pin with its florescent flag. Thinking they were safe to go, the crew departed. Unfortunately, it was not the warning flag they could see but the high visibility markings on the sleeve of the wet weather jacket of the engineer. The mood of the company management was transmitted to the workforce but the way it shaped behaviour was not quite as intended.

An effective learning organisation displays four characteristics (Easterby-Smith & Araujo, 1999):

1. It actively generates information.
2. It integrates that information into the organisational context.
3. The organisation makes a collective attempt to interpret the new data through some form of interaction.
4. Individuals and groups are encouraged to take action based on a shared understanding of the problems raised.

These characteristics overlap Hollnagel's cornerstones of resilience but, of interest, is the suggestion that learning arises from collaborative action at an organisational level with individuals empowered to act based on new knowledge generated through this shared understanding. Organisational learning requires effort, but it seems that, often, the action is an afterthought. Hollnagel proposes that a resilient organisation is anticipatory, and we saw in Chapter 1 that an effective training system should deliver competence designed to cope with potential risks to the operation rather than rehearse past events.

9.11 HOW LEVEL 3 WORKS

The work of Level 3 is to configure assets and develop processes that generate a return on investment (or, in the case of not-for-profit operations such as air ambulance and police, provide a capability for a defined term within the budget). The decisions made at Level 3 and, thus, its defining characteristics, are to do with protecting and accruing capital. The task of the crew is to fulfil the obligation the company has with the customer: they deliver on the contractual promise made by the company to the client. Those operational controls that are put in place (checklists and procedures) fall under the auspices of regulatory oversight. The company must negotiate with the agents of the regulator as part of the obligations that flow from holding an air operators certificate. This chapter has explored other forms of control: those which are implicitly exercised through the discretionary behaviour of the company, facilitated by the uneven power balance between the employees and their employers. Rational management interventions, however, result in unintended responses, and control mechanisms trigger acts of organisational citizenship and resistance.

Feedback is another component of my system that, supposedly, validates the conduct of operations. Feedback to Level 3 is implicit, in the main: production targets are met, aircraft depart and arrive. Other forms of formal feedback include captains' voyage reports, flight data capture and reports from third parties (complaints or reports of violations). Again, the value of feedback is dependent upon organisations having the capacity and mechanisms for receiving, processing and aggregating information into useful knowledge. The examples we have seen here suggest that feedback, in the shape of attendance rates, policy buy-in and reporting systems, represents signals suggestive of dysfunctionality.

Safety is undoubtedly influenced by Level 3 processes to the extent that they support or impede work at the lower levels. Throughout this book, we have seen examples of job designs, resourcing and scheduling (or rostering) that have had an influence on the way crews have been able to do their job. At one extreme, the design of the job has been an impediment. At the other, crews have lacked the physical and mental resource to cope with demands. In a systems context, Level 3 decisions reduce the buffering available either directly through structural elements such as policies and procedures, equipment, maintenance and job design or indirectly through influencing crew capability. A fundamental emergent property of Level 3, I would suggest, is morale.

9.12 CONCLUSION

This chapter has attempted to map out a framework that explores the relationship between legitimate management action and potential worker responses. Once again, we see that the activity is manifested in decisions that shape policies which then feed through to action in the workplace. Policies influence decisions made in the workplace. I have long argued that 'management' is a capability that can be characterised in terms of quantity and quality. Managers need sufficient capacity to cope with the demands of their role but also the skills to effect policies that will achieve the intended, rather than the unintended, goals. Where management is ineffective, the system migrates from one of resilience to one where small disturbances will be catastrophic. Ineffective management facilitates safety drift through an inability or an unwillingness to anticipate outcomes and take appropriate action.

REFERENCES

AAIB. (2004). Bulletin No 1/2004, report reference EW/G2002/10/21 Airbus A320-214, G-OOAR at Kefallinia, Greece.

Anderson, E. (2017). *Private Government.* Princeton University Press. Princeton, NJ.

ATSB. (2019). Fatigue experiences and culture in Australian commercial air transport pilots. Aviation Research Investigation 2015-095 Final – 22 January 2019.

Calder, S. (2002). *No Frills: The Truth behind the Low-Cost Revolution in the Skies.* London: Virgin.

CHIRP. (2018). Issue 128. 4/2018.

Easterly-Smith, M. & Araujo, L. (1999). Organisational learning: Current debates and opportunities. In Easterby-Smith, M., Araujo, L., & Burgoyne, J. (Eds.), *Organisational Learning and the Learning Organisation Developments in Theory and Practice.* (pp 1–21). London: Sage.

Garud, R. & Shapira, Z. (1997). Aligning the residuals. In Shapira, Z. (Ed.), *Organisational Decision Making.* Cambridge: Cambridge University Press.

Gosnell, G., List, J., & Metcalfe, R. (2019). The impact of management practices on employee productivity: A field experiment with airline captains. Centre for Climate Change Economics and Policy Working Paper 296/Grantham Research Institute on Climate Change and the Environment Working Paper 262. London: London School of Economics and Political Science.

Greig, A. (2019). Is Fatigue Under-reported? Unpublished MSc thesis. Coventry University.

Haslbeck, A., Schmidt-Moll, C., & Schubert, E. (2015). Pilots' willingness to report aviation incidents. *ISAP Conference Paper.*

Hodson, R. (2001). *Dignity at Work.* Cambridge: Cambridge University Press.

Hollnagel, E. (2009). The four cornerstones of resilience engineering. In Hollnagel, E. & Nemeth, C.P. (Eds.), *Resilience Engineering Perspectives, Volume 2: Preparation and Restoration.* London: Taylor & Francis.

Hutter, B.M. (2005). 'Ways of Seeing' Understandings of risk in organisational settings. In Hutter, B.M. & Power, M. (Eds.), *Organisational Encounters with Risk.* Cambridge: Cambridge University Press.

Hutter, B.M. & Lloyd-Bostock, S. (2015). *Investigating Reporting Culture amongst Pilots: A Briefing Study.* London: LSE Enterprise.

IATA. (2017). IATA Economics Chart. Week of 23 June 2017.

IATA. (2019). Fuel Fact Sheet. June 2019.

NTSB. (2002). Accident Brief DCA01MA034. Avjet Gulfstream GIII, Aspen CO. 29 March 2001

NTSB. (2006). Runway Overrun and Collision Southwest Airlines Flight 1248 Boeing 737-7H4, N471WN. Chicago Midway International Airport Chicago, IL. NTSB/AAR-07/06. 8 December 2005.

SHK. (2004). Report RL 2005:20e Incident involving aircraft LN-RPLat Gothenburg/ Landvetter Airport, O county, Sweden, on 7 December 2003. Case L-59/03.

Tett, G. (2015). *The Silo Effect*. London: Simon & Schuster.

Vogel, D. (2014). An Analysis of Human Factors Aspects in Operational Fuel Saving. Unpublished PhD thesis. London: City University.

10 Facilitating Aviation – Decision-Making at Level 4

10.1 INTRODUCTION

Level 4 in my hierarchical model comprises those entities that facilitate the conduct of aviation, both private and commercial. They provide some of the tools of production – the most obvious being airports and air navigation services – and also the regulatory frameworks and infrastructure without which modern aviation would be impossible. They may be state-run entities, private companies or some form of private/public enterprise. I have argued that the distinguishing feature of each level in the hierarchy is the mandate it holds to make specific classes of decisions. At Level 4, agencies make decisions about the provision of asset and the granting of permissions to lower-level actors. For example, governments build airports and runways. They also fund satellite systems that allow aircraft to navigate. Pilots cannot fly without a license issued by an authority. Airlines cannot do business without an operator's certificate granted by an authority. A manufacturer cannot sell an aircraft unless it has an airworthiness approval, again, issued by an authority. Organisations at this level do not fly aircraft for gain, but without them, commercial aviation is not possible.

In this chapter, I will elaborate on the role of Level 4 agencies, how they exercise control and, importantly, some of the challenges they face that can give rise to failure. I want to look specifically at regulators and accident investigation agencies. Both are established by state legislation and exist in a symbiotic relationship. While aviation authorities exercise control, investigation agencies act as a feedback channel. Investigators need independence if they are to do their work properly but, while they inhabit the same ecosystem as authorities, they act in an advisory capacity. They lack power, and this has implications. Both create and promulgate rules and procedures, advice and guidance that shape activity, directly and indirectly, at lower levels. In the context of this book, activity at Level 4 creates opportunities for cross-scale interactions in the system, some of which we have seen in previous chapters. The impact of Level 4 on operations, then, is both direct through the need to comply with regulations but also indirect, because of the system dynamics. The key emergent property, I suggest, is trust. Failure on the part of either of these two key actors damages the travelling public's trust in the aviation system and, hence, its economic viability.

I start with a discussion of the nature of regulation and then explore the dynamics of Level 4. As I write, the manufacturer, Boeing, is mired in the fallout from its ill-fated introduction of the B-737MAX. Although the story is ongoing, I will use what we do know about the certification of the aircraft to examine the relationship between authorities and manufacturers. I then explore the ditching of an aircraft off the coast of Norfolk Island, in the Pacific, in 2009 to throw light on the workings of an investigation and how the act of investigation is rarely a search for a simple truth. I concluded by looking at the relationship between Level 4 and the lower levels in my hierarchical model.

10.2 THE NATURE OF REGULATION

I want to start with a very brief look at how regulation has developed in aviation. The first powered, controlled, heavier-than-air manned flight took place on 17 December 1903 at Kitty Hawk, North Carolina. In France, Santos-Dumont flew his aircraft in 1907 while the Wright Brothers travelled to Europe in 1908. Bleriot made the first crossing of the English Channel on 25 July 1909. In less than 6 years, aviation had gone from a 12-second hop to a 30-minute international flight. The subsequent development of aviation has been phenomenal. Prior to 1914 and the outbreak of war, in the United Kingdom, regulation was primarily aimed at keeping people *on the ground* safe. The Board of Trade was responsible for the registration and certification of aircraft and pilots. War is often a crucible for technological development and, by the end of the 1914–1918 war, aircraft capable of carrying passengers for some distance became available as military bombers were decommissioned and converted. After 1918, although regulation remained focused on the airworthiness of structures, the promotion of the growth of air routes connecting the outposts of the empire became important, and we can start to see the emergence of a commercial imperative in regulation in addition to safety. It was originally the British government's intention to delegate oversight to a registry, following the maritime model. The registry would verify the airworthiness of aircraft and, presumably, issue certificates. Up to this point, progress was driven by aircraft manufacturers, hence the emphasis on airworthiness. In 1933, the Gorell Committee was established to look at the broader aspects of aviation (Chaplin, 2011). As part of their work, analysis was undertaken of the 355 aircraft accidents that had occurred between 1926 and 1933. Of interest to us is the fact that in 258 cases (72.67%), 'no part of the aircraft or power plant was in any way at fault'. Specifically, 177 cases (49.85%) were attributed to 'errors of airmanship alone'. The committee concluded that 'entirely the wrong values have been placed on the relative importance of the pilot, the machine and operational activities in arriving at the regulations'. Furthermore, 'the failure of the pilot is by far the most potential source of accidents in flying'. Hobbs (2004) in an analysis of 100 aircraft accidents in Australia between 1921 and 1932 found a similar result, rather undermining the myth that increasing aircraft reliability prompted a need for crew resource management to look at the 'human factor'. It has always been there, we just never looked for it. The publication of the Gorell Committee's report followed that of the investigation into the R101 airship disaster and emphasised the need for a more comprehensive approach to aviation regulation. From the outset, then, it seems that the framing of regulations was problematic. The entity to be controlled – pilot or aircraft – was open to debate, and the initial assumptions about the issues to be guarded against were proving to be ill-found.

The Second World War (1939–1945) further disrupted the development of commercial aviation, but as it was drawing to a close, the Allied Powers began to build the frameworks that would govern the coming peace; one of which was to set up the International Civil Aviation Organisation (ICAO), a branch of the UN, that would set the standards for global aviation. In the United Kingdom, increases in the technological complexity of aircraft, such as the de Havilland Comet, the world's first jet airliner that flew in 1949, and moves towards broader international cooperation, exemplified by the Concorde project, provided an added impetus for collaborative approaches to regulation. The harmonisation of rule-making in Europe has led, ultimately, to the formation of the European Aviation Safety Agency (EASA). Within the United Kingdom, regulation had been undertaken by different branches of government, but an independent authority, answerable to a government minister, was established in 1969 with responsibility for:

- the economic regulation of route licensing
- the control of air traffic
- the regulation of safety

In the USA, aviation developed along different lines to those seen in Europe. The carriage of mail was a significant stimulus in the development of commercial aviation, and legal oversight was the responsibility of the Department of Commerce. The death in an air crash of a famous American Football coach in 1931, followed by that of a US senator in 1935, prompted Franklin D. Roosevelt to sign the Civil Aeronautics Act into law, creating the independent Civil Aeronautics Authority. Accidents have often been significant triggers of regulatory change. For example, in 1919, the collision between two Vickers Vimy aircraft flying in opposite directions, in bad weather, and following the same railway track north of Paris resulted in the introduction of the semicircular flight rule. In June 1956, two passenger aircraft collided over the Grand Canyon, killing all the 128 passengers and crew on board. The accident highlighted the fact that although aviation had grown significantly, oversight had not. In 1958, a law that created the Federal Aviation Agency was passed, with a mandate to provide safe and efficient use of the national airspace. In both the UK and US cases, oversight was originally framed around registration, guarantees of airworthiness and commercial development. The system safety responsibility was a later addition.

10.3 REGULATION AS HIERARCHICAL CONTROL

To illustrate the dynamics of control at this level, I want to revisit the example we looked at in Chapter 8 of the crew of a DHC8-Q400, dealing with a propeller overspeed (SHK, 2007). When seeking certification of an aircraft type, the manufacturer, as part of the approval process, has to provide an aeroplane flight manual (AFM) that covers operational instructions, performance data, the minimum equipment list and checklists for normal and abnormal/emergency conditions. When the aircraft is delivered to the operator, it comes with an aircraft operations manual and the quick reference handbook (QRH), both comprising subsets of the information in the AFM. The operator is required to produce its own operations manual (OM) for issue to

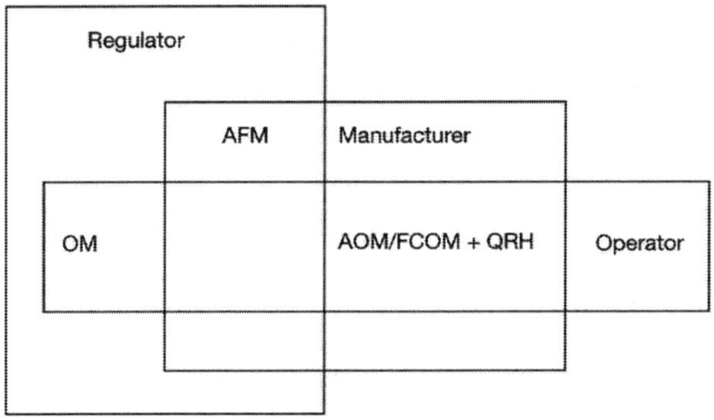

FIGURE 10.1 Document control.

pilots. The authority's flight operations inspector has oversight of the OM. These relationships are illustrated in Figure 10.1

The AFM is approved by the airworthiness branch of the authority issuing the release to service. The QRH is not approved because it is deemed to be derived from the AFM. The QRH is viewed as a version of the information in the AFM, produced for the benefit of the operator. Any changes to the QRH are negotiated between the operator and the manufacturer and, so, are outside of any oversight process.

As a result of the fragmented approach to document control described above, there were differences between the Q400 AFM and QRH. In the event of overspeed, the Q400 AFM describes the items on the checklist as 'actions', whereas they are 'memory items' in the QRH. This simple change has implications for how a crew would enact the items on the checklist. In communication terms, it changed the force of the statements. Ironically, as we have seen, it would have no effect on the outcome in this case, but it illustrates how meaning can change without that being the intention of the original authors of the documents.

The propeller-overspeed checklist also contains the statement 'IF unable to control propeller RPM'. The Swedish accident investigators asked the manufacturer to clarify what that statement meant. The response was:

> The AFM/AOM does not define controllable/uncontrollable with respect to the propeller rpm. Pilot aircraft knowledge and training is assumed to provide the necessary basic knowledge of what constitutes normal and abnormal propeller operation.

The investigation discussed the fact that several other operators had encountered problems with the propeller-overspeed checklist. Some crews had dealt with the event in the way the checklist intended while others had used the checklist only partially or even not at all. Several operators pointed out the lack of clarity. The view of the manufacturer, as reported to the investigators, was:

> We disagree that the procedures are not satisfactory.

This example reveals much about the gaps in regulatory control as much as the control process itself. Control is anchored in certain key points, such as the aircraft design and the AFM, but considerable latitude exists in the process as a whole. Belcher (2002), in an examination of maritime collision regulations, COLREGS, found that some collisions directly resulted from efforts to avoid accidents by actively trying to follow the rules. Unfortunately, as Belcher points out, regulations have 'lacunae' that need to be filled by practice. It is in these spaces that things go wrong. In the example above, the exact meaning of a statement in a checklist is left to the individual pilot to interpret and act upon. The intent of the initial steps in the checklist, discussed in Chapter 8, and their correct application was clearly understood differently by different airlines. These are examples of Belcher's 'lacunae'. In this example, the manner in which control is exerted is through the approval of source documents and the oversight of the translation of documents into local procedures. Divisions of responsibility (airworthiness and flight operations in this case) and delegation (development of the QRH) allow opportunities for control to be weakened. The response of the manufacturer to the accident investigation report also shows how the process of control can be disputed and it is better thought of as a negotiated relationship.

Aviation authorities rely on various forms of communication, some of which have the force of mandatory requirements while others are simply for advice and guidance. For communication to fulfil its function as a control, the content of that communication must be available to those affected by it; it must be understood, capable of being complied with and, finally, enacted at the operational level. Formal communication is only effective to the extent that it is complied with or can be enforced. If we reflect on the Pelee island crash discussed in Chapter 2, Canadian Aviation Regulations prohibit a pilot taking off with ice on the wings of an aircraft, a response to a previous fatal accident (Minister of Supply and Services, 1992). This constraint is fully understood by Canadian pilots, and Georgian Express conducted winter operations training during which this message would have been reinforced. The pilot of the Pelee island aircraft had attended the course. His performance on the day was not influenced by the regulation. In another context, because of the incidence of injuries to personnel at UK airports, the Health and Safety Executive (HSE) issued guidance material in 2000 on the management of aircraft turnarounds. The impact of the guidance material was evaluated (HSE, 2006), and the subsequent report commented that 37.5% of respondents to a questionnaire were not aware of the guidance material while a further 12.5% were aware of the document but had not read it. Fully half of the intended audience had not been affected by the advice. Communication through approved documents is a fundamental aspect of regulatory oversight but, as we have just seen, is not infallible. Regulation, in any context, needs to strike a balance between constraint and facilitation. The manner in which this path is navigated goes to the heart of the problem of oversight. In the next section, I want to explore risk at the level of the regulator.

10.4 REGULATORY FAILURE

Lodge (2015), in discussing the failure of regulation, suggests that authorities differ to the extent that they see their purpose as primarily to avoid harm or to liberalise markets. In a letter to the Secretary of State, the chair of the UK Civil Aviation Authority (CAA) (CAA, 2019), discussing general aviation, commented on the need to be 'proportionate in the exercise of statutory functions' and, later, talks about 'regulating innovatively'. Aviation regulators are no less resource-constrained than other entities. For example, on 25 December 2002, a King Air 100 crashed on approach to Eveleth Airport, Minnesota (NTSB, 2003). The company was overseen by the FAA Minneapolis flight safety district office. The inspector who had oversight of the operator was also responsible for eight similar air taxi companies, 37 crop-spraying operators, one flight school and 12 designated training pilots and examiners according to his terms of reference. Because of budget cuts and a hiring freeze, the office was 50% undermanned. At the time of the accident, the inspector was overseeing 37 charter companies, 155 agricultural companies, 4 heavy-lift helicopter operations, 11 flight schools, 96 designated training pilots and examiners and, finally, 168 non-certified operators. To cope with these resource constraints, there is an increasing trend to move away from prescriptive methods of regulation towards performance-based models with greater delegation to the regulated entity to provide evidence of compliance. Safety management systems (SMSs), international audit standards such as the IATA Operational Safety Audit and new models of training delivery (AQP/ATQP, EBT/CBT) are representative of such trends.

Lodge identifies a number of ways in which regulation can fail. First, a flawed analysis of the current situation will result in inadequate oversight. Interventions may be inappropriate or can fail because of issues around jurisdiction. Second, regulation may fail because an intervention might not be politically feasible. Finally, the statutory basis and, particularly, the resources available to the regulator may be insufficient, as we have just seen. There is no agreed definition of 'regulatory failure', but Hutter and Lloyd-Bostock (2017) suggest that a significant risk to any regulator is that failure represents a challenge to its own legitimacy. In short, if a regulator cannot effectively regulate, why should it exist? Linked to failure is the concept of 'regulatory capture'. Shapiro (2011) identifies two aspects of capture: first, the interests of the regulated entity unduly influence the behaviour of the regulator, and second, the regulator fails to act in the manner intended by the constituting authority, usually the state. Typically, a failure to act would be seen in terms of the manner in which the general public is adequately protected. Capture, it could be argued, is both a state of failure, in that the process is no longer fit for purpose, and also a precursor to catastrophe.

The need for regulation was clearly recognised early on in the development of aviation, and the scope of regulation has both broadened and changed in nature over time. Because the risks to be managed are not a constant but, rather, reflect the environment within which aviation is conducted, the technologies employed and the prevailing social attitudes, regulation is constantly challenged. Similarly, the tension between safety and market liberalisation creates the circumstances in which there is a possibility for failure. In the next section, I want to explore the dynamic tension between oversight, cost/effectiveness and safety by looking at the introduction of SMSs in Canada.

10.5 CHANGE MANAGEMENT AND REGULATORY FAILURE

I mentioned above that responsibility for oversight is increasingly being delegated to the operational level with the authority maintaining control through a 'system of systems' concept. The implementation of the SMS policy is just one example. The experience of implementing SMSs in the USA is summarised in a report from the Government Accountability Office (2014):

> ICAO first mandated SMS worldwide for air traffic service providers in 2001. ICAO later specified that member states should mandate SMS implementation for airports, air carriers, and others by 2009. FAA began SMS implementation in 2005, but FAA officials informed ICAO that the agency and industry would not be able to meet the 2009 deadline. The United States filed a "difference" with ICAO—indicating that it does not yet completely comply with the standard—with the understanding that implementation is under way and that FAA is in the midst of a rule making to require SMS for commercial air carriers and certificated airports. There have been other actions within the United States to encourage SMS implementation. For instance, in 2007, NTSB recommended that FAA require all commercial air carriers to establish an SMS and in 2011 added SMS for all modes of transportation to its Most Wanted List. [The] NTSB identified SMS as one of the most critical changes needed to reduce the number of accidents and save lives.

Transport Canada was, in fact, the first authority to act on the SMS requirement with an initial target of having all aircraft operators that perform maintenance on their own aircraft compliant by May 2005, with a permissible extension to September 2009. This represented 74 of the 2324 approved operators in Canada. The implications of SMSs for the work of the authority are illustrated in Table 10.1. In an audit of aviation safety (Auditor General of Canada, 2008), it was found that Transport Canada had failed to consider the maintenance of the conventional oversight of safety during the transition. Resources had shifted from traditional oversight to SMS activities with no consideration of the implications, such as reduced inspections. The number of inspectors and engineers needed under the new regime had not been identified, and short- and medium-term performance indicators had not been developed. As a result, the Canadian Transportation Safety Board added SMS implementation to its risk 'watch list' in August 2010.

The roll-out of SMS was just one of many changes being managed by Transport Canada. On 1 January 2003, oversight of 'private' operators was delegated to the Canadian Business Aviation Association (CBAA). This class of operators included not only corporate flight departments but also forest fire suppression aircraft, helicopters and the Royal Canadian Mounted Police. The incentive to delegate was to save money on the part of Transport Canada. In order to achieve or renew a license

TABLE 10.1
Implications of SMS (based on Office of Auditor General of Canada)

Role	Traditional Approach	SMS Approach
Inspector	Inspectors are regulatory compliance auditors	Inspectors are system evaluators
Company	The company responds to regulatory requirements	The company proactively manages risks
Department	[The authority] inspects aircraft, records and personnel directly	[The authority] assesses whether companies have effective processes to ensure safety

to operate, companies had to comply with an audit process designed by the CBAA that included elements of the proposed SMS regulations. In addition to the shift to a new model of oversight, then, this segment of the industry was also implementing the new approach to safety management. The CBAA was keen to expand its membership but felt that it did not have the authority to compel the adoption of these new codes and all that it implied in terms of change management. Because of their own internal lack of resources, individual operators could delegate implementation to independent third-party contractors approved by the CBAA. As a result, regulation, and the associated level of control, was being diluted. In fact, the CBAA said that 'some [...] operators may never reach full SMS compliance'.

On 11 November 2007, a Bombardier Global 5000 business jet touched down short of the runway at Fox Harbour, Nova Scotia (TSB, 2009). The aircraft suffered major structural damage, and all the occupants were injured to varying degrees. The subsequent report found shortcomings in the implementation of the company's SMS. The accident report notes that some 320 operators certificates had been issued by the CBAA and 750 audits completed. Unfortunately, the process established required only the audit cover sheet to be submitted to the CBAA and, therefore, it was impossible to assess the quality of the audit and the work done by the auditors. The subcontracting of the audit process to third parties was justified by the CBAA in terms of cost saving. So, ironically, in an attempt to save money, oversight was delegated by Transport Canada to the CBAA who, in turn, delegated to third parties and, in the meantime, effective control was lost.

We discussed forms of safety drift in Chapter 2, and I proposed 'rationalisation' as one mode of drift. If we accept that rationalisation is usually driven by a pressure to find savings and 'efficiencies', then, the specific implementation of SMSs in the Canadian business aviation community is probably a variation on this theme. We can see something similar in the case of the 'girting' of a tug on the River Clyde, in Scotland, coincidentally also in 2007. Girting happens when the vessel being towed overtakes the bow tug and pulls it over, thereby causing it to capsize. The subsequent investigation report (MAIB, 2008) described how, after a previous grounding of a vessel in 2009, the responsible UK Government department pulled together various pieces of guidance and regulation into the 'Port Marine Safety Code' (PMSC), which required Port operators to compile hazard lists and to implement SMSs. Ports were required to submit a statement of compliance with the code, which Clydeport, the company that ran harbour operations on the Clyde, had done. Prior to the introduction of the PMSC, Clydeport was certified by the Lloyds Register under the ISO9001 quality management process. The ISO9001 framework was used as the basis for the PMSC SMS system, and the process was run by a part-time consultant employed by Clydeport. When the PMSC was introduced, the role of 'designated person', who was responsible for providing the duty holder with independent assurance that the SMS was working properly, was undertaken by two members of the Port Board. This responsibility was lost at some point and, by the time of the accident, the ISO9001 quality plan had been nominated as the 'designated person' for the PMSC SMS. A document was now standing in for a person, and that document was overseen by a part-time consultant. Three crews lost their lives in the accident.

These attempts to implement the SMS concept in two very different modes of transport are cases of, primarily, intervention failure, using Lodge's taxonomy, in that the proposed solution and its implementation were inappropriate. The intent of the regulation and the circumstances of its implementation had become disconnected. We can also see that there was an element of design failure in that the resources allocated to regulatory activity were inadequate. The examples illustrate how organisations seek to implement a process, and demonstrate compliance, in a cost-effective manner. Hence, we see subcontracting being put in the place of direct oversight. The nature of control, through regulation and guidance, was not always fully understood by those responsible for its implementation, and the feedback loops, represented by audit and statements of compliance, were unreliable.

Of course, the framing of regulation and its implementation do not directly cause accidents. Instead, modes of regulatory failure combine to create the context for inadequate risk management. We saw earlier the chair of the UK CAA talking about regulation being 'proportionate'. Lodge observes that the regulator's risk tolerance is reflected in the degree to which regulation is deemed to be burdensome. In the United Kingdom, the financial meltdown of 2006–2007 has been blamed on the then government's 'light touch' approach to financial regulation. In the next example, we will also see how 'capture' can shape outcomes, ultimately leading to a state of crisis for the regulator.

10.6 FAILURE, CAPTURE AND CRISIS – REGULATORS, AIRCRAFT MANUFACTURERS AND THE CONSTRUCTION OF SAFETY

On 29 October 2018, a Boeing B-737MAX, operated by Lion Air, crashed 12 minutes after take-off from Jakarta, Indonesia. A second B-737MAX crashed on 10 March 2019 after take-off from Addis Ababa, Ethiopia. In both the accidents, the ability of the pilots to control the behaviour of the aircraft was compromised. The following day, the Chinese Civil Aviation Authority (CAAC) ordered all B-737MAX aircraft to be grounded. This action was quickly followed by a number of other countries. The FAA waited three days before ordering the grounding of the model in the USA. The initial response from the manufacturer, Boeing, was to suggest that the problem lay with the performance of the pilots. The company's CEO, responding to journalists' questions at a press conference after a shareholder's meeting, repeatedly stood behind the 'Swiss Cheese' mantra of any accident being the result of multiple factors in an attempt to deflect any discussion away from the role of Boeing. Soon after, he was sacked.

The Boeing 737 design concept is over 40 years old and, having passed through a dozen iterations, is probably coming to the end of its growth potential. Competition from the younger, more efficient Airbus A320 family of variants prompted Boeing to look at freshening up the B-737. The solution to fit more powerful, fuel-efficient engines created a problem of simple aerodynamics. The additional thrust from the engines, together with the shape of the nacelles, created a nose-up vector which needed to be counteracted under certain conditions. The solution was the Manoeuvring Characteristics Augmentation System (MCAS). This system adjusts the horizontal stabiliser trim to pitch the aircraft's nose down if it detects a potential stall condition developing. The initial MCAS design case was for circumstances of high speed and high angles of attack but it later became clear that it would also need to deal with low-speed situations as well. This had implications for the scale of control input the MCAS would need to make. In addition, MCAS drew data from a single angle of attack (AoA) indicator and, as a result, lacked redundancy. Unfortunately, reminiscent of the SAS MD-82 ATR discussed previously, the designers did not appear to consider either a loss of the AoA input due to mechanical damage or the implications of a repeated triggering of the system. The gap between the planned MCAS, as initially designed, and the actual installed version was not approved by FAA. In the event of an MCAS failure, pilots would only have limited time to detect the condition and to respond.

Intense media scrutiny followed the fatal accidents. A journalist, Dominic Gates at the Seattle Times, wrote (and continues to write) extensive pieces on aspects of the problem: Seattle is Boeing's hometown, and the newspaper won a Pulitzer Prize for its coverage in 2020. A whistle-blower provided information to EASA, and reports circulated of the FAA being put under pressure by the pace of aircraft development and subsequent delays. FAA line managers, rather than technical experts, signed off on reports. The US Senate Committee on Aviation held formal hearings and, finally, on 1 June 2019, the FAA convened an international panel, the Joint Authorities Technical Review (JATR, 2019), to review the flight control certification process.

The B-737MAX quickly became a crisis for the manufacturer, and its response was being managed in a very complex political environment. In 2016, President Donald Trump came to power with an anti 'Big State' agenda. In February 2017, an executive order had been signed, requiring all federal government departments to cut back on 'red tape'. At the same time, Trump had embarked upon a trade war with China, a state keen to expand its domestic aviation industry. The Chinese-manufactured Comac C-919 aircraft had its maiden flight on 5 May 2017 and is a direct competitor of the B-737. The speed with which the Chinese acted against the B-737MAX could be seen against this background. Aboulafia (2019) makes the point that the CAAC has a habit of banning things on a whim, has little experience, has a poor track record and is not free of political interference. In the USA, the debate about the safety of the B-737MAX was, at times, being conducted on the President's Twitter account and, as the largest manufacturing exporter in the USA, the company's woes were estimated to be having a negative impact on the US economy equivalent to a loss of 0.25% of GDP (Greene, 2019).

At the time of writing, work is underway to return the aircraft to the air, and the full story of the B-737MAX is beyond the scope of this chapter: it still has some distance to run. My interest is primarily in what light the event can shed on the nature of regulation and the exercise of control in the context of a system. Two aspects of the B-737MAX story are of specific relevance. The first is the manner in which the certification of MCAS was approached. This has the potential to create a cross-scale effect, manifested in a crew's ability to control the aircraft. The second is the decision not to communicate details of MCAS to customers and, ultimately, pilots. This, clearly, will shape the underpinning knowledge available to crew about the aircraft and its behaviour.

The FAA publishes airworthiness requirements in the Code of Federal Regulations. A new model of aircraft must satisfy the latest iteration of the rules contained in regulations. Because the B-737MAX was deemed to be a modified version of the B-737NG, it was dealt with in a different manner. Under the Changed Product Rule, the 737MAX had to satisfy:

- The regulations in effect when the 737 was originally certified in 1967
- Regulatory amendments in effect when the 737Max certification application was made
- Amendments in effect during the time between original certification and the application for the modification
- Amendments that came into force AFTER the date of application that Boeing elected to comply with
- Special conditions, exemptions and 'equivalent level of safety' findings
- Additional design requirements and conditions

Under this framework, if a change is made, then, both the change and any areas affected by it have to be assessed for compliance. Unaffected areas do not have to be assessed. The JATR report (2019) was compiled by a team comprising representatives from NASA, the FAA and the CAAs of eight countries and the EU. The report's recommendations are in 12 groups, and several of the recommendations reflect the complexity of the certification of modern aircraft. The report observed that regulations were found to be out of date and were not consistent in their interpretation and application. The process of certification can take up to 5 years, but there was no mechanism for testing and updating the original design assumptions. The context of certification, then, was problematic.

In order to reduce the requirement for assessment, Boeing proposed that MCAS – which was not installed on any previous model of 737– was not, in fact, a new system but, instead, was a modification to the existing flight control trim system. For this reason, it was decided not to provide a warning of an MCAS failure as this would be reflected in the existing speed trim fail light. The original design case for the MCAS was based on a specific manoeuvre – the 'wind up' turn. In this context, if necessary, MCAS would reduce the load on the controls by making a slight (0.6°) reduction in back pressure. However, the scope of MCAS was then enlarged to include low-speed situations where a greater input (2.5°) would be needed. The revised design case was not risk assessed. Concern was also raised about the apparent 'single point of failure' represented by the fact that the MCAS was associated with a single AoA indicator. This vulnerability was raised in 2015 but dismissed on the grounds that single points of failure were permissible if the outcome was not likely to be 'catastrophic'. A possible failure of MCAS was not deemed to be catastrophic.

In addition to framing rules, regulators conduct physical oversight. The FAA had delegated oversight of parts of the certification process to Boeing and was exercised by the Boeing Aviation Safety Oversight Office (BASOO). Delegation of responsibility was signed off in law in the 1920s, and by 2004, some 13,400 Designated engineering representatives were acting on behalf of the FAA in the US aviation industry (Downer, 2009). The BASOO comprised 45 FAA employees and 1500 Boeing staff. The JATR team commented that BASOO staff numbers were inadequate and, given that some of the FAA engineers were relatively junior, experience levels were possibly inadequate. In addition, as we saw earlier, there was evidence of pressure being put on engineering unit members. Downer observes that delegation provides access to pools of expertise unavailable from within an agency's own resources and that, in fact, rather than confirming compliance with regulations and standards, agencies such as the FAA put trust in other, more expert, people to do it on their behalf. In the case of the B-737MAX, the trust was misplaced. The JATR report observed that Boeing exerted undue influence over the FAA, which, in turn, was guilty of possibly placing the interests of Boeing above their own responsibility to guarantee safety. It seems, then, that the FAA had a flawed understanding of the situation and, certainly, lacked the resources to maintain robust oversight.

The second point of interest was the way in which information about MCAS was communicated. Commonality between aircraft models and variants is an advantage in that it reduces the need for additional, possibly extensive, customer training. Therefore, it was in Boeing's interest to suggest that the B-737MAX was no different from other variants. The need for training had been raised but was blocked, largely on the grounds that to include training on MCAS and dealing with possible malfunctions was estimated to add US$1 m to the unit price of each aircraft. In the event of a failure, it was assumed that the pilots would detect and respond within 4 seconds, a reaction time considered by Boeing to be a 'longstanding industry assumption'. The issue of pilot responses to a failure was raised in an e-mail that suggested that, while 4 seconds might be a plausible response, anything in excess of 10 seconds would be catastrophic because of an inability to prevent and arrest an overspeed (Senate Committee, 2019). In a recommendation reminiscent of the MD-82 ATR saga in Chapter 3, during certification a decision was made to remove information about MCAS from the draft FCOM. As a result, the FAA Flight Standardisation Board, which makes decisions about training requirements, was not fully aware of MCAS functions and was not in a position to assess training needs. The Boeing AFM does not include all the normal, non-normal and emergency procedures. Most of the operating procedures are in the FCOM, and so, changes could be made without getting FAA approval. Once again, the regulator was not fully aware of the situation over which it was supposed to exercise control.

The example of the B-737MAX illustrates how relationships between agencies can become subverted. Although the FAA has considerable power, its freedom of action can be constrained by economic, political and procedural factors. We saw earlier that regulatory capture reflects a situation where the needs of the body being overseen take priority and the interests of those put at risk by that body's activities – in this case, the general public and flight crews – are overlooked. The B-737MAX saga is developing into a classical example of regulatory capture. However, it is also a crisis for both Boeing and the FAA, a condition I want to explore in more detail later.

Throughout this chapter, we have seen how aviation regulators have sometimes struggled to exercise oversight of safety. Equally, many of the themes raised in Chapter 2 were apparent in the examples we have just looked at: inadequate staffing levels; lack of expertise; fragile regulations; flawed communications processes; inadequate analysis of possible risks. We also see that environmental factors – geopolitics manifested as competition and industrial output as an element of national GDP – can shape decision-making.

10.7 INVESTIGATION AS FEEDBACK

The problems with the B-737MAX were revealed after two fatal accidents. It was the process of investigation that provided the details of what went wrong. Accident investigation agencies occupy an interesting niche. They are mandated under an ICAO requirement, represent a feedback loop within the aviation system but lack the formal authority to effect change. They can simply advise, suggest, recommend and nudge. They lack the power to hold regulators to account. In this section, I look at the history of aviation accident investigation before looking at an investigation in detail.

In 1912, in the United Kingdom, the Royal Aero Club created the Public Safety and Accidents Investigation Committee. A member of this committee became the first Inspector of Accidents for the Royal Flying Corps in 1915. At the end of the Great War (1914–1918), the Accident Investigation Branch was created in the Air Ministry in 1919. With the formation of the CAA in 1972, the renamed Air Accident Investigation Branch (AAIB) was moved to the predecessor of the current Department of Transport, with the chief inspector of the AAIB answering directly to the Secretary of State for Transport. In the USA, the responsibility for investigation initially lay with the Department of Commerce, but after the crashes mentioned above, the wisdom of allowing an agency to, effectively, investigate itself was questioned. An independent investigation 'Air Safety Board' was established in 1938. In 1967, all transportation agencies were brought under the Department of Transportation, and the new NTSB moved with them. Recognising the importance of independence in investigation, the NTSB was established as an independent entity in 1974.

Regulation and investigation have developed in much the same way in different countries, and there are two key characteristics of all investigation agencies: they (usually) have independence from the regulator but they have no power to mandate action; they simply make recommendations, and it is for the authority to act. From a systems perspective, an accident, especially, and some very serious incidents represent a special case of failure given that extreme events are uncommon. During an investigation, the limitations of regulation and oversight are often exposed and, thus, the process provides important feedback.

The demands of an investigation can create tensions between different agencies. A simple example would be the fact that a fatal accident crash site is technically a crime scene until the legal authorities have been satisfied that no illegal act has been committed. The idea of preserving evidence is relevant to both criminal and air accident investigations. In a legal sense, evidence must also be quarantined so that, in the event of prosecution, it can be presented in such a way that no claim can be made that it has been tampered with or manipulated. Investigating an accident requires the evidence to be handled differently. In the event of a fatal accident, a Coroner's Court is usually convened to establish who died and how they died. Again, in an aircraft accident, the coroner often defers to the investigating agency. Any contradictions between these different agencies are dealt with through agreements and memoranda of understanding (MoU). The key issue here is that accident investigation occupies a space between multiple agencies with, often, conflicting requirements. The tensions that arise from this will be explored later.

Reuss (2016) makes the point that accident investigation has historically focused on the technical issues but changes in our understanding of failure mean that the challenge now is to understand the relationship between humans, technology and organisations. He also points out that investigations are increasingly being influenced by the media, the views of family members of victims and, paradoxically, the challenge of cooperation between departments and agencies. Hutter and Lloyd-Bostock (2017) identify two key changes in safety regulation. First, risk regulation is becoming supranational – we have to comply with rules that are negotiated across multiple jurisdictions – and, also, is increasingly driven by non-state actors. They also observe that investigation has shifted away from 'what happened and why' to 'how to anticipate what might happen next and stop it'. On this latter point, society has become increasingly intolerant of failure. As such, a catastrophic loss is seen as an affront, and victim groups demand reparations: someone has to pay. Interestingly, victim groups – an example of a non-state actor – can often be more effective at imposing demands upon regulators than accident investigators are able to. After a Bombardier Dash 8 Q-400 crash at Buffalo, New York, in 2009 (NTSB, 2010), families lobbied the US Congress for, among other things, tighter rules on pilot experience requirements for employment, a factor that had nothing to do with the original accident (Zremski, 2019). Level 5 in my hierarchical systems model – the Environment – includes those factors that influence behaviour in the system but are not, in turn, influenced by those subordinate elements. Aviation authorities are subject to political control and, while acting as an independent body, nonetheless have a political master. Authorities are also exposed to public scrutiny. Equally, investigation agencies are notionally independent, but environmental factors can shape the relationship between regulators and investigators and ultimately, the industry itself, as we will see in the next case study.

10.8 THE PEL AIR WESTWIND DITCHING

The facts of this event are quite straightforward. On 17 November 2009, an IAI Westwind aircraft was dispatched from Sydney to Samoa on an aeromedical retrieval flight (ATSB, 2012). Onboard were the crew of two pilots, a doctor and a flight nurse. The route included a stop at Norfolk Island for refuelling before continuing on to Samoa, where it arrived at about 7 am local time. Therefore, this was a flight through the night. The crew went to a hotel to rest prior to the return flight, scheduled to depart at 1830 local later that day. The crew had less than 12 hours on the ground.

An hour before departure, the captain attempted to file the flight plan for the return journey but discovered that the internet service on the island was not working. He called the company's main office but could not get through to the duty manager. He then called the briefing officer of Airservices Australia, and between them, they created a flight plan based on reversing the one submitted the day before. The pilot was given the latest weather forecast for Norfolk Island, which was acceptable for a landing at the planned time of arrival but the briefing officer did advise of a trend in the forecast from 1500 UTC, but the actual details were not requested. The captain then arranged for the aircraft's main fuel tanks to be filled. The aircraft did have additional wing tip tanks, but these were left empty as the captain deemed that the weather at Norfolk Island did not require additional fuel for a diversion to an alternate. The first officer (FO) was not present during any of the planning but did look at the notes the captain made about the Norfolk Island weather. She was also advised about the fuel. There was a slight delay with loading the passengers, and the aircraft finally got airborne 15 minutes late.

The crew did not update the weather en route and, by the time they arrived at Norfolk Island, the weather was below minima. Four approaches were made before the captain decided that they should ditch the aircraft just off the coast at around 2200 local time. Both the crew and the four passengers were rescued. The aircraft wreckage and flight data recorders were not recovered on cost grounds, a decision partly influenced by the fact that all the witnesses to the event survived.

In the aftermath, the pilot was proclaimed a hero and comparisons were made with the ditching of the A-320 in the Hudson River in January of that year. He was also something of a celebrity, having been a finalist in an Australian reality TV show, 'Bachelor of the Year'. Therefore, media attention was intense. However, in December 2009, barely a month after the crash, the Australian Civil Aviation Safety Authority (CASA) withdrew the pilot in command's license. A preliminary report into the investigation was published by the Australian Transportation Safety Bureau (ATSB) in January 2010 stating that:

> The investigation is continuing and will include further examination and analysis of the:
>
> - meteorological information and its effect on the decision-making and actions of the crew during the flight
> - fuel planning relevant to the flight
> - operational requirements that were relevant to the conduct of the flight
> - crew resource management
> - aeromedical flight classification and dispatch.

Within a matter of weeks, then, the regulator took punitive action despite the fact that the accident investigation was still ongoing. It is a tenet of safety investigation that its purpose is not to assign blame, but regulators do have a responsibility to consider possible safe courses of action after an adverse event. Pending the outcome of an investigation, any response runs the risk of being either premature, wrong or both. Nonetheless, CASA chose to act against the captain, but no action was taken against the FO.

10.9 THE FALLOUT

The final 78-page full ATSB report was published in 2012 (ATSB, 2012). Having stressed that the findings should not be seen as attributing blame to individuals or organisations, the report contained the following findings as to cause:

> Contributing safety factors
>
> - The pilot in command did not plan the flight in accordance with the existing regulatory and operator requirements, precluding a full understanding and management of the potential hazards affecting the flight.

- The flight crew did not source the most recent Norfolk Island Airport forecast or seek and apply other relevant weather and other information at the most relevant stage of the flight to fully inform their decision of whether to continue the flight to the island or to divert to another destination.
- The flight crew's delayed awareness of the deteriorating weather at Norfolk Island combined with incomplete flight planning to influence the decision to continue to the island, rather than divert to a suitable alternate.

Other Safety Factors

- The available guidance on fuel planning and on seeking and applying en route weather updates was too general and increased the risk of inconsistent in-flight fuel management and decisions to divert [minor safety issue].
- Given the forecast in-flight weather, aircraft performance and regulatory requirements, the flight crew departed Apia with less fuel than required for the flight in case of one engine inoperative or depressurised operations.
- The flight crew's advice to Norfolk Island Unicom of the intention to ditch did not include the intended location, resulting in the rescue services initially proceeding to an incorrect search datum and potentially delaying the recovery of any survivors.
- The operator's procedures and flight planning guidance managed risk consistent with regulatory provisions but did not effectively minimise the risks associated with aeromedical operations to remote islands [minor safety issue].

Other key findings

- At the time of flight planning, there were no weather or other requirements that required the nomination of an alternate aerodrome or the carriage of additional fuel to reach an alternate.
- The aircraft carried sufficient fuel for the flight in the case of normal operations.
- [,]

The process for releasing an ATSB report includes an early sight of the text by interested parties before public release. The first draft was sent to those interested parties – the operator, the pilot and CASA – on 26 March 2012. Two rounds of consultation and drafting were completed and the final version was released to the public on 30 August 2012. Publication of the final report triggered something of a furore. A television documentary was broadcast on 3 September 2012 that was very critical of both the ATSB and CASA. The mere fact that the documentary was broadcast so soon after the publication of the report suggests that it had been in production, probably, since the initial draft was released. In particular, the apparent emphasis on the actions of the captain, with little discussion of the role of the operator or CASA, was considered biased. Suspicions had been raised when it emerged that a CASA audit conducted immediately after the crash had not been shared with the ATSB.

Pel Air had been subject to an audit by CASA prior to the accident and deficiencies had been found in a number of areas. Immediately after the crash, a special audit of the company was initiated and report produced on 1 August 2010. An additional review of 'lessons learnt' from a regulatory perspective was commissioned at the same time. CASA oversight procedures called for a variety of checks and audits to be conducted against a schedule. For example, a site inspection was supposed to be completed each year, a systems audit was required every 3 years, check and training and supervisory pilots were to be observed on initial award and subsequent renewal of approvals. Ramp checks and observations of operational flights were to be done on an ad hoc basis.

A system audit was done in 2006 when Pel Air merged its air operator certificate with another company. Audits were conducted again in 2008 and 2009 because of safety concerns. A site inspection was undertaken in 2006 but not again after that. Six-monthly 'Safety Trend Indicator' exercises were completed on 60% of the required occasions between 2006 and 2010. These exercises are supposed to alternate between a desktop telephone interview and then combined with a site inspection. Observation flights were done with management and training pilots, but no observation of line pilots was done. During this period, 31 requests for corrective action were issued, together with one safety alert.

The management report observed that the inspector assigned to Pel Air was not current on the aircraft type although specialist Westwind Inspectors were based at other offices. A team of 16 CASA inspectors descended on Pel Air for the special audit after the ditching, but the audits done in 2006 and 2007 made use of just three CASA staff. The available inspector establishment was reduced by two after a reorganisation in early 2009 and, in addition, the local office was understaffed by two. It was 25% behind on its surveillance targets at the time of the review. The report makes two interesting observations about the behaviour of inspectors. First, the process of signing off on requests for corrective action was based on the operator submitting a plan for *future* remedial action and not necessarily on an inspection of *actual* remedial action. The second observation related to the unwillingness of inspectors to create tension by digging too deep and, therefore, asking too many difficult questions. The Pel Air inspector was considered to have a good relationship with the company and, on that basis, it was assumed that CASA had clear oversight of the operation. The reality was much different. In fact, the inspector did not know that Pel Air was routinely conducting medical flights to Samoa. Oversight of the operation, then, had been suspect for a number of years prior to the accident.

Within days of the broadcast, a Senate Inquiry was convened, which reported in May 2013 (Commonwealth of Australia, 2013). The inquiry report recommended, among other things, that the aircraft wreckage be recovered so that the flight data recorders might be examined. An independent audit of the ATSB procedures by a foreign investigation agency was requested and duly conducted by the Canadian Transportation Safety Board (TSB, 2014). The TSB report was published on 1 December 2014 and was broadly supportive of the ATSB.

The Senate Inquiry drew attention to an earlier coroner's investigation into a fatal accident at Lockhart River on 7 May 2005. The 2 pilots and 13 passengers were killed when the aircraft struck the ground. The Queensland coroner's report (2007) challenged the relationship between CASA and the ATSB and recommended that a new MoU be drawn up between the two organisations. The Pel Air investigation was being conducted as this MoU was being finalised. The TSB audit report observed that a structural reorganisation that increased the independence of the ATSB as an agency was in progress but that, coupled with the fact that the MoU was still in negotiation, the lines of accountability between CASA and the ATSB were blurred. The Senate report suggested that there was collusion between the two agencies in the drafting of the first accident report in order to minimise possible criticism of CASA's role in the oversight of Pel Air, although the Senate report did state that no evidence to this effect could be found; the suggestion was, nonetheless, left in play. One recommendation of the Senate report was to, again, redraft the MoU.

The ATSB report of the ditching was reissued on 23 November 2017 (ATSB, 2017), now running to over 500 pages. In relation to the accident, it stated:

> Contributing factors
>
> Contrary to the consistent practice of the operator's Westwind fleet, the long-distance flight to a remote island aerodrome departed without uploading the maximum possible amount of fuel prior to departure. Had the flight departed with the maximum amount of fuel, it is very likely the aircraft would have had sufficient fuel to divert from the top of descent or to hold at the remote island for a significant period of time.
>
> The captain's pre-flight planning did not include many of the elements needed to reduce the risk of a long-distance flight to a remote island or isolated aerodrome. Limitations included:
>
> - miscalculating the total fuel required for the flight under normal operations
> - not obtaining relevant forecasts of upper-level winds and, in the absence of such forecasts, underestimating the potential headwind component
> - not calculating the additional fuel required to allow for aircraft system failures
> - not obtaining a current aerodrome forecast and NOTAMs for potential alternate aerodromes
> - not calculating a point of no return (PNR).

The original report identified 12 factors associated with the event. The final report identified 30 additional findings. However, at the core of the report, the performance of the captain was still considered fundamental to the outcome.

10.10 POSTSCRIPT TO PEL AIR

On 28 June 2017, a Socata TB-10 Tobago aircraft crashed after take-off, killing the pilot and two passengers (ATSB, 2019). The flight was organised by a charity, providing transport from remote areas for people needing to visit hospital, known as community service flights (CSFs). Volunteer pilots provided their services and the use of their aircraft and, on the company's website, were described as 'Our Heroes'. The 78-year-old pilot of the Tobago was qualified for flight under visual flight rules but did not hold an instrument rating. He had departed Murray Bridge Airport, South Australia, at 0800, arriving in the area of Mount Gambier Airport at around 0940. The weather at the time was overcast at 200 ft agl. Having descended through cloud, the aircraft was seen by observers and caught on CCTV, manoeuvring at low level.

After two attempts, the pilot landed and collected his passengers. Visual flight rules require a non-rated pilot to remain clear of cloud. At the time of take-off, both the horizontal visibility and the presence of complete overcast rendered the flight technically illegal, given the license held by the pilot. Nonetheless, he took off at 1020 and crashed 65 seconds later. The final ATSB report contained seven findings, two relating to the pilot's actions, one relating to CASA's oversight of CSF operations and four relating to the charity that organised the flight.

The ATSB undertook an extensive analysis of accident rates for different types of operations and demonstrated that flights conducted by the charity were significantly more risky than other private flights. Operations conducted by pilots with private licenses were, in turn, more risky than commercial flights. The report drew attention to this fact and also commented on the need for greater risk management and for, possibly, tighter regulation by CASA.

The response was swift in coming. The report was dated 13 August 2019. Within days, the CEO of the charity had appeared in the media claiming that, among other things, 'The ATSB offered no safety recommendations to pilots flying light aircraft in bad weather'; that 'it is regrettable that the Bureau made no relevant safety recommendations, nor gave any guidance whatsoever, to pilots flying in poor weather conditions – the cause of the accident'; that 'its Pel Air (Norfolk Island ditching) all over again – for that particular report the ATSB were found to be grossly incompetent and were ultimately required to redo the report'. In what can only be seen as a significant attempt to deflect criticism, the ATSB was cast in the role of a culprit. Supporters of the charity took out full-page newspaper ads to protest against the findings, and lobbying was successful in achieving yet another Senate inquiry into aspects of the analysis of the accident by the ATSB. The subsequent report (Commonwealth of Australia, 2019) accepted the technical analysis of the event but criticised the ATSB's use of statistics to identify CSF operations as being of higher risk.

After this accident, CASA introduced guidance on the minimum qualifications of pilots and also a slightly more stringent maintenance requirement for inspections at 12 months or 100 in-service hours. Given that most CSF operations were conducted using privately owned aircraft, and it was found that few such aircraft actually accumulated more than 100 hours in 12 months, this was hardly an onerous change, but it drew criticism from the inquiry. In a telling comment, one of the inquiry-participating members stated that 'Angel Flight cannot be disrupted by unnecessary regulation'. Further, he said 'CASA has an *over* regulation issue (my italics)' and that it should 'treat safety as primary, but be mindful of the need to maintain a healthy industry. In the past, CASA regulations have drowned general aviation, and that has to change' (Parliament of Australia, 2019).

This sequence of accidents, from Lockhart River through Norfolk Island to Mount Gambier, illustrates how investigation and regulation increasingly have to negotiate a public discourse that can be politically motivated and contradictory. The ATSB was criticised after Norfolk Island for not saying enough about the organisational factors and the role of the regulator. After Mount Gambier, the ATSB was criticised for going too far into organisational factors and not saying enough about how to stop unqualified pilots attempting to fly in unsuitable conditions. Some of the criticism seemed to reflect a fundamental lack of understanding of the role of the regulator in licensing pilots and the duty of the pilot that flows from holding a license.

10.11 REGULATION, INVESTIGATION, CONTROL AND FEEDBACK

The Pel Air ditching and subsequent report illustrate the political landscape of investigation and regulation. CASA's oversight of smaller operators had been criticised before, not just in the case of the Lockhart River crash, and the challenges of maintaining oversight were seen in Chapter 2 in relation to the Pelee Island and the California Learjet accidents. The state aviation authority has a legal obligation and is, therefore, accountable for the conduct of aviation inside its territory. Accident investigation agencies, similarly, facilitate aviation by determining risk, based on adverse events, and proposing mitigations. Investigation reports represent feedback on the functioning of the system. However, as we have seen, the process of both regulation and investigation is problematic.

Hutter and Lloyd-Bostock (2017) looked at regulators dealing with crises in order to examine regulators, themselves, in crisis. They make the point that a crisis can bc an existential risk for the regulator in that the way in which the crisis is managed can call into question the need for the regulator in the first place. They observe that regulators exist in a social, cultural and political environment: environmental factors in my model. The legitimacy of a regulator flows from the trust that it engenders in those it regulates, the general public affected by its failures and the government to whom it is answerable. Regulatory failure can be seen as a failure to manage risk.

Two aspects of regulatory crisis are of specific interest: accountability and blame. Accountability flows from a failure to anticipate a problem and, as Hutter and Lloyd-Bostock observe, it is often the case that the risk management tools used by the regulator are themselves a source of risk. Aviation authorities typically impose different burdens of compliance on different types of operations, becoming increasingly more complex as the size of aircraft, and numbers of passenger carried, increases. Similarly, the scope of the operation shapes how the investigators approach their task. Air operators, then, exist in an ecosystem of safety with risk as a variable, rather than a constant. The behaviour of regulators reflects what Downer referred to as their 'risk appetite' or, as Hutter and Lloyd-Bostock observe, their view on their role in protecting business interests. This tension was reflected in the comments by the Senate Committee member after the Mount Gambier accident.

Associated with accountability is the issue of blame. In the event of a failure, someone has to pay. Pel Air presented a crisis for the regulator, CASA, when it became, first, a source of public outrage and, second, a cause for a government inquiry. It seems that CASA failed to adequately address risk in the case of smaller operators and, when exposed, were fortunate when the media cast another agency as the villain. Actors attempt to avoid blame by framing an issue in a positive way or by finding ways to transfer the blame onto others. For example, we saw earlier that Boeing suggested that the MCAS is just a part of the standard aircraft trim system and that the lack of AoA redundancy did not represent a 'single point of failure' situation. They also claimed that many of the assumptions they used in relation to pilot reaction times were 'industry standard' and were 'the same as those used by the FAA'. The initial attempt to blame poorly-trained or inexperienced pilots for the two MCAS-related crashes was being maintained, albeit less explicitly, by the new CEO of Boeing as late as March 2020 (Kritroeff & Gelles, 2020). The discussion of the accidents in Australia illustrates multiple attempts to deflect blame onto the ATSB.

I have proposed that the emergent property at this level is trust. Both the political masters and the travelling public (and society more broadly) must have trust in the aircraft they travel in and the system that supports the conduct of aviation. A breach of trust can have serious implications. For example, in September and October 2007, the Scandinavian airline, SAS, experienced four landing gear failures on its DHC8-Q400 fleet. The last one was on 27 October 2007, and the next day the airline decided to suspend its use of the aircraft and replace the entire fleet. The technical failure could not be attributed to the airline and other operators had, similarly, suffered undercarriage failures. The drastic and significant decision to replace the whole fleet was the result of a loss of public confidence. Passengers, when making a booking, were asking what aircraft type was used on the route. When told it was the Dash 8, they chose to make other arrangements (personal communication). Public opinion represents an environmental – Level 5 – factor. In this specific instance, the public lost faith in an aircraft type, and the operator paid the price. One goal of Level 4 actors is to prevent a breach of trust in their ability to guarantee safety.

In the context of the aviation system, power resides with Level 4 agencies even if their ability to effectively exercise that power is problematic. For example, a review of regulation in Australia (Australian Government, 2014) observed that 'relationships between industry and the CASA have, in many cases, become adversarial'. Insufficient resources, both physically and intellectual (as in the detailed understanding of technical aspects of the technology being overseen), is a consistent motif running through any examination of the functioning of regulators as is the problem of information asymmetries: regulators often are unaware of the actual status of an entity being overseen. Feedback from the point of production to those that of frame regulation is indirect and usually lagging behind operational developments. Because investigation agencies typically only look at fatal accidents or serious events, their findings typically address significant disconnects between intended system functioning and its actual enactment. These challenges increase opportunities for well-intentioned regulation to result in cross-scale interactions that result in failure at the operational level.

10.12 CONCLUSION

The significance of Level 4 is that it both shapes the performance of the aviation system and also gives it legitimacy. Without rules for guidance and permissions for action, the system would be chaotic. But, we need to consider the implications of the discussion in this chapter for individual competence. I suggest that there are two key areas of interest: what is said and what is not said. Of course, knowing the body of policy and procedures, technical information and checklists is fundamental to any job and would typically be associated with the concept of 'airmanship'. However, because this body of knowledge is an act of communication, it requires interpretation. It is not enough to read the rule, we have to be able to understand it and be able to apply it across a range of situations, not all of which might have been considered in the original framing of the rule. I suggest that 'comprehension and application' represent a subset of the competence of communication. Belcher commented that regulations are incomplete; they have gaps: 'lacunae'. So, another competence is how to navigate the 'not said'. Filling in the gaps is a form of decision-making in that a course of action must be determined based on the intent of the 'what is said'. An understanding of the dynamics of Level 4 should induce a healthy scepticism in relation to the output from regulators and manufacturers and, thus, a more robust approach to 'working within the rules'.

REFERENCES

Aboulafia, R. (2019). 'Boeing's Crisis Strengthens Beijing's Hand'. Foreign Policy March 2019.

Auditor General of Canada. (2008). Chapter 3 – Oversight of Air Transportation Safety – Transport Canada. Report of the Auditor General of Canada, May 2008.

ATSB. (2012). Transport safety report. Ditching – Israel Aircraft Industries Westwind 1124A, VH-NGA 5Km SW of Norfolk Island Airport. Final Report No: AO-2009-072. 18 November 2009

ATSB. (2017). Fuel planning event, weather related event and ditching involving Israel Aircraft Industries Westwind 1124A, VH-NGA. 6.4 Km SW of Norfolk Island Airport. 18 November 2009. Final Report No: AO-2009-072 (Reopened) Final -23 November 2017.

ATSB. (2019). Collision with terrain involving SOCATA TB-10 Tobago, VH-YTM, near Mount Gambier Airport, South Australia. 28 June 2017. Aviation Occurrence Investigation AO-2017-069 Final – 13 August 2019.

Belcher, P. (2002). A sociological interpretation of the COLREGS. *Journal of Navigation*, 55(02), 213–224.

CAA. (2019). Letter Hutton to Schapps.

Commonwealth of Australia. (2013). Aviation accident investigations. Rural and Regional Affairs and Transport References Committee. 23 May 2013.

Commonwealth of Australia. (2019). Performance of the Australian Transportation Safety Bureau. Rural and Regional Affairs and Transport Legislation Committee.

Chaplin, J.C. (2011). Safety regulation - The first 100 Years. *Journal of Aeronautical History*. Paper 2011/3.

Downer, J. (2009). Watching the watchmaker: On regulating the social in lieu of the technical. LSE CARR Discussion Paper 54; June 2009.

Government Accountability Office Report No: GAO-14-16. Aviation safety. June 2014.

Greene, M. (2019). Boeing is dragging down the US Economy. *Financial Times* (21 October 19).

HSE. (2006). Aircraft turnround: The impact of HSG209 and emerging good practice. Research Report 428 prepared by BOMEL Limited for the Health and Safety Executive 2006.

Hobbs, A. (2004). Human factors: The last frontier of aviation safety? *International Journal of Aviation Psychology*, 14(4), 335–341.

Hutter, B.M. & Lloyd-Bostock, S. (2017). *Regulatory Crisis*. Cambridge: Cambridge University Press.

JATR. (2019). Boeing 737 MAX Flight Control System: Observations, findings and recommendations. Submitted to the Associate Administrator for Aviation Safety, U.S. Federal Aviation Administration. 11 October 2019.

Kritroeff, N. & Gelles, D. (2020). 'It's More than I Imagined': Boeing's new CEO confronts its challenges. *New York Times* (5 March 20).

Lodge, M. (2015). Managing regulatory failure. *LSE Knowledge Exchange HEIF*, 5.

MAIB. (2008). Girting and capsize of tug Flying Phantom while towing bulk carrier Red Jasmine with 1 person injured and loss of 3 lives. Accident Investigation Report No: 17/2008.

Minister of Supply and Services. (1992). Commission of inquiry into the Air Ontario Crash at Dryden, Ontario.

NTSB. (2003) Aircraft Accident Report Number AAR-03-03: Control and Impact With Terrain, Aviation Charter, Inc. Raytheon (Beechcraft) King Air A100, N41BE Eveleth, Minnesota October 25, 2002.

NTSB. (2010). AAR-10/01. Loss of control on approach. Colgan Air, Inc. Operating as Continental Connection Flight 3407 Bombardier DHC-8-400, N200WQ. Clarence Centre, New York. 12 February 2009.

Queensland Courts. (2007). Office of the State Coroner. Findings of the Inquest into the Lockhart River air crash.

Parliament of Australia. (2019). Press Release. 7 October 2019. An inquiry into the performance of the Australian Transport Safety Bureau, and in particular its report on the June 2017 crash of a flight conducted on behalf of Angel Flight Australia,

Reuss, J. (2016). Independence does not mean isolation. *ISASI Journal.* pp 24–29. July–September 2016.

Senate Committee. (2019). US Senate Committee on Transportation, Aviation Sub-committee. Transcript of hearing on 30 October 2019.

Shapiro, S.A. (2011). The complexity of regulatory capture. *Roger Williams University Law Review*, 17 pp 221–257.

SHK. (2007). Report No: RL 2007:12e. Aircraft accident to LN-RDA at Kalmar Airport, H county, Sweden, on 6 April 2006. Case L-08/06.

TSB. (2009). Aviation Investigation Report No: A07A0134. Touchdown Short of Runway. Jetport Inc. Bombardier BD-700-1A11 (Global 5000) C-GXPR. Fox Harbour Aerodrome, Nova Scotia. 11 November 2007.

TSB. (2014). Independent review of the Australian Transportation Safety Bureau's Investigation methodologies and processes. Quebec, Canada.

Zremski, J. (2019). Congress urged to water down Flight 3407 pilot training rules. *The Buffalo.*

11 Training for Competence

11.1 INTRODUCTION

This book set out to address the issue of what 'crew resource management' (CRM) means in a systems context and how that translates into a set of behaviours necessary for an individual to function in that system in a competent manner. Three questions remain to be answered: what to train, how to train it and how to measure success. I am writing this chapter at a very peculiar time in the history of aviation. After years of growth and with the prospect of the Asian and, especially, Chinese markets offering enormous potential, the Covid-19 pandemic has almost brought the world to a halt. Aircraft are being parked, staff laid off and the prognosis is that it will take years for the industry to recover. At the same time, concern about global warming, and aviation's environmental impact, is causing some to ask if we should even attempt to return to 'business as usual'. Instead, perhaps the pandemic should be seen as an opportunity to rethink the way the global economy works and the future role of air transport. Any thorough analysis of the problem should look at existing processes, including training, and question their validity for the future shape of the industry.

In this chapter, I will set out a framework of candidate competencies that flow from the systems model outlined in the book. I will illustrate how the competencies can be used to develop training curricula and then look at the methods available to effect behaviour change. We need to recognise that a training intervention is just one approach to bringing about change and I will consider alternative methods. Finally, truly resilient systems are based on robust skill sets. The fundamental goal of training is to bring about a change in behaviour or to support the maintenance of effective behaviours. This chapter offers some solutions to these problems while, in the next, we will look at measurement.

11.2 THE TRAINING PROBLEM

Before moving on to the issues of what to train, how to train it and, then, measuring performance, I want to just return to the core question that arose in Chapter 1: what is training supposed to achieve? I suggested that a good starting point was Wood's view of a competence model as being an understanding of the variability and uncertainty faced by an operator – the demand side – coupled with an understanding of how the strategies, plans and countermeasures provided by the organisation – the supply side – will handle these. This book has elaborated on aspects of these two questions. In terms of variability and uncertainty, when we looked at the idea of threats we saw that these were, on the one hand, attributes of the operating environment that require a response from crew but, on the other hand, by their very nature they can induce anxiety and increase workload, thereby impairing crew in proportion to an individual's susceptibility. Figure 11.1 illustrates environmental variability. It shows the distribution of safety reports recording bird strikes submitted by one airline across a year. Most bird strikes are

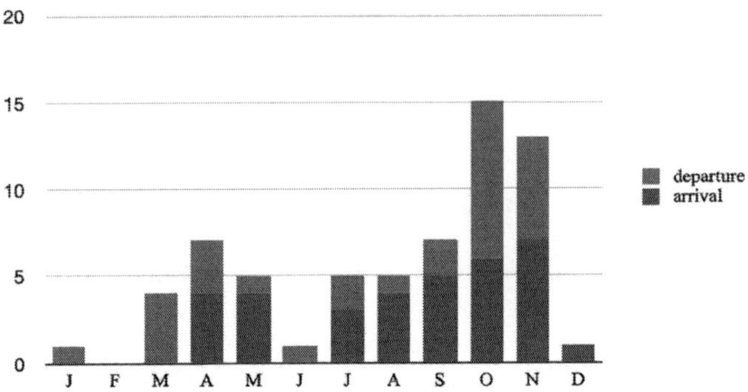

FIGURE 11.1 Bird strike reports over 12 months.

inconsequential but some can cause severe damage to an engine or, in the case of the Miracle on the Hudson, can even cripple an aircraft. The demand side of the equation requires that crew must cope not just with the range of environmental conditions encompassed by the operation but also the implications, manifested as stress and fatigue. Non-ergodicity dictates that crew will have to be able to constantly modify plans in real time. Because of cross-scale interactions, we know that crew may well be presented with device configurations and technical systems behaviours that have lain dormant or were not fully communicated. The crew will have to cope with conflicting goals, especially when working across organisational boundaries. On the supply side, we know that interfaces will be opaque, that policies and procedures will be underspecified and that the organisational context will trigger frustration and resistance.

A competence model will have to map onto the demand side while navigating the deficiencies in the supply side. This suggests that competencies will include skills associated with action and also with analysis and interpretation. In Chapter 1, we looked at the key aspects of ID and I suggested that a problem we faced was that the conventional tool kit was predicated on having a task to analyse. Having attempted to explore performance in the context of a system, we can now attempt to derive some 'competencies' that support that performance. We do have one issue to address before moving on: what constitutes 'competence' in relation to CRM? The approach taken throughout this book is that effective performance is goal-directed and comprises a suite of skills underpinned by robust knowledge structures. There is no separate entity called 'CRM'. With that in mind, in the next section I want to outline a competence framework.

11.3 ASSESSMENT FRAMEWORKS AND 'COMPETENCE'

Once the concept of assessing 'CRM' skills was accepted, the best mode of sampling was considered to be observation on the flight deck and, to that end, various marker frameworks were proposed as tools for capturing performance in the workplace. There are three methods commonly available for developing assessment schedules. The first method looks at past failure and attempts to identify behaviours that would have reduced the probability of an undesired outcome. Probably the longest-standing

TABLE 11.1
Extract from University of Texas Behavioural Markers Rating Scale

SOP briefing	The required briefing was interactive and operationally thorough	Concise, not rushed, and met the requirements Bottom lines were established
Plans stated	Operational plans and decisions were communicated and acknowledged	Shared understanding about plans Everybody on the same page

aviation marker framework, the NASA-UT Crew Effectiveness Markers, represents this approach. It formed the basis of the subsequent University of Texas Behavioural Markers Rating Scale. Table 11.1 illustrates two of the 13 suggested markers, each of which was rated on a four-point scale (Helmreich et al., 1990)

The next approach is the 'expert committee' model. This method involves panels of subject matter experts debating what acceptable behaviour looks like. The European NOTECHS scheme (Flin et al., 2003) and the more recent EBT/CBT framework (EASA, 2019) exemplify this approach and is, possibly, the most common method used. Examples are at Tables 11.2 and 11.3.

The third approach is to conduct structured interviews with operating crew, using tools such as critical incident and repertory grid, to seek examples of behaviour both good and bad. Having assembled a range of statements, subject matter experts then undertake a 'card sort' activity' (MacLeod, 2005). The set of statements are assigned to categories according to similarity or relatedness. An example is in Table 11.4.

In previous chapters I have used data from a number of line operations safety audit (LOSA) surveys. A content analysis of the narrative reports written by observers is another source of information about crew performance. Table 11.5 illustrates a structure that emerged from 423 comments extracted from reports. Only the most significant clusters are included.

TABLE 11.2
Categories and Elements of NOTECHS

Category	Elements
1. Cooperation	Team-building and maintaining Considering others Supporting others Conflict-solving
2. Leadership and managerial skills	Use of authority and assertiveness Providing and maintaining standards planning and coordination Workload management
3. Situation awareness	Awareness of aircraft systems Awareness of external environment Awareness of time
4. Decision-making	Problem definition and diagnosis Option generation Risk assessment and option selection outcome review

TABLE 11.3
Extract from EASA Competency Framework

Application of Knowledge (KNO)

Description: Demonstrates knowledge and understanding of relevant information, operating instructions, aircraft systems and the operating environment

1. Demonstrates practical and applicable knowledge of limitations and systems and their interaction
2. Demonstrates required knowledge of published operating instructions
3. Demonstrates knowledge of the physical environment, the air traffic environment including routings, weather, airports and the operational infrastructure
4. Demonstrates appropriate knowledge of applicable legislation.
5. Knows where to source required information
6. Demonstrates a positive interest in acquiring knowledge
7. Is able to apply knowledge effectively

Application of Procedures and Compliance with Regulations (PRO)

Description: Identifies and applies appropriate procedures in accordance with published operating instructions and applicable regulations

1. Identifies where to find procedures and regulations
2. Applies relevant operating instructions, procedures and techniques in a timely manner
3. Follows SOPs unless a higher degree of safety dictates an appropriate deviation
4. Operates aircraft systems and associated equipment correctly
5. Monitors aircraft systems status
6. Complies with applicable regulations
7. Applies relevant procedural knowledge

TABLE 11.4
Interview-based Marker Construction

Task Management

This dimension relates to the conduct of the work. It includes the consistent and appropriate use of checklists and procedures, making effective use of time. It also includes the avoidance of distraction and maintaining the bigger picture of things happening in and around the aircraft. There are three critical clusters:

- Use of SOPs (understanding and interpretation of SOPs, their application, attention to detail, diligence)
- Creates solutions, adapts procedures and improvises as demanded by situation. Able to work within resource constraints. Manages time by prioritising tasks
- Makes an appreciation of risk and considers implications of actions.

Positive indicators include:

A consistent, but flexible, use of SOPs. Monitors the conduct of work during busy periods and positively verifies that tasks have been completed. Maintains an even tempo of work (no unnecessary haste or urgency). Actively develops a mental picture of what to expect during the next stage of flight (e.g. through verbalisation). Anticipates and thinks ahead. Is aware of time available/remaining,

Negative indicators include:

Too strict an adherence to or rigid application of SOPs. Spending too much time out-of-the-loop on admin tasks. Rushing or delaying actions unnecessarily.

TABLE 11.5
Emergent Structure from LOSA Narratives (% of comments)

Systems management (10.6%)
(Wrong selection, left in wrong position, inappropriate selection, hesitation, confusing systems behaviour)
Application of procedures (13.2%)
(Acting from memory, acts with no clearance, failing to act, incorrect actions, unaware of need to act, inconsistent actions)
Communication (21.5%)
(Explanation, clarification, intervention style)
Task management (22%)
(Anticipation, clarification, prior requests forgotten, workload, prioritisation, path management and control)

The approaches illustrated differ in that the first three – the UT, NOTECHs and EBT frameworks – reflect the views of those involved in their construction. They are heavily influenced by the opinions of the collaborators and this is sometimes apparent in the language used in the descriptions. The latter two are rooted in the observation of workplace performance and the structure emerges later. One fundamental weakness of these approaches is that they were applied to the problem of developing a process of performance assessment, not to elucidate a comprehensive competence model. They were designed to sample, not to catalogue. By default, then, they will be incomplete in that their coverage of performance will not be comprehensive. Also, their formulation will reflect their intended purpose.

A further problem is that the frameworks described were designed to be used in any context. No consideration was given to whether different categories of aviation demand different competencies. For example, as we have seen, business aviation pilots have to spend time managing the expectations of their clients as well as flying the aircraft. Freighter crews have to deal with complex operational issues, often to do with the cargo they are carrying, while at remote destinations with little operational support. Police and emergency medical support helicopters often have to make decisions under intense pressure. While it is probable that there is a core of common performance across all types of aviation, we also need to address the complete operational envelope, to match training to the risk profile and to the operational demands faced by the user. In the next section, I want to explore what competence requirements emerge from looking at aviation from a systems perspective.

11.4 A SYSTEMS VIEW OF COMPETENCE

A competence, in the way it is used in training, is simply a construct that describes a class of behaviour that is deemed significant in the performance of work. The conventional approach to describing a competence is to identify both the specific behaviours needed to do the job and also any underpinning knowledge that supports performance but it is important to remember that 'behaviour' and 'knowledge' are of a different order. For example, to write this book, the performance is the way I express my ideas and formulate my arguments but in order to do so, I also need some knowledge of syntax, grammar and vocabulary and of the use of a word processing package on a computer. Comprehension, integration of data, manipulation of information and the creation of new ideas might be considered 'competence' but knowledge about MS Word is not.

It is possible to differentiate between knowledge that is essential for a specific behaviour and that which underpins performance generally. We can trace a line from an understanding of climate, through knowledge of local weather, to sources of current weather information and how to interpret a weather report until, finally, we get to decision-making in relation to fuel loads based on the probability of holding at the destination because of weather. The underpinning knowledge around climate and weather is generic but supports the specific skill of processing meteorological reports. This, in turn, facilitates effective decision-making, which is probably a competence. The fundamental question is, what background information supports performance?

We also need to recognise that information has a half-life in that, over time, research will render previously agreed knowledge incorrect. The knowledge base that pilots draw upon is refreshed in cycles, each of a different duration. Some fundamental information might survive a career. Policies will change over a shorter cycle while procedures can be quite unstable. The other side of the equation is the fact that information stored in memory is not stable. We forget and memories become unreliable. When we check a specific element of a policy in a document we find that it is not quite as remembered. Neither the knowledge base that supports operations nor the knowledge held in memory is stable but maintenance of both is effortful. Knowledge, then, needs constant housekeeping but recognition of this fact and taking the necessary action possibly reflects an attitude to both the individual's role and to the job.

Throughout this book, I have suggested, possibly by implication, that mastery of technology and familiarity with policies and procedures is the foundation upon which competent performance is built. In simple terms, for each component or system on the aircraft we need to know, or have knowledge of:

What it does
How it works
How to work it
What it interacts with
How to deal with malfunctions
How to operate with degraded or non-functioning equipment
What additional risks are associated with reduced capability

For policies, procedures and checklists I need to know what information is available, where to find it, what its scope is, what is *not* covered and how to use it.

If systems knowledge is the foundation then the next step is to consider how it supports the management of a task. We now need to integrate this knowledge into the operational context and, therefore, our goal in relation to aircraft control might be to:

Configure and operate systems to the required standard
Maintain a valid conceptual understanding of what automation is trying to achieve at each stage of the trajectory

Use appropriate and optimal automation modes
Evaluate situations to confirm congruence between actual and desired conditions
Project how interventions will affect the trajectory
Communicate to sustain crew understanding of aircraft status and goal accomplishment

From this analysis, we can start to identify what skills can be aggregated into competence and what aspects represent underpinning knowledge. A top-level conceptual structure such as this allows us to audit in-service training to see if there is a need for specific training interventions but these performance elements would need to be further analysed in terms of behavioural objectives for use in ab initio training development.

In this section, I have simply tried to clarify the relationship between underpinning knowledge and competence. The application of a systems concept to the task of aviation suggests that crew have to function in three specific domains. First, we can identify a space bounded by the normal variability of the operation and that of the performance of human actors. This is the domain of 'normal operations'. Next, there is a point where the crew are having to cope with contingencies that require additional effort. This could simply be events that are non-routine but, nonetheless, normal (Kluge, 2014) that require physical reference to checklists, documents or third parties to verify processes or it could be anomalies that require clarification before action. This class of activities will add to cognitive load, have additional risks attached to them but should be with the expected capability of the crew to resolve. The final area of interest is the space where the operation is approaching the system boundary. Buffering capacity is limited and crew typically have to shift to a new mode of functioning in order to restore margins. Here, we might expect higher rates of failure among crew.

11.4.1 Competence as 'Normal Work'

We start with normal operations, a template for which might be:

Operational Readiness (pre-flight)
Planning for the anticipated task
(Load, fuel, weather, minimum equipment list (MEL)/DDG) (quantities, requirements and implications)
Planning for probable contingencies
(Route, destination, MEL/DDG) (range of possible encounters and implications)

Trajectory Management (in-flight)
Goal definition
Goal state convergence
(Next Goal State, Current Status and Interventions required)
Synchronising activity
(Communication (intra-, inter-), coordination of effort, conflict resolution)
Current 'most likely alternative course of action'
(Update 'what-ifs')

From this framework, we can derive a set of generic behaviours that revolve around routine planning, organising and validation. Planning involves both resourcing requirements for the expected task (planned fuel, payload, information relating to next goal state) but also anticipating change (weather) and preparedness for contingencies (implications of a deferred defect, possible en route events). Organising involves coordination with third parties (refuelling, loading, passenger boarding, maintenance, air traffic control (ATC)) and task allocation within the team. Validation is the process of establishing if goal constraints have been met. It involves processing third-party inputs (load sheets, weight and balance, flight plans, fuel receipts and ATC clearances), seeking information from aircraft systems and confirming agreement within the crew. Validation is similar to controlling in an active sense but also includes the passive checking and confirmation of current goal status.

Much of this activity flows from the application of fundamental skills, such as checking flight plans and load sheets, calculating the aircraft's weight and balance, interpreting weather reports and relevant NOTAMs. The skills are the enactment of knowledge gained through training and they facilitate the functioning of the system by supporting decision-making on the part of the crew. Communication has not been addressed specifically in this discussion and its ambiguous role has been already noted. I argued in Chapter 8 that communication is the driving force of the system. Although a clear relationship between underpinning knowledge and performance can easily be established in some areas, it is not clear if the same relationship exists for decision-making and communication. For example, what underpinning knowledge about decision-making supports better decision-making? Given that competencies are simply constructs, the final structure we arrive at is simply one that should be useful in some way. The structure should clarify the requirement for training inputs as well as inform decisions about assessment although, as we will see in the next chapter, only a subset of competence is probably amenable to direct measurement through observation. At this stage, given that this chapter looks at training, it is probably more beneficial to have a broader, rather than a narrower, focus. Therefore, I suggest that the fundamental competencies that support normal operations are:

- Planning
- Organising
- Validating
- Deciding
- Communicating

11.4.2 Competence as Management of Anomalies – Performance in Transitional States

The next space we need to explore is that part of the system envelope where the operation has departed from what I have called normal. This could be a procedure or a manoeuvre not often rehearsed, a system malfunction that is not an emergency or an unexpected change to the flight plan. These are infrequent routines that still rely on standard operating procedures (SOPs), in the main. Because these are infrequent actions, they are less automatic in their execution, which means they are more susceptible to interruptions and distractions. They need more attentional control, thus resulting in higher mental workload and there is lower remaining mental capacity for coping with any additional events. Many of the examples examined in the previous chapters fall into this category.

Kluge (2014) identifies a number of SOP-driven action sequences we must manage when dealing with anomalies. They comprise: fixed (must do), contingent (depends), parallel (do at the same time) or free (can do whenever) sequences. Control of these action sequences make demands upon the cognitive capacity of the operator and, interestingly, also map onto Hockey's tripartite structure of things we must do, things we should do and things we want to do, discussed in Chapter 4. The implication being that psychological fatigue will impinge upon anomaly management.

Kluge's work was mainly based on an analysis of complex technical systems such as power plants. She offers the following generic task analysis of the skills needed to manage anomalies:

- Cope with higher mental workload
- Clarify the current status of a situation before the control actions are taken
- Interpret warning and indicators correctly based on systems knowledge, integrate separate indications and account for all data
- Select an action based on how the system is controlled, select correct SOP or checklist
- Plan and prepare the control activity (SOP, procedure, intervention)
- Monitor the system behaviour so that processes remain safe and within predefined limits based on systems knowledge
- Describe the event at a higher level based on process and safety
- Develop cause and effect relationships, estimate the state of 'hidden' variables
- Execute actions based on operators mental model
- Check whether the system reacts in the intended direction
- Make sense of unexpected phenomena occurring as the task progresses

When dealing with contingencies, it appears that four clusters of behaviours are significant:

- Coping with increased cognitive demand
- Discrepancy clarification
  - (Causal analysis (why is this happening?), options available (what can we do?), actions required (how do we do it?))
- Implications of change
  - (Revised risk profile (new risks?), operational constraints (new limitations on what we can do), changes to capability (remaining functionality))
- Referencing observed behaviours to expectations

A characteristic of performance in this domain is the need for crew to be able to function under greater cognitive load, partly resulting from elevated stress, which suggests that training should support the development of coping capacity, and I will return to this later. A consistent theme in previous chapters has been the fact that the system margin is a zone where discrepant signals are present. Furthermore, buffering reflects the system's capacity to cope with alternative modes of action. By default, this rather assumes that we are capable of estimating an intervention's probability of success – its efficacy – in the margins remaining. Kluge's analysis supports this view. In terms of a competence model, the behaviours described here suggest that we might consider 'analysis' as a candidate. The other behaviours seem to be elaborations of the competencies associated with normal operations.

11.4.3 Competence in a Crisis – Performance at the Boundary

Throughout this book I have tried to avoid, as far as possible, any reference to catastrophic, fatal accidents unless the specific case allowed me to reflect on normal operations. In this section, we are moving towards the periphery of the normal envelope. These boundary states will probably involve damaged or degraded systems, extreme weather events or some significant challenge to the crew. In all probability, we are now beyond the company's designed coping strategies and success will probably be predicated on skill, ingenuity, luck or all three. Abnormal or emergency situations, these boundary conditions, are typically unfamiliar, unplanned and unexpected. SOPs do not always work. Quite often, causes are not immediately apparent nor are they comprehensible. One of the problems is that, in emergency situations, we often have multiple goals but it is not always clear how these goals are to be achieved. Uncertainty is the theme of such situations, as is intense stress. This is the area of 'knowledge-based' behaviour – fundamental knowledge (systems and procedures) is used to create rules that can be used to drive behaviour and improvisation is often required. On 4 November 2010, a Qantas Airbus A-380 departing from Singapore suffered a major engine failure (ATSB, 2013). The significant damage to the aircraft and its systems resulted in the crew having to deal with in incessant cascade of alarms and failure indications. After trying to cope using the standard approach taught in the simulator, the captain realised that they faced an impossible task. He then decided that rather than trying to deal with what was wrong with the aircraft, it would be more beneficial to establish what systems were still working and to manage the situation with the assets remaining. The aircraft landed safely back at Singapore but the outcome required a radically different approach to emergency management.

Again, Kluge's work offers a task analysis for dealing with abnormal and emergency situations. The defining characteristics of these situations appear to be increased ambiguity and elevated stress. While time may be a factor, it is more often significant simply because fundamental processes consume more of it. Therefore, the tasks Kluge has identified reflect the need to seek clarity and increase control. She proposes three significant clusters around diagnosis, communication and group functioning:

Diagnosis
Establish deviations from a normal state
Collect information about cues
Establish abnormal cues based on the mental model
Compare assumptions about cause and effect relations among cues

Use critical thinking
Apply team knowledge (task, team, process and goal-related)

Communication
Use proper phraseology
Pays attention to completeness of standard reports (listening)
Seek information/clarification and check understanding
Communicate information to the team
Exchange information and comprehensions to establish a shared understanding of the problem
Formulate and communicate hypothesis about cause and effect relationships

Group functioning
Resolve opposing interpretations based on team conflict resolution
States appropriate priorities
Update situation periodically
Monitor and support others
Provide guidance and suggestions to others
Apply stress-coping strategies

Of interest is the role of a more formal approach to communication. In Chapter 8, we saw that modes of speech change under conditions of increased task demand as the social dynamic changes. More rigorous attention to structure, completeness and distribution of communication acts is important. We also see that additional resource must be allocated to maintaining the structure and functioning of the team. Collaboration emerges as a possible candidate for our list while the need for coping is reinforced.

11.4.4 An Outline Competence Model

This exploration of performance can be summarised in an eight-factor model of competence (Table 11.6). Each competence can be elaborated as a set of behaviours, activities and, in most cases, underpinning knowledge. It is almost possible to make a case for 'deciding' and 'communicating' to be supra-ordinate competence, the former being an outcome of the actions of other competencies while the latter being a facilitating process. Fundamentally, the model describes the behaviours associated with goal-directed performance in systems model. I make no reference to the manual control of an aircraft because I consider that to be a skill. It supports performance and is a reversionary mode should other technical systems fail but it serves no purpose in itself (unless the pilot is engaged in recreational flying). That said, the model is not a framework for assessment and, undoubtedly, the evaluation of flying skills is important. We will look at assessment in the next chapter.

11.5 NON-TRAINING INTERVENTIONS

Before looking at the design of training to support competence development, I want to consider alternative approaches to behaviour change. Early versions of CRM

TABLE 11.6
Proposed Competence Model

Competence	Supporting Activities
Analysing	Causal analysis; risk appraisal; establish the gap between observed and expected; establish abnormal cues based on mental model; compare assumptions about cause and effect relations among cues
Planning	Identify options; establish operational constraints; clarify remaining capability/ functionality; planning for contingencies
Organising	Identify actions required; establish resources required; implementing contingency plan
Validating	Referencing observed behaviours to expectations; establish deviations from normal state; use critical thinking
Deciding	Validate rule set; identify information requirements; validate efficacy of option; establish time requirements
Communicating	Use proper phraseology: pay attention to completeness of standard reports; seek information/clarification/ check understanding; exchange information and comprehensions to establish a shared understanding of the problem; formulate and communicate hypothesis about cause and effect relationships
Collaborating	Monitor, support others, provide guidance and suggestions; states appropriate priorities; update situation periodically; resolve opposing interpretations based on team conflict resolution
Coping	Create space and time; control stress

training requirements included a reference to 'organisational factors', although the term was typically not elaborated on. The latest EASA guidance includes a reference to 'the operator's safety and organisational culture' (EASA, 2017). Again, there is no clarification of what such training is supposed to cover. In Chapter 9, I explored some aspects of the relationship between the organisation and front-line crew. The key message of the chapter was that the legitimate actions of management can generate unintended consequences. From a systems perspective, business models create situations where work processes do not map onto human physiology. The evidence suggests that this mismatch can impair performance and shape risk tolerance, which means that buffering is reduced. Furthermore, differences between individuals in terms of their need for recovery, well-being and LWB results in sickness/absence that places a further burden on the remaining available workforce.

Table 11.7 shows the distribution of pilots listed as long-term sick (LTS) by age compared with the distribution of pilots in each age group. Pilots on LTS have been off work for longer than 19 days and, therefore, need their aviation medical certificate to be reissued. As an aside, the administration of a pilot's return to work is itself an organisational problem than can increase lost productivity. The group included individuals suffering from cancer, other long-term illness, musculoskeletal problems, sleep disorders and so on. On a daily basis, in this airline, approximately 300 pilots were unfit for duty of whom one-third were on LTS. Over a 5-year period (2009–2013), LTS events rose from 131 per 1000 pilots to 171 per 1000 pilots while days lost to LTS rose from 9.2/pilot to 14.92/pilot. The data suggests that some age groups are over-represented.

TABLE 11.7
Distribution of Pilots on LTS ($n=2800$)

Age	% Pilots	% LTS
<25	2.9	0
<30	14.6	9.7
<35	11.8	2.4
<40	13.9	8.7
<45	19.9	25.61
<50	17	14.6
<55	9	18.2
<60	7.3	14.6
<65	3.2	4.8

Table 11.8 looks at the distribution of pilots appearing in a monthly list of those taking more fuel than required by the flight plan. These two tables together suggest that age can affect both availability for work and operational effectiveness. In Chapter 1, I discussed age effects in railway workers in relation to reduced risk perception and over-estimating the efficacy of actions. In the broader context, this data suggests that the demands made on staff, as reflected in contracts of employment and work schedules, might benefit from being tailored to accommodate the effects of age. Organisational design at a strategic level, therefore, may be one route to shaping behaviour at the operational level.

In Chapter 9 we saw that the adoption of fuel efficiency policies was variable and reflected a number of factors, one of which was possibly perceived increased effort. In the study by Gosnell et al. (2019), airline captains were advised that their company was participating in a project that looked at fuel efficiency. Pilots were assigned to one of four groups: a control group; those provided with information about personal performance; those who were set targets; those who were offered a financial incentive (a charitable donation if targets were met). Interestingly, simply being told of the project improved behaviour with the control group making the largest contribution

TABLE 11.8
Excess Fuel Uplift by Age of Captain

Age	% Total CNs	% on List
>60	8.2	19.6
55–59	17.72	13.7
50–54	16.72	13.7
45–49	24.9	29.4
40–44	25.3	21.56
35–39	6	1.9
<34	0.3	0

to fuel efficiency during the project. Providing information about performance and setting improvement targets also had an effect. The trial demonstrated that behavioural modifications could result in a potential annualised saving of US$9.15 million at 2014 fuel prices plus the associated reduction in CO_2 emissions. Importantly, one spillover effect was an improvement in job satisfaction for those captains included in the experimental groups. In a post-trial review, of the 335 participating captains, approximately half welcomed a continuation of such management strategies while just 4% preferred a return to the pre-study status quo. This framework of interventions could be adapted to other behavioural change targets but significant in its approach is the fact that it involved active engagement of crew, target setting and feedback.

It is not my intention to discuss airline management, generally, here. However, from a broader systems safety perspective, it does appear that management activity is a significant factor. From a training perspective, where stress and fatigue, for example, are included in syllabi, there is a danger that the disconnect between the classroom message and the lived experience of the crew will provoke cynicism in trainees. We do have an issue with allocating ownership of the problem in that fatigue management is a shared endeavour between both the company and the employee. That said, alignment between training and operational reality is essential if the investment is to be beneficial. Organisational design is clearly one area where operational safety and efficiency can be influenced by means other than a training intervention.

Changes at the level of the organisation are clearly strategic, the impact of which would take time to work through but technological developments offer other, more immediate, non-training interventions. The use of flight data in safety has a long tradition but advances in data collection, accessibility, processing and visualisation offer considerable potential for non-training interventions. The biggest single barrier to progress in this field is the long-standing paranoia in the industry about data falling into the wrong hands. Pilots do not trust management not to abuse data while airlines fear information being made public. A case could be made for the data generated during a flight to be 'owned' by the crew operating that flight. A similar case is being made that internet companies that benefit from access to personal information, browsing and purchasing behaviour should somehow recognise the rights of the individuals who created that data, either through stronger privacy rules or through a usage charge, possibly through corporate taxation. While it is recognised that smarter use of flight data can benefit the broader safety and operational efficiency agenda, it could also be used to develop individual pilot competence.

Post-event visualisation is now commonplace and many safety departments will permit pilots to request that a flight be made available for viewing if the individual had some concerns about performance. However, accepting that reflection and self-analysis are important in refining mental models, the routine access to personal information should be the next step in the use of flight data. One such approach is CEFA Aviation's AMS tool (Figure 11.2). Using cloud-based technology and with powerful confidentiality firewalls in place, it allows pilots to access data from their own flights (de Courville, 2019). It uses a conventional visualisation interface but pilots can look at the approach and landing flight path, controls and displays to see how the event was managed. Displays could be enhanced to direct attention to aspects of control that could be indicators of proficiency. For example, Ebbatson (2009; Ebbatson et al., 2010)

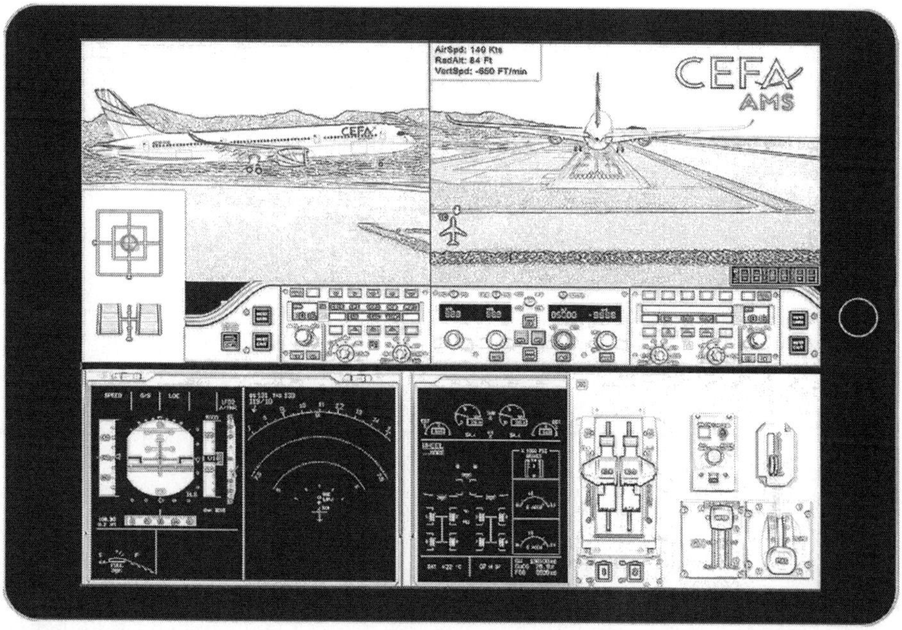

FIGURE 11.2 Flight data feedback. (© 2020 CEFA Aviation.)

studied manual handing skills and found that tracking of speed, together with control inputs (magnitude and frequency) were indicative of recency and, thus, skill decay.

There are now machine learning (ML) algorithms that can make use of large data sets. The application of artificial intelligence and ML to flight data has tended to focus on anomalous events but some use has been made of these approaches to look at normal operations. Li (2013) and her colleagues (Li et al., 2015) used an algorithm to cluster flight data to look for significant patterns and also to detect anomalies. The algorithm can present, first, density-based relationships between data values which shows the normal distribution of a parameter. It can then, second, identify any flight that differs from the normal distribution in a statistically significant way. These outliers are typically those flights that would trigger a flight data monitoring (FDM) alert. In fact, in trials, the algorithm has been shown to be more accurate at detecting anomalous flights than conventional systems. However, my interest here is in how flight data could be used to provide non-training interventions.

Figure 11.3 shows the data for, in this case, pitch angle on final approach. The data was collected from 300 arrivals made by Airbus A-350 aircraft. The central darker zone represents 50% of the most tightly clustered data. The outer, lighter zones represent 20% of the data at each margin of the distribution. Because this is a statistical technique, there will always be some loss of data at each stage of the algorithm. Pitch angle is a parameter that is monitored in the FDM programme and by overlaying the trigger threshold, it is possible to illustrate how close the normal operation approaches an arbitrary boundary: the FDM programme is a surveillance mechanism and triggers are set at values that warrant attention but might not necessarily equate to a system boundary. In this case, it would be possible to set actual upper and lower limits. At the upper limit, the pitch angle on touchdown would result in a tail strike while, at the lower limit, in effect, the aircraft would not be flaring adequately and the result would be a hard landing. We have now established the buffering capacity of the system in relation to this one parameter.

The algorithm can also look at a specific flight and Figure 11.3 also shows the trace of a single flight superimposed on the general distribution. We can now let a pilot look at their performance compared to the normal distribution. Of course, a single parameter is unlikely to fully explain an event and to understand what happened on this approach we would need to look at, say, turbulence as measured in vertical acceleration and possible wind components (head/tail or crosswind). If a parameter is captured in flight data, it can be presented using this technology. Smarter use of data coupled with the use of leading indicators offers the prospect of intelligent tutor tools that can offer feedback to pilots post-flight based on actual data drawn from their own performance. There are, nonetheless, some problems with this concept. Data simply captures inputs to the aircraft and subsequent configurations and responses. It cannot capture either motive – the reason why an input was made – or the collaborative aspects of trajectory management.

One aspect of collaboration, of course, is communication. Looze et al. (2014) looked at the prosodic characteristics of speech. Prosody comprises intonation, tone, stress, and rhythm. Tools already exist that can evaluate cockpit communication based on the prosodic components such as pitch and cadence. Chapter 8 (Excerpt 8.2) discussed pauses and overlaps in conversation and prosodic analysis is capable of detecting these features (Figure 11.4).

Graphical analysis of speech can point to episodes during crew interaction where performance was suboptimal (time 06:30 in Figure 11.5). Semantic analysis is the

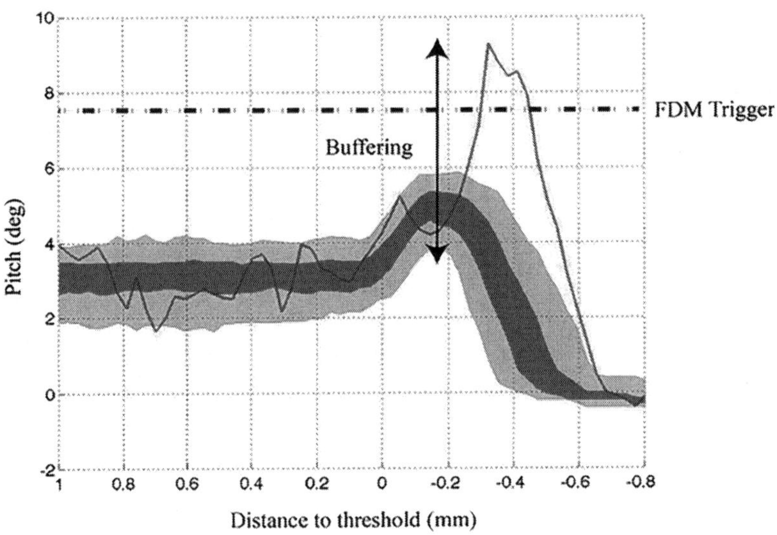

FIGURE 11.3 Cluster analysis of flight data. (Courtesy Dr. Lishuai Li.)

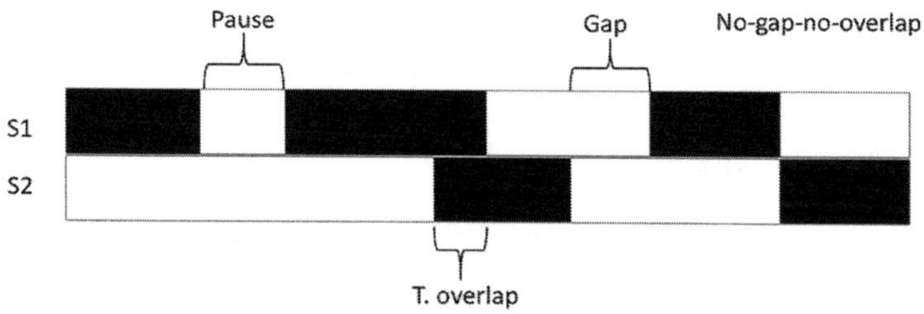

FIGURE 11.4 Pauses and overlaps in crew communication. (© 2020 Vocavio Technologies.)

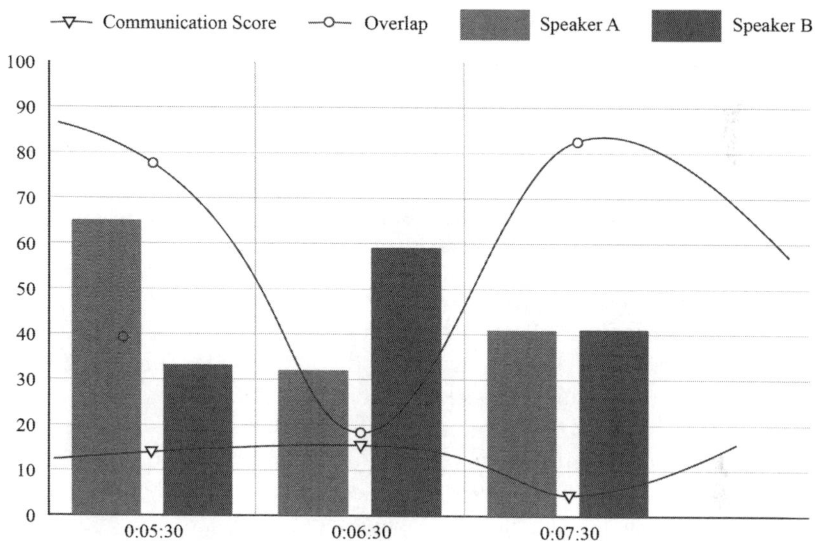

FIGURE 11.5 Prosodic analysis of pilot communication. (© 2020 Vocavio Technologies.)

next stage and the Vocavio system is already capable of evaluating SOP compliance based on detecting keywords. The functional communication framework outlined in Chapter 8 points towards semantic triggers that could be used to identify types of acts based on crew dialogue.

The use of flight data and communications analysis are forms of operational feedback. They offer opportunities for reflection on performance and represent approaches that could be adapted to create individual targets and structured feedback for crew. They can also be adapted to provide opportunities to develop knowledge structures. Martin et al. (2011) used 'what-if' scenarios to develop crew skills. Candidates were presented with scenarios and asked to reflect on possible solutions to problems. A post-trial evaluation found that pilots felt that they benefitted from this type of activity. By linking repertoires of problem scenarios to samples from flight data and crew conversation, it would be possible to afford rich opportunities for reflection.

There is a long tradition of social learning in aviation, often called 'hangar tales'. The sharing of experiences often allows a fine-grained approach to problems that might be beyond the scope of a classroom treatment. Learning through 'stories' has proven to be powerful, affords space for reflection and clearly support the development of mental models through aggregation and compilation. Consider this LOSA observer's report:

> Instruction from ATC: Reduce speed 270kts, expedite passing FL170. The captain (PF) initially reduced the speed target to 270kts, extended the speed brake and, after watching a slow deceleration to 270kts, then decided to increase the speed target to over 300kts in order to expedite to FL170. The Vertical deviation was showing the aircraft as 5000' high on profile and the aircraft was above the 3degree path. The PF then selected V/S 1000fpm, actually reducing the rate of descent. This was when ATC issued track miles to touchdown of 38nm. The PF decided to expedite the descent at this stage and extended the speed brake, triggering an EICAS SPEEDBRAKE message. FLCH was then selected. ATC then issued 210 kts, shortly followed by "Direct [waypoint], via [waypoint] cleared ILS 25R". Approaching 5500' the PF reduced speed to 180kts despite ATC requesting to maintain speed 210kts, as per the previous clearance. 180 kts was also 5 kts below the minimum speed for the current configuration (Flap5 185 kts). The PM did not pick up on this. As the speed reduced below the minimum flap speed the PF called for the next stage of flap (Flap10). Shortly prior to [waypoint], ATC then requested a spot wind check from us and then told us to reduce speed 180kts until 7nm. We had already reduced from 210kts to 180kts.

Crew familiar with the aircraft type and the local area would identify with the challenge of descent profile management, balancing the needs of slowing down while losing height, of remembering the current configuration of the aircraft, which automation mode is best for achieving a specific goal, when ATC clearances contain conditional instructions and so on. Stories extend our experience base. By reflecting on what others have encountered we can elaborate on our own experience of operations in a similar or same environment. We can reflect on similarities between our experience of the same event and that of the storyteller and establish causal factors. We can contrast their responses with our own and evaluate the efficacy of different interventions. Stories are often engaging and, so, the learning effect can be more powerful.

In Chapter 9 we looked at issues around safety reporting and the barriers to participation. Dijkstra (2015) describes an app-based approach to allowing crew to share stories in a context outside of formal safety reporting. Stories submitted are amenable to analysis but, importantly, because the narratives are usually richer than safety reports, the approach has the potential to tap into what constitutes expertise and, importantly, competence. An informal learning system built on a storytelling concept would be adaptive in that narratives would change over time to reflect crew coping with emergent threats, unexpected events arising from procedure changes or other external disturbances. Records, stored in a database structure, could be cross-referenced and aggregated into key themes. Where beneficial, links can be made to policy and procedures but also, if relevant, to external sources such as significant research or core texts.

The methods discussed here under the broader label of organisational and personal development represent an indirect approach to building system resilience. Unfortunately, providing smarter feedback and sharing stories do not meet the demands of a structured, standardised approach to crew training. However, these approaches do address many of the significant issues discussed in the book at source while offering solutions that support the concepts of autonomy and motivation.

11.6 ARE STRESS AND FATIGUE A SPECIAL CASE?

Coping is one of the candidate competencies identified earlier. In Chapter 4 we explored stress and fatigue in some detail and it is clear that an organisation design intervention along the lines discussed in the previous section could moderate the problem. I have also suggested that ab initio training will be radically different to in-service training and airlines will need to adopt different models of training analysis and delivery in order to be properly performance focussed. Coping would appear to be one such competence that needs a structured, phased approach.

In the context of the earlier discussion, coping includes dealing with instantaneous stress (startle), reducing cognitive load, creating capacity and reinforcing control. But in addition to stress management, coping capacity is shaped by endemic fatigue state, general health, fitness and nutrition. The current interest in pilot mental health post-Germanwings has seen the growth of initiatives such as pilot peer-support mechanisms but the risk is that these might appear to offer a level of mitigation but actually simply further shift ownership of a part of the problem away from the company. Broader lifestyle and wellness campaigns in companies are undoubtedly beneficial but participation is often on a voluntary basis. A true personal readiness programme must engender a recognition of need on the part of the individual while exercising a degree of compulsion. There is a thin line between responsible support on the part of organisations and intrusion and this relationship needs careful navigation. There also appears to be much to suggest that an approach that addresses stress and fatigue as an attitudinal goal, perhaps using methods drawn from behavioural economics, is warranted.

One fundamental question, possibly implied but not so far addressed in this book, is where does organisational responsibility for risk end. Many years ago, I read that Swiss Railways ran a scheme that paid for employees' skis and bindings to be maintained at the start of the winter season. The company found that the cost was less than the lost productivity arising from employees off work with lower leg injuries. Ownership of risk extended beyond the workplace. Similarly, having once broken down on the motorway, the recovery service mechanic donned his hi-viz jacket as he got out of his vehicle, telling me that in the event of an accident his family would get better compensation if he has taken the correct steps to protect himself. Of course, the jacket would not save his life if struck by a juggernaut on a motorway but wearing the jacket was deemed to reduce risk. Contrast this the aviation experience after a captain was struck and killed by a vehicle while doing the pre-flight inspection. The wearing of hi-viz jackets became mandatory when walking on the ramp and a failure to don a jacket was punishable with a fine.

Issues around sleep hygiene and stress-coping are clearly targets for ab initio training but coping, in the sense of dealing with operation disturbances – threats – is the domain of in-service training. Research in a number of different domains suggests that coping can be developed through task-related training. Physiological stress, as measured by the secretion of hormones epinephrine and norepinephrine, can be reduced through intensive simulator training. Landman et al. (2017) suggest that coping can be enhanced during training by affording pilots the opportunity to better frame situations. In this context, a frame is simply the instantiation of the individual's mental model held at that moment. Reframing involves adjusting the mental model to better match the current circumstances. In effect, the individual needs to establish a new goal state in relation to the disturbance and to derive the constraint set that now applies. A failure to reframe will result in a failure to fully understand the relevance and meaning of the available information. By providing a greater variety of scenarios with different levels of unpredictability, pilots can develop richer constraint sets which allow for faster switching between goal states. In another study (Landman et al., 2020) it was found that a simple mnemonic could mitigate startle. They trained pilots to use a four-item procedure: (1) Calm down: take a deep breath, sit up straight and relax shoulders and hands. (2) Observe: call out the basic flight parameters. (3) Outline: formulate a hypothesis about the problem. (4) Lead: formulate and execute a plan of action (Landman et al., 2020). Item 1 derives from relaxation therapies and is a well-established coping technique. The remaining three steps are all elements of the other candidate competencies: analysis, planning and deciding. By verbalising parameters, an act of communication, control is exerted over the information gathering process but it also ensures that all crew have the same data upon which to base future action. Although developed in relation to a specific research question – the management of 'startle' – we can see an overlap here with the management of threats generally. Those individuals prone to see threats as existential would benefit from using the techniques listed here. This suggests that, in fact, a holistic approach that addresses problems in a broader generic, competence-based sense might serve better than a symptom-directed strategy.

Because of the significance of coping to the safe conduct of flight, I argue that it warrants being considered a competence. There are clearly behaviours that support the maintenance or re-establishment of emotional control and cognitive clarity under conditions of stress and uncertainty. There is underpinning knowledge that supports performance but because of the complex nature of the contributing factors, I would argue that an approach to developing coping requires more than just a training intervention. Ab initio pilots can be prepared for the world of work and, therefore, be provided with foundational knowledge. Appropriate attitudes towards self-management can be inculcated. But the true impact of the workplace and out of work factors will only be experienced once in employment. Effective organisational design and support structures will mitigate some of the sources of debilitation. In-service development should come through exposure in a training environment but there is a clear divide between allowing crew to rehearse the skills of reframing outlined earlier and then applying those skills under conditions of stress. This suggests that we need a multi-phased approach that makes use of different media, appropriate to the pilot's stage of professional development.

11.7 DEVELOPING TRAINING INTERVENTIONS

To summarise my position, a competent performance in a systems context requires an individual to apply classes of functions (competencies), enacted through specific activities and supported by underpinning knowledge, across a range of contexts. These functions need to be applied under conditions of increasing complexity, ambiguity and cognitive demand depending upon the operation's proximity to the system boundary. As we have seen, an effective training regime must develop competence in relation to operational demands. If there is a mismatch between training provision and operational need then there is an elevated risk of crew responses to anomalies lacking efficacy. By implication, training ought to be informed by the safety management system, given that this is usually the source of information about operational risk. In this section, I want to consider how training can better support the development and maintenance of competence.

Conventional CRM training contains content drawn from social and cognitive psychology. Applied research in the field of aviation has grown enormously since CRM first emerged but it is questionable that this growing body of knowledge stands in relation to competence in the same way that the fundamentals of, say, meteorology might. At the ab initio level, there is possibly the space to deal with some of this content in depth but in-service, recurrent training is under such pressure that a true cost/benefit should be demonstrated before resources are allocated to interventions that simply address academic or underpinning knowledge. I do not wish to imply that knowledge of fundamental research is of no value but I suggest that this falls under the heading of self-development. Broadly speaking, conventional approaches to CRM training can be classified as:

Centralised (instructor-led, typically in a classroom)
Distributed (trainee-led, exploiting text or digital media, including video or programmed content (CBT))
Experiential (typically in a simulated environment of either low- (desktop, part-task trainer) or high-fidelity (full-mission simulation))

Developments in technology offer opportunities for training to be more distributed while affording collaborative, high-fidelity experiences. Networked, remotely located training devices and operational equipment that can jointly participate in exercises have been commonplace in the military for many years. In aviation, training applications that allow crew to rehearse cockpit drills using a virtual reality interface while, themselves, each being at remote, networked locations are entering service. Technology is now capable and affordable while the software need to develop training applications is more accessible than in the past. Exploiting this technology does require access to a training design capability.

If competence is essentially about performance, it follows that training should provide opportunities to, first, act and then allow for reflection. This premise suggests that training should be experiential whenever possible. Figure 11.6 shows a template design for a forced-choice event that formed part of a recurrent simulator session. On this occasion, the overall scenario was a standard company route with the aircraft on the return leg to home base. The route passed close by a third company destination. Fidelity was, therefore, high. The trigger event was a depressurisation, chosen because it happened to be a mandatory item to be covered during this cycle. The forced descent resulting from the technical problem meant that there was now

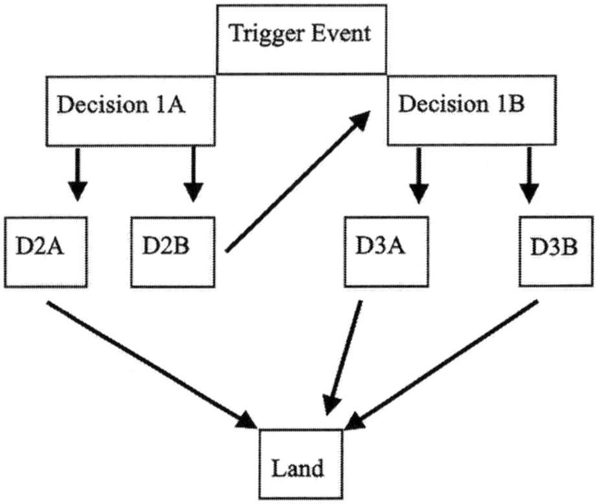

FIGURE 11.6 Game decision tree.

insufficient fuel to complete the remainder of the flight and so a diversion was necessary. There were two available airports. The first was the company destination and the second was still acceptable, if a little further away. The first steps in the design of the activity, then, are to select a plausible storyline (routine company sector), establish a trigger event which forces a response and then offer plausible alternatives. The goal is to design an activity that has face validity, which means that the scenario is sufficiently plausible such trainees can readily buy-in to the activity. We are trying to capitalise on learner motivation.

The next step is to define the attributes of the destinations. Runway direction, length, available navigation aids and surrounding topography were, in this case, set. The variables that can be manipulated are usually associated with the weather. Precipitation, visibility, wind strength and direction, implications for braking action can all be controlled. In this scenario, values were set such that no single destination or runway was obvious but all conditions were plausible, again in an attempt to establish face validity. The intent was to create situations that trainees would find recognisable based on past experience. Having created the scenario. the next step was to consult a group of SMEs. Management pilots were asked to step through the decision points and state the advantages and disadvantages of each choice. They were also asked to state what their preferred choice would be. The decision points in Figure 11.5 are described in Table 11.9.

The responses from the management pilots were aggregated and the most frequently cited reasons were used to create a table that could be used in the subsequent exercise debrief. Interestingly, the results were fed back to the management pilots and some were surprised that their peers, first, sometimes opted for a different outcome and, second, offered justifications that differed from theirs. Even subject matter experts do not always agree. Table 11.10 contains an example of the decision point strengths and weaknesses collected from the management pilots.

TABLE 11.9
Game Plan Decision Points

Decision Point	Action
1A	Land at Destination A
1B	Land at Destination B
2A	Execute go-around and second approach to Destination A
2B	Divert to Destination B
3A	Make approach to Runway 1
3B	Make approach to Runway 2

TABLE 11.10
Example of Decision Factors

Decision 1	Go to Destination 1	Not Go to Destination1
Really should consider	Online port so faster service recovery Crew more familiar than with Destination 2 Closer than Destination 2	Weather close to minima Probability of a missed approach
Might also consider	Better able to deal with medical issues arising from depressurisation. No terrain issues Autoland capability Fuel availability Long runway	No ILS on reciprocal runway Time available for preparation due to proximity Landing performance (wet runway with tailwind)

The planning described here was then used to construct a simulator profile. Rules were established to ensure that the exercise always ended with a successful landing: we did not want crews to fail as that would undermine the training value. The debrief was then conducted using the SME framework so that crew could, first, declare their own decision-making process and then compare with the thoughts of an expert group. However, it was made clear that there was no correct solution. The point was to explore decision-making. Because of the way the exercise was constructed, crew had to trade-off options (into-wind runway with no instrument landing system (ILS) v ILS with tailwind; descend over terrain in marginal weather v descend over sea), prioritise (needs of injured passengers v probability of successful approach) and consider future risk (fuel remaining if unable to land on the first approach). They also had to fly the simulator and deal with the procedural activity associated with the task. The framework is representative of a LOFT scenario although possibly with some additional effort applied to designing the decision points. It also represents a template that can be used to map competence training goals onto technologies.

The choice of trigger event is important. In the exercise we have just looked at, the goal was to explore decision-making and not simply to deal with the malfunction. An over-emphasis on the technical aspects of the problem would have detracted from the point of the exercise. However, the nature of the trigger event is significant in that it can shape the exercise. The next exercise I want to look at explored how crews trade-off activities. The trigger was an ambiguous report of a smell in the cabin, which prompted the use of the 'smoke and fumes' checklist. The exercise forced the crew to balance the management of normal and non-normal checklists while deciding about possible diversions and still control the aircraft trajectory. The trigger event was sufficiently vague but one that would have been familiar to crew. The aim of the exercise, in this case, was to explore task management. What was different this time is that the training medium used was a desktop simulation.

Figure 11.7 illustrates the materials used to create the scenario. Crews of a captain, FO and SO were provided with all relevant company documentation, charts for destinations, port pages and weather information. The aircraft was position on an airway en route to home base. Each hexagon on the grid represented a unit of distance and rules were developed that resulted in the activity running in almost real time. We removed the need for vertical navigation but crew were responsible for speed control and heading. Three alternate airfields were available, all with different characteristics. Because of the dynamic nature of the exercise, the respective advantages of each destination changed as their positions relative to the aircraft changed.

The exercise ended when the crew had either landed at a destination and taken the appropriate action (parked the aircraft or evacuated) or had decided to continue en route. The class then reconvened and each crew described, first, their actions and then explained their rationale. Each class comprised three sets of crew and so it was possible to compare different performances. It was not uncommon to have three very different outcomes. Despite the low fidelity of the activity, it was still possible to see the effects of stress on performance and by having crews comprising different ranks, it was possible to see the social dynamic at work.

The design concept illustrated in Figure 11.5 applies equally to a flight simulator or a desktop scenario. The low-fidelity approach allows for the exploration of competence while the high-fidelity device allows for the competence to be rehearsed under pressure. Effective training should exploit the strengths of available media according to the goal of the specific training intervention. Table 11.11 lists the different available media, together with their relative advantages.

In Chapter 1, I said that training design typically starts with a clear understanding of the desired performance outcome. A competence-based approach to CRM training is no different and, in this case, the outcome is performance defined by the competence model we have developed. We have seen that training is delivered in two different contexts: ab initio and in-service. In theory, the competence framework will be a constant but the range of operational contingencies to be managed by ab initio trainees is likely to be fairly limited. We may also decide that ab initio trainees should be exposed to more underpinning knowledge about CRM concepts. In-service training must fit crews to cope with an escalating range of contexts. We can now start to explore how different training media can be configured to support competence. There are two important questions that must be addressed from the outset: what level of fidelity is appropriate and does the activity need to be dynamic in terms of the role of time in accomplishing the task? Whereas fidelity may be important in terms of training transfer, for example in the case of aircraft manual

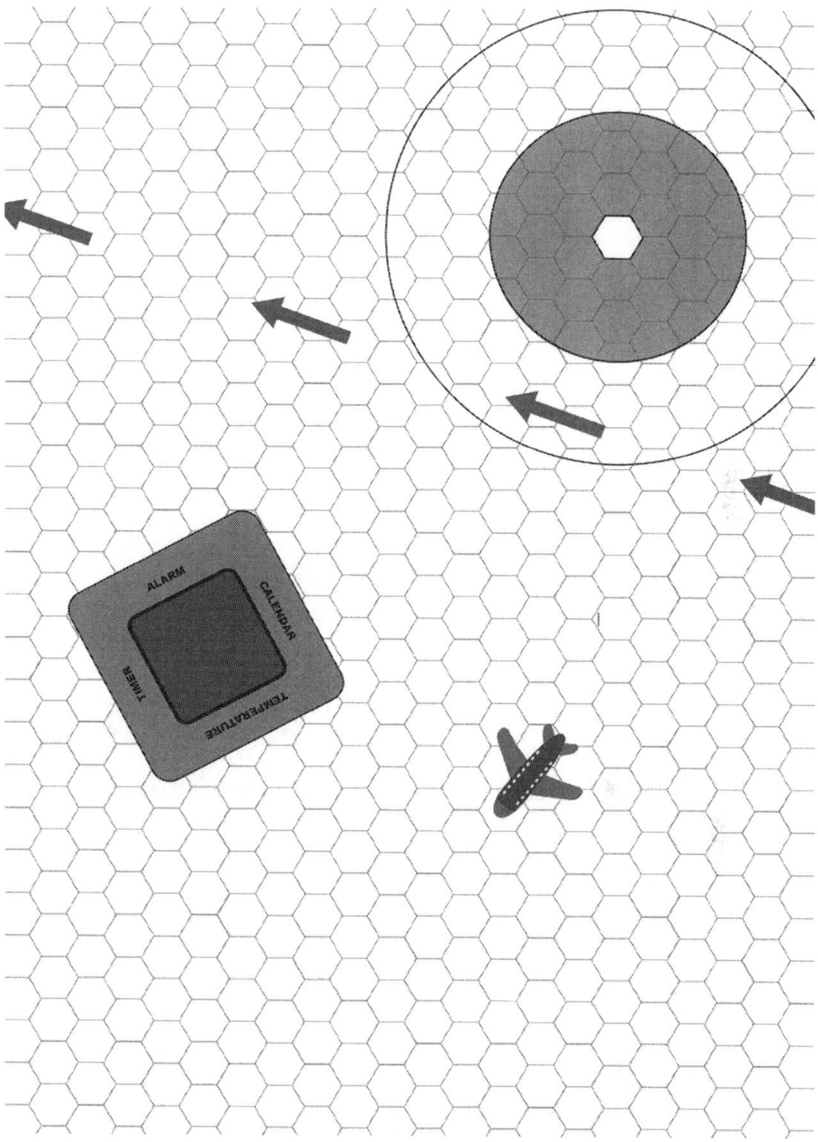

FIGURE 11.7 Desktop game. (Courtesy: Captain Mark Sigson.)

handling, we have already seen that the simple visualisation of a task can still promote learning. Low fidelity actually removes complexity and allow trainees to focus better on the specific learning goal. I will limit this discussion to airline in-service training and I want to conclude this review of training methods by returning to the proposed list of candidate competencies developed in Table 11.6 and to see which of the approaches we have been considering might be used.

TABLE 11.11
Training Media

Media	Key Characteristics
Case studies and problem scenarios	Low cost, no dynamic properties, low fidelity
Desktop exercise	Low cost, low fidelity, very flexible, gameplay can introduce dynamic dimension, able to run multiple crews simultaneously and compare performance
Tablet/iPad[a]	Moderate cost, less flexible, requires IT support and investment in technology, can support group activity
Part-task trainer	Higher fidelity, allows systems management rehearsal, limited group size although some configurations allow for observers, limited scenario repertoire
Fixed-base trainer/simulator	Highest fidelity, can incorporate flight path control, systems management and malfunctions, group size limited, requires scenario-building skills
VR	High-fidelity, distributed, can involve multiple participants, requires investment and IT skills, reconfigurable

[a] My thanks to Dr. Nick Dahlstrom at Emirates for information on the in-house development of this solution.

11.8 MAPPING COMPETENCIES ONTO TRAINING METHODS

I have offered a view of a system as a set of nested decision-making entities and I suggest that competencies, in aviation, are functions that support effective decision-making. Competence such as 'coping' functions by mitigating the impact of external factors. Communication serves to facilitate other functions. Analysing, planning and organising are precursors to action while validating is a mode of control. Finally, collaboration is the process of social interface management. The elements I have identified clearly have an internal structure and performance will be the product of multiple competencies.

The particular issues associated with developing coping have been dealt with in some detail in a previous section. From a training perspective, we need to accommodate opportunities to reflect on how other competencies support coping by developing resistance through performance under stress, always mindful of the fact that too much stress would be counter-productive. A good training model is Vygotsky's 'zone of proximal development' (ZPD) which describes the gap between actual and potential development (Vygotsky, 1978). When applied to training, a ZPD design would present a challenge or a demand that is just beyond what the trainee was last capable of achieving. The gap is always within the expected capability of trainees and success always results in achieving a new standard of performance. As a result, training is motivating.

Analysing and planning require opportunities to identify and clarify goals, establish constraints, consider implications of changing circumstances, verify the legality of options, integrate information sources and plan courses of action. Much of this can be done using text-based case studies developed from a safety or investigation report, a LOSA narrative or based on an SME scenario-building activity, similar to the 'what-if' approach explored by Martin et al. The case should comprise sufficient context to allow trainees to rapidly grasp the problem and provide access to all relevant information. It would help if the information was presented the same way as would be encountered on the aircraft (graphical representations of displays rather than textual lists of parameters). Trainees should be allowed to interrogate the scenario to the extent that it would be possible in real life (to simulate fault-finding, for example). All necessary documentation and support (i.e. access to maintenance or operations) must be available. One weakness with this approach is that some trainees struggle to enter into the activity because of the low fidelity but if the scenario is well-designed this can be overcome. For increased engagement, scenarios could be built using computer game construction programmes and deliver via tablets or smartphones. Both the text-based and IT approaches mean that training can be delivered in a distributed mode.

Training for organising requires opportunities to identify activities associated with achieving a goal, clarifying resource needs, sequencing actions and adapting plans to changing circumstances. It requires trainees to consider the implications of existing priorities in relation to goals. Much of this can be adequately addressed through group discussions (a form of desktop exercise), desktop simulations or interactive games. However, we saw in Chapter 3 that task switching will be affected by cognitive load. Changes to goal states will require task switching as part of organising to meet the revised constraint set. Therefore, there is some benefit in delivering some training in more dynamic environments.

Validating involves comparing available signals and data with expected values. Signals can relate to the task, the technology or the environment and they can indicate the current status or suggest a future condition. Training, then, needs to be dynamic and be able to present action over different time scales. The need to be able to present discordant signals linked to a managed task suggests that a high-fidelity situation might be beneficial, whether it is delivered in a conventional simulator or in a VR environment.

Training for deciding can be delivered through desktop exercises, interactive games delivered electronically and, of course, full-mission simulation. There are three contexts in which deciding should be rehearsed. The first is represented by the forced-choice situations illustrated earlier. As we have seen, these can be implemented through both desktop and high-fidelity simulations. The second context is represented by scenarios where change is subtle and innocuous. The decision point in such situations can be easily missed. Finally, we need to consider situations where problem-solving activity precedes a decision. I have argued that problem-solving is a form of 'deciding about deciding', requiring candidates to clarify the situation, specify information requirements and create a possible response. Problem-solving, then, is built upon analysis and planning. Levels of difficulty can be set by the ambiguity of the situation and the time available for resolution. These scenario characteristics will shape the training media required.

Communicating and collaborating clearly require shared tasks with desktop exercises, simulators and distributed VR environments all offering potential vehicles.

TABLE 11.12
Mapping Competencies onto Training Methods

Competence	Media
Analysing	Case study, desktop simulation, tablet-delivered game e.g. Dispatch problems involving deferred defects into destinations with particular constraints (terrain, weather). Presentation of anomalous system indications related to obscure interactions requiring fault-finding.
Planning	Case study, desktop simulation, tablet-delivered game e.g. Managing diversions to an alternate
Organising	Desktop simulation, group exercise, high-fidelity simulation, VR
Validating	Tablet-delivered simulation, high-fidelity simulation, VR e.g. Establishing system status based on anomalous indications, dynamic signal detection
Deciding	Desktop simulation, tablet-delivered games, high-fidelity simulation, VR e.g. Go/no go, diversion, trajectory management with degraded systems
Communicating	Desktop simulation, group exercise, high-fidelity simulation, VR e.g. Activities that allow for verbalising comprehension of non-normal system status; apply structured communications frameworks (briefings)
Collaborating	Desktop simulation, group exercises, high-fidelity simulation, VR e.g. Multiplayer exercises requiring agreement over solutions, negotiation with third parties
Coping	See Section 11.6

Scenarios should be designed in such a way that participants engage in parallel tasks which require occasional interaction in order to achieve their personal goals. Getting participants to role-play the 'other' side, for example getting pilots to act as the line engineer during a turn round simulation, can provide powerful insights. Table 11.12 attempts to map competencies onto media.

11.9 CONCLUSION

Competence is an ability to perform in a particular field of endeavour to a standard that suggests mastery, that an individual can perform in a reliable manner, that contingencies can be dealt with and, ultimately, that inherent risks will be managed. Training for competence assumes that the organisation's understanding of the range and extent of hazard exposure is accurate and that the ingredients of that skilled performance are understood. Competencies are abstract artefacts that describe classes of performance, each one contributing to the broader ability of individuals to sustain control over work processes. In this chapter, I have offered a framework of competencies derived from an analysis of operator performance in a systems model. The vast majority of work is involved in achieving task goals in a space defined by, on the one hand, the expected range of perturbations that arise from a non-ergodic world and, on the other hand, from the normal variability expected of human performance.

But competence must also extend to operations where system buffering is constrained and those situations where performance approaches a system boundary.

Training for competence requires careful analysis of need and then a mapping of competence onto appropriate media. Historically, CRM has been delivered in an academic sense with real-world performance disconnected from classroom inputs. This chapter has taken a step towards a more integrated approach with training being an attempt to offer task-centred activities that are constructed around specific competencies.

REFERENCES

ATSB. (2013). Transport Safety Report No: AO-2010-089. In-flight un-contained engine failure Airbus A380-842, VH-OQA. Final – 27 June 2013.

de Courville, B. (2019). Breaking airlines flight data monitoring barriers: A pilot's perspective. *Paper Presented at IASS Conference*. Taipei.

Dijkstra, A. (2015). Understanding resilience in flight operations. *Resilience Engineering Association Symposium*. Lisbon, Portugal.

EASA. (2017). Crew resource management (CRM) training NPA 2014-2017, RMT.0411 (OPS.094)-26 June 2014.

EASA. (2019). Appendix to Opinion No 08/2019 (A) (RMT.0599). Rationale behind the proposed amendments to the implementing rules.

Ebbatson, M. (2009). The Loss of Manual Flying Skills in Pilots of Highly Automated Airliners. Unpublished PhD thesis. Cranfield: Cranfield University.

Ebbatson, M., Harris, D., Huddlestone, J., & Sears, R. (2010). The relationship between manual handling performance and recent flying experience in air transport pilots. *Ergonomics*, 53(2), 268–277.

Flin, R., Martin, L., Goeters, K., Hoermann, J., Amalberti, R., Valot, C. & Nijhuis, H. (2003) Development of the NOTECHS (Non-Technical Skills) system for assessing pilots' CRM skills. *Human Factors and Aerospace Safety*, 3, 95–117.

Gosnell, G., List, J., & Metcalfe, R. (2019). The impact of management practices on employee productivity: A field experiment with airline captains. Centre for Climate Change Economics and Policy Working Paper 296/Grantham Research Institute on Climate Change and the Environment Working Paper 262. London: London School of Economics and Political Science.

Helmreich, R., Wilhelm, J., Gregorich, S. & Chidester, T. (1990). Preliminary results from the evaluation of Cockpit Resource Management training: Performance ratings of flight crews. *Aviation, Space, and Environmental Medicine*, 61, 576–579

Kluge, A. (2014). *The Acquisition of Knowledge and Skills for Taskwork and Teamwork to Control Complex Technical Systems*. Dordrecht: Springer.

Landman, A., Groen, E.L., van Paassen, M.M., Bronkhorst, A.W., & Mulder, M. (2017). Dealing with unexpected events on the flight deck: A conceptual model of startle and surprise. *Human Factors*, 59(8), 1161–1172.

Landman, A., van Middelaar, S.H, Groen, E.L., van Paassen, M.M., Bronkhorst, A.W., & Mulder, M. (2020). The effectiveness of a mnemonic-type startle and surprise management procedure for pilots. *The International Journal of Aerospace Psychology*, 30(3–4), 104–118.

Li, L. (2013). Anomaly Detection in Airline Routine Operations Using Flight Data Recorder Data. Unpublished PhD thesis. MIT.

Li, L., Das, S., Hansman, R.J., Palacios, R., & Srivastava, A.N. (2015). Analysis of flight data using clustering techniques for detecting abnormal operations. *Journal of Aerospace Information Systems*, 12(9), 587–598.

Looze, C.D., Scherer, S., Vaughan, B., & Campbell, N. (2014). Investigating automatic measurements of prosodic accommodation and its dynamics in social interaction. *Speech Communication*, 58, 11–34.

MacLeod, N. (2005). *Building Safe Systems in Aviation*. Ashgate.

Martin, W., Murray, P., & Bates, P. (2011). What would you do if....? Improving pilot performance during unexpected events through in-flight scenario discussions. *Aeronautica*, 1(1), 8–22.

Vygotsky, L.S. (1978). *Mind in Society: The Development of Higher Psychological Processes*. Cambridge, MA: Harvard University Press.

12 Assessment of Performance

12.1 INTRODUCTION

In this final chapter, I want to look at some of the issues associated with assessing competence. In the previous chapter, I looked at some examples of crew resource management (CRM) assessment frameworks in order to explore the idea of competence as implemented in these tools. I made the point that a competence model is not an assessment tool but, by using these frameworks as a starting point, it raised the question of what aspects of behaviour could be captured. The approaches illustrated were all based on the idea that performance needs to be observed in the workplace and that the assessment tool is simply a framework for capturing significant samples of behaviour. Observable outputs are, of course, simply the result of hidden cognitive processes and, thus, a comprehensive appraisal of overall competence is probably not possible through workplace observation alone. We may need a different approach.

In this chapter, I look at the problem of assessment from the perspective of reliability: the data collected must stand up to scrutiny or the conclusions we draw will not necessarily guarantee that crew are fit for purpose. I will look at the main sources of unreliability: the assessment tool, the method of grading, the assessors, the mechanics of the process and the context of assessment. We will look at some of the statistical methods we can use to validate data collection tools. Before we get into the detail, there is a fundamental issue that needs to be addressed: what are the differing needs of the various stakeholders in the process?

12.2 WHY ASSESS?

Ultimately, we assess performance because the regulations in force in our jurisdiction require us to. However, this simple demand creates quite a complex problem and the answers to the question 'why assess?' could shape both what we assess and what we use the data for. There are four key stakeholders in competence assessment. First, the individual being assessed wants to know of their progress in training or their performance against a standard, which is usually some organisationally-mandated benchmark of 'acceptable' proficiency. Second, the employer, the airline, wants to know if crews are fit for purpose. Can all crew be used for all possible tasks or are there any constraints on their use? Constraints represent an enforced under-utilisation of an asset and, therefore, foregone returns. Third, the training system wants to know if the training inputs are achieving the goal of producing a competent workforce. Finally, the regulator wants to know if pilots are legal to fly. These four requirements are usually met by a single process: observation on a check ride or in a simulator event using an assessment scheme that is probably not optimised for any specific stakeholder. A failure to recognise that these differences exist in the first place will result in schemes that serve no real purpose other than to satisfy the very first requirement mentioned above: we do it because we have to.

There are four commonly used terms in assessment. Summative assessment refers to progress against an agreed standard. For example, an Instrument Rating is usually a profile flown to specified limits laid down by the Regulator. Formative assessment refers to measurement that supports individual personal progress. We may want to establish the extent to which a trainee has mastered a particular aircraft system, for example. A diagnostic test might be set at the start of training to establish the extent of a trainee's prior knowledge and, therefore, what additional training they need. Criterion-referenced assessment uses an external benchmark against which to compare a score. In the case of the Instrument Rating, it must be flown to standards laid down in regulations. Norm-referenced assessment looks at performance against a reference group. A prospective command candidate could be assessed against a group of recently upgraded captains. The implications that flow from this discussion are that, as well as establishing a benchmark of acceptable performance, we need to consider if the benchmark needs to accommodate differences in training and experience. For example, should new-hire first officers (FOs) be assessed against the same standard as senior captains? We also need to consider how these differences affect the feedback we give to candidates. Do we simply offer an absolute result (pass or fail) or is it better to frame feedback in developmental terms.

Competence assessment is a subjective process in that we are asking one person to express an opinion about another person. More accurately, in an aviation context, it is an act of discrimination. We are choosing between those that are acceptable and those that need additional attention. To make these discriminations, we must first gather evidence against a set of constructs or 'markers' and then we must assign a value to the performance.

12.3 WHAT TO ASSESS?

We saw in Chapter 1 that the 'competence' movement in the 1980s benefited from the drive towards vocational workplace accreditation. The significant proportion of the workforce in most countries had left school with few, if any, formal qualifications but, nonetheless, had developed to be highly skilled workers. How can you recognise that ability and communicate it in such a way that potential employers understood what a job applicant was capable of? Accreditation of Prior Learning (APL) was one approach used in the United Kingdom. Trained vocational assessors would look at a portfolio of work and decide if it met the standard for the award of a qualification. Implicit in this approach is that a candidate could provide artefacts that represented evidence of the breadth of their experience and the complexity of work undertaken. The evidence would be workplace-specific, but it would be set against a generic competence framework.

Exam scores, equally, represent evidence of achievement but, in a competence-based approach, success can only be confirmed through demonstration of competence in the real world, not the exam hall. A competency-based method of assessment requires more valid evidence: a high school certificate in mathematics is not proof of an ability to manage inventory in a warehouse. It follows, then, that assessment requires the individual to demonstrate their ability.

TABLE 12.1
Suggested Marker Framework

Marker	Description
Manual handling	Control of aircraft flight path through the physical manipulation of controls
Application of procedures	Interpretation, application and enactment of policies, procedures and checklists
Management of systems	Configuration of aircraft systems, utilities, automation and navigation
Communication	Quality and effectiveness of intra- and inter-crew communication
Task management	Control of work in space and time

Dekker et al. (2010) raised concerns about the epistemological confidence we can legitimately have in the constructs we use to create knowledge of the world. In Table 11.6, I offered a set of 'competencies' which, I suggested, describe the fundamental processes that underpin competent performance. While they might offer useful targets for training interventions, I have not presented any scientific evidence that any of the items exist in any tangible sense. The fact that they are useful does not make them truths. If we now scale up the problem to that of assessment, it is highly unlikely that we would be able to observe all of these constructs in any meaningful way in the workplace. In Table 12.1, I offer a framework of behaviours that are observable and represent the outputs from the competence set. Its purpose is to facilitate assessment by providing a tool that can be used in a consistent manner.

Weber (2016) compared the performance of assessors in two airlines, one of which did not have a framework to guide the process. Differences were observed between the grades awarded with the raters of the airline who were not using a framework generally giving lower scores, failing pilots more often than the assessors of the airline with guidance and, finally, there was a larger spread of the scores awarded. Using a standardised methodology, then, supports a more consistent approach to assessment.

A poorly designed assessment tool results in raters falling back on their own native world view and then shaping their assessments to fit the marker scheme. Not only do the markers require adequate framing, but the boundaries between them also need to be clear. Given that we are trying to tease out samples from a stream of behaviour, we need to be able to communicate what units of performance to select into which category. If the clusters are too large, then there will probably be overlap between categories. As a result, two assessors are likely to put the same performance into different categories or the same observer might be inconsistent across observations, leading to unreliability. For example, in Table 11.3, 'Application of Knowledge' includes the statement 'knows where to source required information' while 'Application of Procedures' includes 'identifies where to find procedures and regulations'. It is difficult to see how these behaviours differ. Weber et al. (2013) explored the reasoning behind the scores awarded by assessors. It was found that some assessors gave the same score but for different reasons, whereas others gave a different score but used the same reasons. The degree to which the assessment category is open to interpretation will influence variance between observers. On the other hand, if the categories are too tightly defined, then the number of markers we need to capture the full range of performance will increase. Beaubien et al. (2004) suggested that fewer markers are preferable as the effort needed to assess against too many categories degrades assessor performance.

Another problem we need to recognise, as was mentioned earlier, is that not all of the ingredients of a competent performance are 'observable' in the workplace. If we consider the APL approach, flight documentation (weight and balance, fuel slips, completed flight plans etc.) is an artefact that captures underpinning knowledge. Flight data and air traffic control (ATC) transmissions also represent evidence about proficiency. A full exploration of 'competence' would require multiple modes of sampling including performance, underpinning knowledge and other associated skills. Because of the effort such an approach would involve, the default mode has been to draw conclusions from samples of behaviour collected in the simulator or on the flight deck.

An observational framework is no more than a tool through which to capture aspects of underlying competence. To have confidence in the data gathered, we need to strike a balance between a comprehensive coverage and a manageable system. The solution is to identify a few clusters of behaviour that are significant in terms of overall operational effectiveness and clearly map onto the competence model. They must be observable and be described in such a way that the marker can be used in a consistent and standard way by a cohort of assessors. Figure 12.1 illustrates how competencies map on to markers. The marker, in effect, is an aggregate performance. It provides the starting point for analysis in that a poor performance in each marker category can be decomposed into the relevant competence components.

12.4 ASSIGNING A VALUE TO THE PERFORMANCE

Having gathered the evidence, the next step in assessment is to assign a value to the performance. Evaluating pilot performance is not the same as measuring their height or weight. As soon as we start talking about a 'scale', we automatically think in terms of intervals of standard, knowable magnitude. My subject who is 1.82 m tall is exactly 10 cm taller than someone who stands 1.72 m tall. They are twice as tall as someone who is only 91 cm. The intervals on the scale are of a fixed magnitude, and

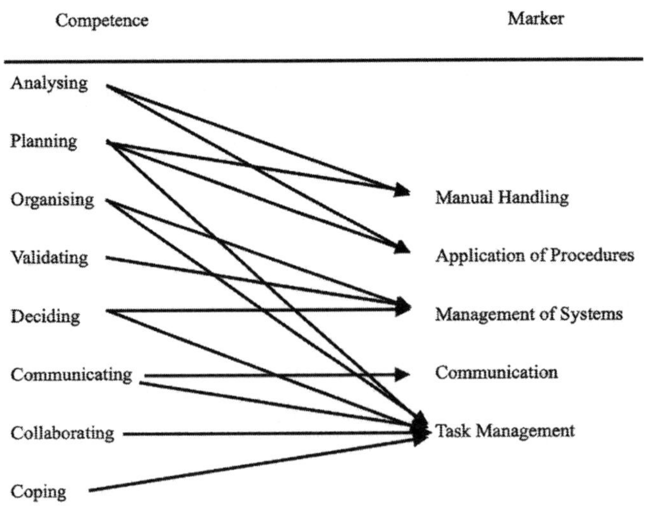

FIGURE 12.1 Mapping markers on to competencies.

the distance between intervals is known. When we grade a pilot's competence, we are actually assigning that pilot to a category of performance. The categories are arranged in a sequence and may have a numerical label, but it is not a scale in the same sense as a ruler is a scale. The non-technical skills (NOTECHS) markers initially used a five-point grading scale:

Very Poor: Poor: Acceptable: Good: Very Good

The problem with this approach is that we have neither a clear understanding of the magnitude of each interval nor the distance between intervals. Is an 'acceptable' performance three times better than a 'very poor' performance, for example?

The problems we face at this stage are how many intervals do we want and how to describe them? If we go back to the question of stakeholders, the aviation authority might just want two intervals: compliant with all requirements or not compliant. The training system might want a finer-grained analysis to identify specific classes of endemic weakness so that the training system can be tweaked. The airline might want to know who can be trusted to do the job, who needs watching and who needs further training. Each interval on a scale represents data about performance. The problem is that with more intervals we suffer an increasing loss of standardisation, or consistency, between assessors. An acceptable compromise in terms of capturing the most information for the least loss of consistency is four intervals (Flin, personal communication). Unfortunately, settling on the number of intervals does not answer what the intervals represent. That decision will flow from your answer to why you want to assess in the first place. Table 12.2 offers a suggested grade scale for use in assessing operational line pilots. A grade scale for use in initial training might need more intervals or different category descriptions. I have deliberately chosen not to

TABLE 12.2
Example Grade Scale

Category of Performance	Description of Performance
Unfit	Pilots in this group, when under observation, breach a safety or regulatory limitation. The performance can be linked to a published standard.
Needs input	Pilots in this group triggered a need for an intervention. It might have been a requirement to repeat a manoeuvre, or it may be something that can be solved by a debrief after an event. The assessor must be able to identify a specific flaw, weakness or deficiency that triggered the need to intervene. A failure to act would result in an operational risk persisting in the system.
Operational standard	Most pilots would fall into this group. The range of observed competence would be broad, but the assessor would be confident that the pilots do not represent a risk.
Resilient	Pilots in this group are clearly in control of the operation, are rarely taken by surprise and typically anticipate problems. Anticipation is a key indicator of this standard of performance, but solutions to operational problems will also be qualitatively better than the 'operational standard'.

assign letters or numbers to each of the intervals simply because to do so reinforces the misperception that assessment is measurement, rather than categorisation.

It should be immediately apparent that the intervals are not of the same size. Given that we are typically assessing line pilots who hold a type rating and many of whom will have years of experience, the 'unfit' interval ought to represent a small proportion of the distribution of performance. The 'unfit' grade is probably the only interval that can be considered 'objective' because it should be anchored to a regulation or a procedure (making it both summative and criterion-referenced). Conversely, the 'operational standard' interval will capture the bulk of the pilots in a company and, thus, will be very broad. The boundaries between the first three intervals are quite clear, but my experience is that the boundary between 'operational standard' and 'resilient' can be problematic. To understand why, we need to look at the sources of unreliability in assessment that flow from the act of assigning a value to performance.

A common problem in all assessment systems is that of 'central tendency': everyone is average. In cases where the scale has five intervals, everyone gets '3'. If it is a seven-point scale, then everyone gets '4'. Attempts to combat this using even-numbered scales simply offset the 'centre': everyone is still a '3'. Central tendency is often a manifestation of a lack of a clear performance standard. If I am not sure of the expected standard, then I cannot go wrong with 'average'. Equally, if I give a low score, then I have to justify to the trainee why I viewed the performance the way I did and that can be uncomfortable. If I give a high score, then I might have to justify my decision to management. So, a '3' is the easy way out. Another problem is 'scale clipping'. In this case, assessors adopt the attitude that no pilot is bad enough to warrant a '1' or else we would not have given them a job. At the other end of the scale, 'there is always room for improvement' and, so, the best you can get is a '4' on a five-point scale. Grade manipulation can sometimes result from experience levels or a belief that pilot ability should improve over time. In the case of experience, it is only natural that a new-hire pilot might not be as competent as an experienced line pilot (summative v formative assessment). The idea that pilots should get better with experience, logically, means that more than four intervals would be needed to capture performance in a career airline. Status can shape grading, and I have seen examples where management pilots always score 'straight 5s' on their assessment, training captains get all '4's and the rest of the pilots fight it out for the remaining numbers. Following on from this idea is the problem of, for want of a better word, vanity. I have seen situations where very senior captains expect top scores not just because of their performance (which is, more often than not, good) but simply because of their position in the airline. I have heard the argument made to retain a 'top score' on the scale simply as a way of rewarding longevity. That said, I have also seen high, but unrealistic, scores awarded to trainees as a way of motivating them. The grading of performance is a social act, but the more the social factors intrude in the process, the less valuable the activity becomes.

Poorly designed grade scales will affect reliability in that the data collected across time will be inconsistent. We need to establish what level of sensitivity – the number of intervals – will provide the data needed to satisfy the measurement goals (the reason for assessing). The intervals or categories must be defined in meaningful terms, using agreed benchmarks if possible. The reasons for categorising performance must be made clear to the pilot group, and the assessors must be disciplined in their use of the system.

12.5 ASSESSORS AS SOURCES OF BIAS

In the last section, we saw examples of how the act of grading a performance can become inconsistent. The attitude of the rater to the process of performance appraisal is just one of many forms of bias that affect the reliability of the system. A fundamental problem is that, being a social animal, I generally like people who are like me. On first encounter, I start sizing someone up. If I find something about them I like, I start to build a positive reaction to them. Then confirmation bias kicks in. The more I see what I like, the stronger the positive evaluation. The problem is that I now discount any signals that might be telling me that my initial assessment was wrong. Goldin and Rouse (2000) found that the per cent of female musicians in the five highest-ranked orchestras in the USA increased from 6% in 1970 to 21% in 1993 after 'blind' auditions were introduced. Candidates played behind a screen and decisions were made on skill alone. 'Unconscious bias' reflects the fact that we cannot help but make judgements about others and the ability to rapidly categorise strangers, while possibly conferring fitness an evolutionary sense, creates problems in the modern world. Airlines, like any organisation, are social entities and people will get reputations. Crew reading this will identify with that sensation of getting their flying roster and, once they see who they have been paired with, will already have an expectation of what that particular day will be like. When an individual arrives for assessment, the groundwork has often been prepared, unconsciously, in advance.

Another source of bias is what is known as the 'horns or halo' effect. If someone creates a bad impression at the start or arrives with a poor reputation, then everything they do will be bad. We are primed to expect the worst. The halo effect is the opposite. Someone who has a good reputation is seen in a positive light and even a poor performance is explained away: 'they were having a bad day'. Primacy and recency describe attributes of memory. In an encounter, our first and last experiences tend to be more memorable, whereas the events in between can blend into a generic memory of what happened. The problem with memory is that an evidence-based assessment scheme requires assessors to attend to the performance and capture samples that can be categorised and evaluated. Reliance on memory will distort the evaluation.

We all have expectations that we bring to the workplace. I have played a game on CRM courses where I divide the class into captains and FOs and get each group to imagine that they are approaching the desk in Dispatch and they are about to join up with their other crew member. What do they expect from that person? The lists can be illuminating, but the activity reflects the fact that we are primed to expect certain responses in specific settings and a failure to meet expectations creates a negative impression. Individuals are activating their mental script for the situation 'arrive at Dispatch' and discrepancies create dissonance. The problem is that these expectations are not a formal specification, they reflect personal preference and, as such, we all differ. Where a power differential exists, as in an assessment situation, the preferences of the rater become the standard: individual bias drives interpretation. One final problem we have to contend with is that attitudes change over time. Usually, assessors become harder to please. As one trainer put it to me: 'I just couldn't face seeing the same mistakes being made over and over again'.

These various forms of bias can be covered in an assessor training course, but awareness of these issues is not mitigation. Returning to Unconscious Bias, evidence suggests that specific training courses are of questionable value (see Atewologun, Cornish & Tresh, 2018 for a review). In fact, some studies show that training merely provides a way to rationalise, and deny, biased behaviour. The fact that giving information about bias does not remove bias has long been recognised.

Assessors need to be recalibrated at intervals, and their work needs to be checked and standardised. A failure to recognise this will result in unreliable data being generated.

12.6 ESTABLISHING RELIABILITY AND VALIDITY

In any domain, testing regimes must meet the requirements of reliability and validity. Although earlier I used the term 'reliability' in a generic sense in relation to the data collection, in measurement it has a stricter meaning and refers to the stability of a testing instrument over time. In short, two observations or samples of the same item taken at different times should return the same score unless the target has changed in some way between sampling. In a quality system, measurement devices, such as scales or chronometers, must be subject to independent testing to verify that they are consistently accurate. This is reliability. Validity on the other hand, is the degree to which the evidence gathered actually reflects the underlying construct being measured. For example, if I am interested in height, I can take a tape measure and check the height of an individual. The value I get will have a strong relationship with the 'real' height of that individual, within tolerable limits. If the measure shows 1.82 m, then their height is probably 1.82 m. Now, let us assume I want to compare a group of comedians on a scale of 'funniness'. I watch each comedian in turn for a period of, say, 15 minutes in a room containing 50 people. Using the decibel app on my smartphone I track the noise from the audience and I record the peak decibel value recorded during the 15 minutes. My hypothesis is that the funniest comedian will promote the loudest laughs. I think the flaws in my approach are clear. Decibels would not be a valid measure of comedic competence, whereas height taken with a tape measure would be. The behaviours we identify as candidates for a measurement must pass the test of validity. What we observe must relate to the underlying construct. This book has shown that human behaviour is messy and does not fall into neat categories. Furthermore, as has been said, outputs flow from hidden internal processes. The shorter the distance between the observed behaviours and the attribute being measured, the less need there is for interpretation and the data will be more valid.

McMullan et al. (2020) conducted a large-scale, systematic review of observational tools used in hospital operating room contexts in order to report on their psychometric properties. Thirty-one tools were identified in the literature, most of which were derived from the original aviation NOTECHS scheme (Table 11.2) but with modified grade scales of between 3 and 8 intervals. Reliability was established for three of the tools using a test–retest method. Various aspects of validity were also assessed. In our case, the range and complexity of statistical testing needed to establish the psychometric properties of a marker scheme are probably unnecessary, but a familiarity with some basic concepts is appropriate.

To illustrate the problem, consider Table 12.3. Two experienced training captains were asked to assess ten newly graduated student pilots to estimate their probability of successfully passing the company's type rating course. The candidates had two attempts at flying a short profile in a flight simulator. To avoid training effect, each assessor saw half the group on their first attempt to fly the profile and half on their second attempt. Assessors were asked to assess the probability of success on a five-point scale:

1. highly unlikely
2. unlikely
3. not sure
4. likely
5. highly likely

The simplest method of calculating the inter-rater agreement is to look at those occasions where both raters gave the same score and then dividing the number of agreed scores by the total number of ratings per individual. The assessors agreed on four out of ten occasions (0.4), suggesting that the process was no better than chance.

Assessment of performance comprises two separate activities. First, assessors allocate performance elements to a category, or marker. Second, assessors evaluate overall performance in each marker and assign the individuals to another category, or grade. In the example, we have just looked at the training captains were assigning candidates to categories – likelihood of success – not grading the performance against a benchmark. We need to be sure that the cadre of assessors is consistent in both allocating observations to markers and then in assigning a value to the performance.

There are two areas of interest to us: inter-rater agreement or reliability (IRR) and inter-rater variability. In fact, these terms are used flexibly in the literature, but I will differentiate between them by using the former when assessors are looking at the same candidate and the latter in cases where multiple assessors are grading a range of candidates. If we look at IRR first, many of the most common statistical tests

TABLE 12.3
Assessor Scores

	Assessor 1	Assessor 2
Captain		
A	2	3
B	3	4
C	3	3
D	3	3
FO		
A	3	2
B	3	3
C	3	2
D	2	3

available were developed in a specific context, such as clinicians looking at cancerous tissue samples and assigning them to a limited number of categories based on the tumour type. Quite often, tests simply compared two raters and the choice was dichotomous: 'is this an example of x, yes or no?'.

The simplest way to test IRR is the method used in the example above: how often to assessors assign the same score to an individual? To increase reliability, tests also need to accommodate measurement error. Any score assigned to a performance includes a component that reflects the true score and a component that reflects measurement error, one example of which is 'luck'. The fact that two raters assign the same score to a performance could simply be chance, especially when a grade scale only has a few intervals. If we consider the case of an exam using multiple-choice questions (MCQs) each with four responses, there is a 25% chance of getting the right answer simply by guessing. It is for this reason that many MCQ exams have penalty scoring for wrong answers. Commonly used tests that allow for measurement error include Cohen's and Fleiss' kappa. IRR is an important tool used during the initial design of psychometric tests to explore their accuracy or discriminatory powers.

The simple calculation of agreement illustrated here is probably adequate for use on an assessor training course to focus attention on group consistency. If we assume that the grade awarded in an exercise is a reliable surrogate for the nature of the evidence collected by the trainee assessor, the output from an IRR calculation can be used as a starting point for a conversation about observational performance. The score is a catalyst for discussion.

To determine the accuracy of the performance of a group of assessors we need a 'gold standard' against which the performance of raters can be compared. While this is feasible in some domains, in assessing competence in an operational aviation context it is remarkably difficult to achieve. Consider the data in Table 12.4, which shows the distribution of grades by six assessors observing the same candidate.

On the four-point scale depicted in Table 12.1, the upper two categories are acceptable – above the cut-off – while the lower two require management effort. From an organisational perspective, we want to minimise the risk of assessing someone as competent when they are not (a false-positive error). Equally, we want to prevent unnecessary retraining if the candidate is, in fact, of an acceptable standard (a false negative error). The absence of a 'gold standard' places an additional burden on the design of the grade scale. Agreement between assessors, in the sense we have been discussing, would not be an acceptable metric simply because, although there might be congruence, they could simply all be wrong! On an initial assessor course, again, data presented as in Table 12.4 highlights any lack of standardisation and can be used to promote a discussion.

TABLE 12.4
Grade Calibration

	A	B	C	D	E	F
Above cut-off	3	2	4	3	3	2
Below cut-off	1	2	0	1	1	2

Finally, once we have trained and standardised our cadre of assessors, we now have to check on the level of IRA. For this, we use the statistic R*wg* (Smith-Crowe et al., 2014). For each marker, we calculate the statistic using the formula 1 − (S × 2/(A2 − 1/12), where S × 2 is the observed variance of the single item, in this case, the scores per marker, and A is the number of response options. A result of >0.8 shows an acceptable level of agreement. This test should be applied to the output from a final standardisation exercise conducted during an initial assessor training course.

All of the examples used so far have involved small groups of assessors looking at the same performance in a training context. To monitor the health of the operational system, we need a test that can cope with multiple assessors looking at different candidates with each candidate being graded by more than one assessor. The training record system must be capable of generating a table of scores awarded by each assessor for the specific candidates. The result can be analysed using Gwet's AC2 statistic (Gwet, 2014).

I want to end this section with a slight caveat. When we assess performance and assign a sample of behaviour to a marker, we are creating a piece of categorical data. I am simply saying that I have *x* samples of the behaviour we have chosen to call category *y*. Because we have labelled the categories, calling them markers, the data are classed as nominal. Similarly, a grade is also categorical data but now we can consider it to be 'ordinal' because the category we have used is on a progression from low to high. The categories themselves have no specific value; they simply represent ways of ordering data. As such, they represent types of non-parametric data. Because we are dealing with non-parametric data, mathematical operations, such as averaging the grade scores to produce a single value for an event, are, strictly speaking, illegal. It is commonplace to see parametric statistical techniques, such as calculating means and standard deviations, applied to competence assessment scores. In the same way that Dekker et al. questioned epistemological overconfidence, I would guard against too much faith in misapplied statistics in assessment.

12.7 TRAINING ASSESSORS

Baker and Dismukes (2003) addressed some of the problems associated with training assessors. Quite often the introduction of assessment using markers represents a culture change, and it can also be quite intimidating for both the assessor and the trainee. The first step is to ensure that both sides of the process understand, and buy in to, the concept of assessment. Assessors must have a thorough understanding of the marker framework in terms of what it represents and where the boundaries between categories lie. They need to understand the grade scale and its use. Finally, they need to understand the process to follow in gathering and evaluating evidence.

Video clips of crews are probably the best media for calibration and standardisation but can be challenging to create. Having crews rehearse non-normal events in a simulator can often provide ample evidence to observe. However, normal performance, especially on line checks, is often routine and, therefore, apparently lacking in events of note. However, assessing 'normal' work is still a valid training goal. Smartphones are capable of capturing adequate video material on the flight deck, but audio recording can sometimes be an issue. Where assessors are from different aircraft types, unfamiliar technical and procedural details can sometimes be a distraction. Video segments of around 5-minute duration are usually adequate.

Training involves giving candidates exposure to examples of workplace performance so that they can identify specific episodes or samples of behaviour. A recommended approach is to, first, simply let assessors watch a video clip and then discuss what they saw. Quite often, trainees will initially offer a generic overview of the performance: they will broadly summarise what they saw in qualitative terms. The goal, however, is to identify specific instances of behaviour. Show the clip a second time but now get the class to cite the episodes they observed. Having gathered the examples on a flip chart or whiteboard, the next step is to get the class to categorise their evidence against the marker framework. Where an episode possibly falls into more than one category, get delegates to explaining their rationale. As a rule of thumb, an episode should be assigned to the marker for which it represents 'best evidence'. That is to say, the most significant impact of the action observed tends to fall into one specific category. The activity is then repeated with a different video but now each trainee collects evidence and categorises before, then, sharing with their colleagues. Observing a performance and capturing evidence is a skill in its own right. Experienced trainers can often correctly evaluate the performance of a crew but struggle to support their conclusion with evidence.

Having collected and categorised the evidence, the next step is to assign a value to the performance. Calibration involves getting groups of assessors to discuss episodes and develop a consensus view about the most appropriate grade. Initially, it will be useful to look at the videos used for the initial observation task as the class will already be familiar with the performance. The class will also have developed a richer set of samples against each marker based on the group discussion. Initially, get the class to assign a grade to just one or two markers rather than attempt the whole suite at this stage. In any case, if manual handling is one of the categories, it will be very difficult to assess from a video. Each trainee assigns a grade and writes it on a post-it label. These are gathered in and then displayed on a whiteboard or flip chart. At this point, the spread of scores can often be surprising. Table 12.5 contains scores for ten trainees observing a video of a captain.

For one of the markers graded, trainees giving the highest and lowest scores are asked to cite their evidence and explain their reason for assigning the performance to the particular category. At this stage, their views are not challenged. The activity can

TABLE 12.5
Allocation of Grades

	A	B	C	D	E	F	G	H	I	J
AP	3	3	3	3	3	3	3	2	4	2
SM	4	2	3	3	3	3	4	3	3	2
TM	4	2	3	3	4	3	4	2	3	2
C	3	2	3	3	4	3	3	2	4	2

AP = application of procedures; SM = systems management; TM = task management; C = communication.

then be repeated with another marker with the discussion being led by the facilitator. This time, trainees can be invited to change their grading based on the views of their colleagues. Finally, using a new, unfamiliar video clip, the class has to gather the evidence, grade and then arrive at a consensus view of the quality of the performance. The exercise is not complete until all agree on the correct score. Garrison et al. (2006), looking at IRR in coders of transcripts, found that agreement improved from 43% to 80% after a similar 'negotiated agreement' activity. The training course should finish with a final standardisation exercise where assessors have an independent opportunity to demonstrate competence in assessment. If possible, scores awarded by the group should be analysed using the R*wg* statistic.

Because of the issue with attitude shift – which could be either hardening and softening over time – periodic recalibration should be done at an annual workshop, where groups of assessors review a performance (Baker & Dismukes, 2002). Trainers assess and share their conclusions about the evidence and the grade awarded. A periodic review of assessor performance by inspection of grades awarded is also recommended.

Rater bias, leading to unreliable assessment, is a significant and ongoing challenge. Tests of IRR will identify the presence of an issue but not, necessarily the cause. The stability of the assessment system across time can also be tracked statistically but, again, this will not identify the cause of any discernible drift. For example, a change in pilot recruitment demographics may influence the data or refreshing of the rater cadre, resulting in a drop in the group's shared expertise, could influence performance. From a systems safety perspective, meaningful assessment is a guide to resilience in that it points to the likely efficacy of crew interventions. Reflecting on the reasons for assessing, data gathered should also point to gaps between the operational risk profile and the capability of crews to cope. An assessment system lacking validity and reliability is an operational risk.

12.8 CONDUCTING THE ASSESSMENT

A fundamental problem with assessment is that the assessor is not invisible. The assessor changes the dynamic of the situation. Called 'observer effect', a suggested advantage of the line operations safety audit (LOSA) methodology is that, because the observer is deliberately unobtrusive, the performance of the crew is closer to 'real life'. Another problem is that assessors are sometimes involved in the conduct of the exercise. This is more obvious in the simulator context where the trainer also manages the scenario, often provides external inputs such as ATC but still has to assess performance. I have witnessed debriefings unravel because the trainer was distracted when a specific action was completed by the crew and a comment about a procedural omission becomes a contest over truth. For line checks, it sometimes happens that the assessor forms part of the operating crew and this is certainly the case with line flying under supervision for new hires and upgrades. Combining operating with assessing increases the risk of poor data collection. The mechanics of the assessment situation, then, militate against the process.

A recognised framework for the conduct of assessment is captured in the mnemonic ORCE. First, we need to 'observe'. Assessment is an evidence-based process, and the evidence is the observed behaviour captured during the event. Of course, we need to 'record' the evidence. Because of the problems of primacy and recency, a failure to record will result in a skewed evaluation. These first two stages are conducted during the event, obviously, but once the session is over, we need to 'classify' the evidence. This involves looking at the samples of behaviour and assigning them to the relevant marker. The final step is to 'evaluate', which is to assign a value to the performance based on the grade scale.

Evaluating the evidence will be shaped by the design of the assessment system. There are three broad approaches to evaluation. The first can be described as the 'global' approach involving a single grade for each marker across the whole event. Another approach would be to segment the profile and collect grades for each phase (taxi, take off and climb, approach and land). A third, tailored, method is to constrain the data collected to the most important behaviour for each phase. For example, a complex non-normal scenario might only be assessed against markers for, say, problem solving and communication.

Another method of evaluation is known as the VENN method (EASA, 2019). This approach suggests that an individual is graded according to:

How well the candidate demonstrated the required behaviours (quality)?
How often the behaviours were demonstrated (frequency)?
How many behaviours were demonstrated (quantity)?
What was the outcome (result)?

The VENN approach uses a marker framework with specific exemplars of observable behaviour for each marker (see Table 11.3). There are problems with this method (see Roth & Mavin, 2015 for a discussion of some of the issues). First, it is impossible to elaborate all of the behavioural events that are representative of a specific marker. It is the role of the assessor, after training and standardisation, to use their expertise to analyse the performance. Competence for a line pilot in productive service is typically assessed during a line check or in a simulator. The problem we face here is that, under normal line conditions, a crew might not have an opportunity to fully demonstrate their capability. The situation simply did not demand any great virtuoso performance. The frequency and quantity criteria are, thus, fallacies.

Another problem is that elaborations of markers using exemplars encourage assessors to use the marker as a checklist. Performance is graded according to the number of exemplars observed. To a degree, we are entering a stalled loop here in that assessors are encouraged to look for frequency and quantity is a part of the recommended process, but the fact that the marker is simply a subset of all possible exemplars means that the process is arbitrary and constrained.

Gontar and Hoermann (2015) found that 'non-technical skills are rated more reliably under high workload conditions than under low workload conditions, and social aspects of non-technical skills are rated more reliably than cognitive aspects'. The stress of high workload exposes fundamental competence and the demands of a line check might simply be insufficient to expose weaknesses in crew competence. Equally, whereas the social aspects of performance are tangible, the cognitive aspects are not. This brings us back to the question about how best to ensure that we have a full understanding of competence.

Data collection opportunities should be reviewed in order to take full advantage of their potential. Rather than line checks being passively driven by events on the day, they could be designed to target specific competencies, especially analysis and planning. The simulator activity typically comprises a set of standard non-normal and emergency events as part of license renewal, but it may also include a LOFT event. The problem is that the complexity of the LOFT scenario may vary from one training cycle to another and may be conducted differently according to the trainer assigned. Equally, the scenario might not tap into all the competences we have decided are of interest. The assessment situation, then, can constrain the data we collect. The fact that the assessment framework might not map onto the data collection opportunities means that training managers must have the skills to recognise the problem and find ways to mitigate, either through effective design of LOFT scenarios or through innovative data collection methods, such as self-assessment. Over an annual simulator training cycle, LOFT scenarios can be coordinated to provide coverage across all competencies. Trainers, including simulator instructors, are chosen because of their skill and breadth of experience. The training event should be sufficiently standardised to ensure that the data captured is trustworthy but still affords space for the trainers to share their expertise. This can be a challenge. The conduct of assessment, however, must be accomplished to the same standard by all raters. The context of assessment, then, is the final source of bias in the data collection process.

12.9 CONCLUSION

This book has been a reflection on the performance of airline pilots conducting work. However, the intention has been to consider the implications for training. Having established a model of performance, how do we achieve that goal through dedicated training interventions? This chapter has dealt with the problem of establishing the level of competence demonstrated by the crew. Measurement is of fundamental importance simply because the consequence of poor performance is potentially catastrophic. The problem we face is that failure is usually seen through a 'safety' prism when, in fact, outcomes are the product of crew applying maladapted solutions to operational problems. A safety paradigm is possibly too constraining.

Such is the nature of work that it is probably too messy to neatly fit into a psychometric. We need ways to categorise and assign values to performance. The problems associated with these demands are well understood. Airline training departments need to understand the implications of these problems for their task of confirming the fitness of the crew to undertake operations.

Validity and reliability are legitimate concerns even if they might not be statistically demonstrable in the aviation training context. They force us to constantly question the performance of the assessment system. The mandatory training and checking events represent a sampling regime and, therefore, should be optimised to gather the data needed to establish that the training system is fit for purpose. Finally, the vulnerability of the operation depends upon crew with robust skill sets. Effective assessment is an essential part of building a resilient aviation system.

REFERENCES

Atewologun, D., Cornish, T., & Tresh, F. (2018). Unconscious bias training: An assessment of the evidence for effectiveness. *Equality and Human Rights Commission*. Research report 113.

Beaubien, J.M., Baker, D.P., & Salvaggio, A.N. (2004). Improving the construct validity of line operational simulation ratings: Lessons learned from the assessment centre. *International Journal of Aviation Psychology*, 14, 1–17.

Baker, D.P. & Dismukes, R.K. (2002). A framework for understanding crew performance assessment issues. *The International Journal of Aviation Psychology*, 12(3), 205–222.

Baker, D.P. & Dismukes, R.K. (2003). A gold standards approach to training: Instructors to evaluate crew performance. NASA/TM-2003-212809.

Dekker, S.W., Nyce, J.M., van Winsen, R., & Henriqson, E. (Spring 2010). Epistemological self-confidence in human factors research. *Journal of Cognitive Engineering and Decision Making*, 4(1), 27–38.

EASA. (2019). Appendix to Opinion No 08/2019 (A) (RMT.0599). Rationale behind the proposed amendments to the implementing rules.

Flin, R. & Martin, L. (2001). Behavioural markers for crew resource management: A review of current practice. *The International Journal of Aviation Psychology*, 11(1), 95–118.

Garrison, D.R., Cleveland-Innes, M., Koole, M., & Kappelman, J. (2006). Revisiting methodological issues in transcript analysis: Negotiated coding and reliability. *Internet and Higher Education*, 9, 1–8.

Goldin, C. & Rouse, C. (2000). Orchestrating impartiality: The impact of "Blind" auditions on female musicians. *The American Economic Review*, 90(4), 715–741.

Gontar, P. & Hoermann, H-J. (2015). Reliability of instructor pilots' non-technical skills rating. *18th International Symposium on Aviation Psychology* (pp. 366–371). Dayton, OH.

Gwet, K.L. (2014). *Handbook of Inter-Rater Reliability: The Definitive Guide to Measuring the Extent of Agreement among Raters*, 4th edition. Advanced Analytics LLC, Gaithersburg, MD.

Helmreich, R., Wilhelm, J., Kello, J., Taggart, W., & Butler, R. (1990). *Reinforcing and Evaluating Crew Resource Management: Evaluator/LOS Instructor Reference Manual* (Tech. Manual 90-2). Austin: NASA/University of Texas.

McMullan, R.D., Urwin, R., Sunderland, N., & Westbrook, J. (2020). Observational tools that quantify nontechnical skills in the operating room: A systematic review. *Journal of Surgical Research*, (247), 306–322.

Roth, W. & Mavin, T. (2015). Peer assessment of aviation performance: Inconsistent for good reasons. *Cognitive Science*, 39, 405–433.

Smith-Crowe, K., Burke, M.J., Cohen, A., & Doveh, E. (2014, March). Statistical significance criteria for the rWG and average deviation inter-rater agreement indices. *Journal of Applied Psychology*, 99(2), 239–261.

Weber, D. (2016). Judging airline pilots' performance with and without an assessment model: A comparison study of the scoring of raters from two different airlines. *Journal of Aviation/Aerospace Education & Research*, 25(2). pp. 39–70.

Weber, D.E., Roth, W.-M., Mavin, T.J., & Dekker, S.W.A. (2013). Should we pursue inter-rater reliability or diversity? An empirical study of pilot performance assessment. *Aviation in Focus–Journal of Aeronautical Sciences*, 4(2), 33–57.

Index

T

U

V

W